# Micromechanics of Fracture in Generalized Spaces

*In memory of my father*

# Micromechanics of Fracture in Generalized Spaces

**Ihar Miklashevich**

ELSEVIER

AMSTERDAM • BOSTON • HEIDELBERG • LONDON
NEW YORK • OXFORD • PARIS • SAN DIEGO
SAN FRANCISCO • SINGAPORE • SYDNEY • TOKYO

Academic Press is an imprint of Elsevier

ACADEMIC PRESS

Academic Press is an imprint of Elsevier
84 Theobald's Road, London WC1X 8RR, UK
30 Corporate Drive, Suite 400, Burlington, MA 01803, USA
525 B Street, Suite 1900, San Diego, California 92101-4495, USA

This book is printed on acid-free paper ∞



**British Library Cataloguing in Publication Data**
A catalogue record for this book is available from the British Library

ISBN: 978 0 08 045318 7

For information on all Academic Press publications
visit our web site at http://books.elsevier.com

Printed and bound in Hungary
08 09 10 10 9 8 7 6 5 4 3 2 1

# Contents

# Preface to English Edition

Another monograph about—and one should ponder a while here. Fracture?—Or generalized spaces? What is it, and how is this all linked together? If we try to restrict the subject of this book by some narrow topic, it will not be quite fair. A brief look at the list of contents will show that the problems not traditional for the fracture theory are considered in it. At the same time, a considerable number of issues traditionally referred to the fracture theory are not mentioned or mentioned briefly.

This choice is connected both with the author's scientific interests, and with a certain methodology position. To our mind, initially, mechanics was developed as a part of physics in a broad meaning of this notion, i.e., as a method of studying the real world in all its integrity. However, gradually, both the integrity and reality were substituted by a sophisticated mathematical apparatus and computer mathematical models, which now 'keep living' independently of the objects of study themselves. Now, it is probably senseless to pity the situation that it is hard to find researchers who work equally fruitfully in several areas of natural sciences—from mathematics to biology, and from medicine to paleontology—like the scientific titans of the past. This phenomenon has both objective and subjective explanations and reasons. It seems (subjectively to the author) that this is explained first of all with the loss by researchers of presentation about the cooperative, synergetic development of the world. Therefore, any attempts to restore this multilevel and multiply connected representation look reasonable and justified.

This book was written with these particular presentations in mind. The concept of *geometrization* was chosen, first of all, because of its intuitive obviousness; the energy-based approach—because of its fundamentality; the mathematical apparatus is the required minimum to get the idea of the real processes. When obtaining most of the results, we aimed at a finished analytical form. It is always possible to build a more complicated model of a phenomenon, but it is important to get the **first working** model. Therefore, many provisions and issues are marked in the book but not developed in detail, and many more topics are dealt with synoptically. This is why the English version is about a third larger than the original Russian one. It allowed inclusion of new materials, correction of mistakes and partially reprocessing the old materials with accounts of new presentations.

The bibliography is an important part of the book. In view of the contemporary publication practices, the compilation of an all-inclusive bibliography in any technical area is an impossible task. Nonetheless, the bibliography is intended to be comprehensive in the sense that it includes entries which describe the research results on essentially all the aspects of fracture. In most of the cases, references to Russian-language literature were replaced by the available translations into English. Thus, most of the references should be available in a reasonably complete technical library. All the references cited in the text are included in the bibliography, and many additional references are included as well. Some judgment was required in the selection of references for citation in the text, and I have done my best to accurately identify sources for key steps in the evolution of ideas.

## Introduction: selection from preface to the first edition

### *Collective effects in mechanics of deformed bodies*

Deformation and fracture are an integrated multi-level process of changing the structure and form-building of solids. Already the definitions of the basic notions of the mechanics of a deformed solid, for example, an ideally elastic and ideally rigid body, contain the presentation of the continuum and its properties. Therefore, analysis of collective effects arising at deformation is an actual problem. The hierarchical nature of deformation and fracture is the reflection of the collective nature of the destruction process. Strictly speaking, any processes observed at deformation can be referred to as collective effects, since at the microscopic level, the processes of macroscopic deforming are connected with deformation and break of interatomic bonds (change of interatomic space). This change cannot take place as an isolated process, but always only as a correlated interaction of many structural elements of the body. However, elaboration of a consecutive microscopic theory of such interaction *ab initio* does not look feasible so far. Such a non-contradictory description should comprise both the continual theory of dislocations, and the *geometrized* deformation theory, based on a stochastic approach. The continual dislocation theory is an attempt to describe, in the physical field level, the defect structure of the material. The *geometrized* deformation theory binds the deformation together with the continuum property (of an abstract non-deformed body), in which deformation is taking place. Such geometrized theory can be built on the basis of presentation of deformation as of a motion in some abstract space. The properties of such motion and space are discussed in the calibration deformation theory. Correlation of both approaches is possible by means of presenting these processes as the ones which are implemented within the united hierarchical space in different structural layers.

As the final result, both the continual and 'geometric' approaches to deformation should be coordinated with fluctuation processes in fundamental atomic–molecular and quantum levels, since these levels in particular statistically justify the equations of the continual theory.

### *Generalized mechanics of the continuum*

The necessity of simultaneous consideration of the processes which take place at different structural levels is explained by a well-known fact: in the process of mental 'splitting' of the materials used in engineering into smaller and smaller volumes, we arrive to a *qualitative* change of properties at some level. This limit does not always clearly exist; however, for example, in crystalline bodies it is expressed rather clearly. Therefore, for structurally sensitive materials, strictly speaking, the methodology of continuum is not applicable. Nevertheless, it is admissible to transfer the methods of the continuum mechanics—which deals with the mechanical behavior of the matter in the macrolevel—to the microlevel, the level of media with microstructure. This method is called the method of continuous approximation. Probably the first scientists to knowingly offer the theory of media with microstructure were the Cosserat brothers in 1907. The methods of continuous approximation operate with such notions like *polarity, non-locality, configuration complexity and correlation function*. The science area which studies the behavior of materials with microstructure through using the methods of continuous approximation is called the *generalized mechanics of the continuum*.

## *Micromechanics and physics*

*The micromechanics of the solid body* considers macromechanical properties of structural materials, including polycrystals, from microscopic positions. To create new materials and to make their efficient processing, it is not enough to understand and be able to analyze the properties of microstructure of the material. It is necessary to define the links of the required macroscopic characteristics of the material with the microscopic characteristics of the structure, and to be able to reproduce the preset macroscopic properties. The approach to the defective structure, developed in the present monograph, from the physical viewpoint and related with the use of the physical apparatus of describing field structures, enables one to approach closer to a solution of this problem. It seems that the ideology of modern physics has penetrated insufficiently deep so far into the conceptual and 'technical' mechanism of fracture mechanics, although the required axiomatic basis is in place. A joint consideration of deformation and fracture processes with account of the influence of the defect distribution on the metric structure of the deformed material on the basis of the calibration field theory makes one of the 'super-problems' of the modern fracture theory. We would like to hope that this book is a certain step towards its solution. The monograph is based both on original researches of the author, and on critical analysis of modern literature.

The plan of narration is as follows. Chapter one gives a brief analysis of the existing presentations about the hierarchical structure of deformation on the basis of the general system theory. The fracture fractality is considered as a special case of the hierarchical structure of deformation. Certain issues are also dealt with concerning the mathematical structure of the space associated with fracture.

Chapter two, which is of mostly auxiliary and review character, considers certain mathematical presentations of the manifold and generalized space theory. The chapter justifies the necessity to use the spaces, more general than the Euclidean one, to describe deformation and fracture. It is suggested to use the Finsler space as one of the simplest ones. A description of a solid as a field structure (a structure that has a potential in its every point) is considered. The chapter considers group transformations and the procedure of covariant differentiation in abstract spaces, as well as basic geometrical characteristics, and introduces the tensors of torsion, curvature and segment curvature (the first, second and third tensors of Cartan curvature). The procedure is considered of associating geometry with deformation processes and connection of geometrical characteristics of space with the structure of real deformed medium. A possibility is shown to consider the crack propagation processes as motion in the Finsler space.

Chapter three is dedicated to crack propagation in a defective medium. Special attention is given to the issues of violation of symmetrical properties of the continuum at deformation and fracture of solids. A general variation approach is considered to crack propagation in a solid. A possibility has been studied to build a point symmetry group connected with deformation on the basis of analysis of the classic stress–deformation diagram. Restrictions on the elements of this group have been obtained.

Chapter four applies the general formalism to analyze crack propagation in a real medium. An equation of a crack trajectory is obtained and its stability studied. Crack propagation in media with different mechanical properties and in media with a sharp border is considered. Conditions are defined for the onset of stochastic modes of crack propagation, and phase portraits are built of the crack propagation process.

Chapter five is dedicated to the theory of fractals and the use thereof in fracture. The initial sections consider the general mathematic theory of fractals. It is specified in application to the problems of fracture mechanics. A provision is justified that the fractal properties of the crack and fracture surface are a direct consequence of extreme principles implemented in nature. It is shown that fracture has fractal properties in all structural levels. Dependence has been found of the fractal dimension of the fracture surface on the defective structure of the medium. A physical justification is offered of the onset of the fractal structure of fracture.

Because of well-known difficulties of recent years with modern scientific literature and poor accessibility both of well-known monographs and of modern studies, the material related to narration of non-traditional sections of the fracture theory is presented as explicitly as possible and in a closed form. However, the mathematical strictness was not an end in itself, and wherever possible, physical justifications of conclusions and provisions have been presented. It allows the reader to do, at a preliminary acquaintance with the book, without any additional literature. Along with that, many rather important issues have not been covered in their entirety, both because of limited space and because they fall outside of the main context of the book. Partially, these issues are specified in footnotes, and partially, are briefly dealt within the main text. This mentioning may move certain readers to a more detailed elaboration of the mentioned problems. As a rule, the chapters have summarizing remarks. These remarks are intended to draw attention once again to the issues touched on in the respective sections, and, possibly, to set problems for the future.

A rather specific selection of the material for the book, dedicated to fracture, requires a basic acquaintance with the course of mathematic and tensor analysis in the usual university scope, and also acquaintance with the ideology of the quantum field theory and group theory. A competent reader may easily skip certain auxiliary mathematical aspects. It is thought that the book will be generally interesting and helpful both to non-specialists—as a tool of introduction into the subject, and for specialists in fracture theory—as the basis for further studies. Certain specific and more complicated issues are marked by an asterisk, and can also be skipped at initial reading.

*Acknowledgements*

This book is not the fruit of my individual work only. My friends and colleagues have helped me a lot with their friendly advice and sincere interest in the researches carried out. Without this sort of participation and multiple discussions, the book would have hardly appeared in the form as it is. It is not practically feasible to enumerate all the contributors.

I would like to mention especially the German Service of Academic Exchange (DAAD), which supported my work in the universities of Braunschweig and Stuttgart. This support made my fruitful work in libraries possible.

Associate Professor V. V. Barkaline (BNTU, Minsk, Belarus), whose contribution is referenced in the respective sections of the monograph, in fact contributed much more to the appearance of this book. He was the main appraiser of the whole monograph, making many sections of the text more correct and highlighting a number of mistakes and discrepancies.

Professor I. Kunin (Houston, Texas) exhibited his invariable interest in discussing many issues specified in the monograph, and I express my sincere gratitude to him for that.

I would like to express my special thanks to the employees of Elsevier, whose kind proposal to publish the book had promoted materialization of ideas, and to mark the self-sacrificing work of my translator Mr. Leonid Schukin who has managed to overcome multiple difficulties of various computer programs and specific terminology and, I hope, has clearly translated the scientific Russian into scientific English.

Certain topics of this monograph were developed under the grants of the Belarusian National Fund of Fundamental Studies and within the Republic's Scientific-Technical Programs 'Mechanics' and 'Nanomaterials'.

# List of Basic Definitions and Abbreviations

EPR — Energy production ratio
MHST — Multilevel hierarchical system theory
SIF — Stress intensity factor
$\gamma$ – Surface energy density
$R_{i,j}$ – Distortion tensor
$e_{ijk}$ – Antisymmetric tensor of Levi–Civita
$\omega$ – Rotation or elastic rotations vector
$W$ – Energy density
$\sigma_{ij}$ – Stress tensor
$\mathcal{M}$ – Elementary manifold
$\mathcal{M}_n$ – Elementary manifold of dimension $n$
$\mathfrak{T}(\mathcal{M}) = \bigcup_{x \in \mathcal{M}} \Omega_n$ – Tangent fibering
$\Omega_n$ – Tangent affine space
$(\varpi)$ – Manifold of admissible coordinate systems
$A^i_{i'}$ – Transition matrix between coordinate systems
$\partial\Omega$ – Connected domain border
$\Omega$ – Open connected domain
$\mathbf{u}$ – Displacement vector field of the body points when transiting from the initial ideal state into the state with a crack
$T_\alpha$ – Fissuring tensor (for multiple cracks) or crack tensor
$A$, $a$, $\varepsilon$ – Crack length
$\mathbf{e}^T$, $\mathbf{e}$, $\mathbf{e}^p$ – Tensors of full, elastic and plastic deformation, respectively
$\sigma$ – Applied stress
$\mu$ – Shear modulus
$\nu$ – Poisson ratio
$n_i$ – Direction cosine of external normal to crack surface
$u_j$ – Displacement of crack edges
$x$ – Direction of crack propagation
$\eta$ – Incompatibility tensor
$R^l_{ij,k}$ – Curvature tensor
$H(x, y)$ – Hamiltonian function
$F(x, \dot{x})$ – Metric function, Lagrangian
$x^i$, $\dot{x}^i$, $y_i$ – Independent variables
$T_n$ – Linear vector space of dimension $n$
$P(x^i)$ – Manifold point
${}^f g_{ij}$ – Metric tensor of Finsler space
${}^r g_{ij}$ – Metric tensor of Riemannian space

${}^{c}g_{ij}$ – Metric tensor induced by connectivity
$s$ – Curve length parameter
$x^{\mu}$ – Current coordinates
$dl^{(r)}$ – Length element of Riemannian space
$(r)$ – Index at variables of Riemannian space
$t$ – Natural parameter of the curve
$S$ – curvature
$R$ – Torsion
$K$ – Segment curvature
$\Gamma^{*}_{kjl}$ – Symmetric connectivity coefficients of Finsler space
$\delta_{ij}$ – Kronecker symbol
$\xi^{l}$ – Field connected with defects
; – Covariant $\delta$-differentiation
$C^{i}_{k.j}$ – Tensor connected with change of metric tensor along the chosen directions, Cartan torsion tensor
$S^{j}_{i.kl}$ – First Cartan curvature tensor
$A^{j}_{i.k} = F(x, \dot{x})C^{j}_{i.k}$
$\tilde{\Gamma}^{\mu}_{\sigma\lambda}$ – Connectivity object without torsion
$D$ – Fractal crack dimension
$\Delta^{ij}$ – Metric deviation
$S$ – Entropy
$G$ – Full group of translations and rotations of a solid
$\partial_{\beta}$ – Unit vector of current configuration
$\partial_{\alpha}$ – Partial derivative by $k$-component of parameter $x$
$\partial_{\alpha|}$ – Partial Derivative by $k$-component of parameter $y$
$L$, $\mathcal{L}$ – Lagrangian, Lagrangian density
$\Gamma^{*\beta}_{\delta\alpha}$ – Connectivity coefficients of Finsler space
$\Gamma^{\beta}_{\delta\alpha}$ – Space connectivity coefficients; connectivity coefficients of the tangent Riemannian space
$D$ – Covariant derivative
$F_{\delta}$ – $\delta$-parallel body from $F$
$\mathbb{R}^{n}$ – $n$-measure space of real numbers
$\zeta$ – Roughness exponent

# List of Figures

# 1 Deformation Models of Solids: Description

## Contents



### 1.1 Description of hierarchy systems[1]

The complexity of the deformation and fracture processes in solids and a rather unsatisfactory theoretical interpretation of many accompanying effects have caused and are causing researchers to correct already known results and to put forward new concepts that more adequately describe real processes. Therefore, we see today an active elaboration and study of our presentations about the hierarchical and synergetic mechanism of plastic deformation and fracture.

#### *1.1.1 General description of hierarchy structures*

Currently, in the study of physical processes in matter in the condensed state, the statistical paradigm is the basic one, where macroscopic properties of a physical body are interpreted based on statistical averaging of microscopic behavior of its component particles. In this case, a physical body is presented to be a set of a small number of statistical collectives of particles, similar or identical in their physical characteristics with known interaction laws, which are quantum mechanical in the general case [155].

Despite huge achievements of statistical methods, a sphere of phenomena exists, where, on the one hand, the identified collectives of elements comprising a condensed body include heterogeneous objects, differing in their characteristics, and, on the other hand, each of such objects is notable for its individual behavior and cannot be exhaustingly described by a set of parameters, like the charge, which define the interaction laws of these objects. First of all, these are classic mesoscopic objects, which have currently become important due to the development of nanotechnologies. Besides, the quantum mechanics of the condensed state also has the problem of non-local correlations that is

[1] This section was written jointly with V. V. Barkaline

related to description of integral systems and integral properties of macroscopic objects already in the chemical level [290].

In this context, the authors find it important to develop an alternative approach to the statistical one that is naturally based on the general systems theory [207, 208, 365]. This new approach is based on the mathematical definition of a system as a relation that is set at the input and output objects of the system, the role of which is usually performed by certain preset multitudes $\mathfrak{X}$ and $\mathfrak{Y}$. In this case, the above relation is set as $\mathfrak{S} \subseteq \mathfrak{X} \otimes \mathfrak{Y}$, where $\mathfrak{S}$ is a system. At this approach, the existence of a system means the presence of sufficient grounds to eliminate certain implementations of 'input–output' pairs. In the absence of such grounds, the pair should be included into the system. It corresponds to complex behavior, at which not one but several outputs may correspond to a given input. Right because of this complexity, the systems theory provides an alternative to the statistical approach, which does not require introducing any statistical ensembles to explain the multiversion behavior of the systems.

The systems characterized by one-to-one mapping of the input and output objects are called ***functional***. We have already noted that general systems are not functional. However, any general system can be described as a functional one by setting one more multitude—system state object $\mathfrak{C}$, such that there exists one-to-one mapping $R : \mathfrak{X} \otimes \mathfrak{C} \rightarrow \mathfrak{Y}$, called global system response $\mathfrak{S}$ [208].

From the stand-point of the general systems theory, a general approach to describing an integral system, which interacts with its environment, is ***decomposition***, that is, presentation in the form of an organized manifold of interacting objects, each of them being a system by itself. Decomposition is not defined uniquely and depends on the whole body of knowledge about the system and dynamics of its environment. The ***multilevel decomposition***, has proved to be most fruitful in various sciences, in which system elements are distributed by the levels according to their spatiotemporal scales and intensity of interactions. As applied to physical systems, this decomposition corresponds to really observed structural levels of system organization within a certain experimentally fixed environment. In this case, one level hosts subsystems, comparable in their specific spatial parameters, life time and interaction forces. Usually, the knowledge system about the given system assigns to each such level a particular special theory or a manifold of mathematic models of the elements and interactions of this level. An integral description of the whole system is reached by a definite interaction between subsystems of different levels, and by establishment of interaction channels of the levels with the system environment.

An essential restriction of multilevel decomposition as applied to rather complicated systems is a rigid (strict) fixing of the selected level manifold, since it is assumed that inter-level connections are too weak to cause a radical reconstruction of the level system related to appearance or destruction of them. Besides, all the subsystems are interacting only through their input and output objects, while the status objects of these subsystems are isolated from each other and can be controlled only by the variables of the system environment.

A way to overcome the above restriction has been offered within the development of the multilevel hierarchical systems theory (MHST) [207]. Its appearance lies within the methodology of natural hierarchies [97], detected at a comprehensive view at the natural evolution of biological, social and technical system, and of the knowledge system.

The MHST was devised as a general system theory of decision-making in the conditions of uncertainty as a preset aim of the system. At the first stage, MHST was classified by

Table 1.1
Classification of MHST.

| Level name | Level description |
|---|---|
| Stratum | Description or abstraction level |
| Layer | Complexity level of decision made |
| Echelon | Organization level |

the type of the problems solved by the level and their complexity, and then, several level types were introduced into consideration in accordance with the aims of the system [207] (see Table 1.1). ***Strata*** appear because a complex system, by its definition, cannot be described at the same time clearly and precisely, so, these are complementary descriptions according to N. Bohr [209]. A compromise between a potential simplicity of description and need to account for multiple behavioral peculiarities of the system (multiplicity of 'input–output' combinations) is searched in hierarchical construction of the system status object. The system is set by a family of models, each of them describing behavior of the system from the viewpoint of a certain abstraction level. The following peculiarities are general for stratified systems [207]:

- Selection of the stratum, in the terms of which the system is described, depends on the knowledge and aims of the observer.
- Descriptions of system functioning at different levels are not related with each other in the general case. The principles and laws used to characterize the system in the given stratum, in the general case, cannot be deduced from the principles used at other strata.
- Each strata has its own set of terms, variables, concepts, laws and principles, which make up the language of this stratum. The stratum languages form a hierarchy with semantic relations between any neighboring members of the hierarchy.
- There exists an asymmetric dependence between the system functioning conditions in different strata. The requirements for functioning of the system in a particular stratum play the role of restrictions of the behavior in the lower strata.

A formal definition of a stratified system implies the following. It is assumed that the input and output objects of a certain system $\mathfrak{S} \subseteq \mathfrak{X} \otimes \mathfrak{Y}$ are represented in the form of Cartesian products:

$$\mathfrak{X} = \mathfrak{X}_1 \otimes \mathfrak{X}_2 \otimes \ldots \otimes \mathfrak{X}_n; \tag{1.1.1}$$

$$\mathfrak{Y} = \mathfrak{Y}_1 \otimes \mathfrak{Y}_2 \otimes \ldots \otimes \mathfrak{X}_n. \tag{1.1.2}$$

Then, each couple $(\mathfrak{X}_i, \mathfrak{Y}_i)$, $1 \leq i \leq n$, is assigned to particular stratum. The $i$-th stratum of system $\mathfrak{S}$ is the system $\mathfrak{S}_i$, which may be presented in the form of global reaction $R_i : \mathfrak{X}_i \otimes \mathfrak{C}_i \to \mathfrak{Y}_i$, while the neighboring strata are connected by informational $\mathfrak{I}_i$ and control $\mathfrak{M}_i$ mappings of the type:

$$\mathfrak{M}_n : \mathfrak{Y}_n \to \mathfrak{C}_{n-1};$$

$$\mathfrak{I}_i : \mathfrak{Y}_i \to \mathfrak{C}_{i+1}; \quad \mathfrak{M}_i : \mathfrak{Y}_i \to \mathfrak{C}_{i-1}, \quad 1 < i < n;$$

$$\mathfrak{I}_1 : \mathfrak{Y}_1 \to \mathfrak{C}_2.$$

Thus, the response of the given stratum to the input stimulus from this very stratum may spread over all the strata. If the response remains within the stratum of the input stimulus, the system is named ***completely stratified***.

The status objects for physical systems usually represent certain internal variables of the layer, which uniquely define the system's output. For temporary systems, all the objects represent multitudes of time functions, and for them the states in a given moment of time define not only the output at this moment, but also the future states (at the preset input) [234]. In this way, the cause-and-effect relations are represented in the system. The general systems theory is sufficiently expressive, however, to represent the time and logic dependencies of a more general type than the physical causality. Whereby, the causal 'input–output' system is replaced by a system of decision-making that is formally defined in the following way: a system $\mathfrak{S} \subseteq \mathfrak{X} \otimes \mathfrak{Y}$ is called ***a decision-making system***, if a family of problems has been set $Q_x, x \in \mathfrak{X}$, with a multitude of solutions $\mathbf{Z}$, and a mapping has been set $T : \mathfrak{Z} \to \mathfrak{Y}$, such that the couple $(x, y) \in \mathfrak{S}$ then and only then, when $z$ is solution $Q_x$ and $y = T(\mathfrak{z})$.

***Layers*** appear in systems of making complex decisions as a result of decomposition into a family of subproblems such that the solution of any subproblem of this family fixes certain parameters of the next subproblem. In this case, the last subproblem of the family becomes completely defined, and it may be solved in accordance with the functioning aim of the system-referent, for which the last subproblem defines the control stimulus. This hierarchy is called a layer hierarchy of decision-making [97]. The most important feature of this hierarchy is a capability to find an acceptable solution in the conditions of uncertainty. For this purpose, a decision-making system should implement at least three layers: self-organization, training and selection (Fig. 1.1).

A certain global aim is put in front of the system-referent $S \subseteq \mathfrak{X} \otimes \mathfrak{Y}$, for example, a controlled object, in the form of restrictions on permissible states, and, respectively, on output $\mathfrak{Y}$. These restrictions are set by solution $z$ taken out of a certain multitude of possible solutions $\mathfrak{Z}$. The system-referent is subject to the effect of non-controlled factors characterized by a certain multitude $\mathfrak{U}$.

If multitude $\mathfrak{U}$ is a single-point one, solution $z \in \mathfrak{Z}$ is generated only by the *layer of selection* as a solution of an optimization problem: by using the preset output function $P : \mathfrak{Z} \otimes \mathfrak{X} \otimes \mathfrak{U} \to \mathfrak{Y}$ and preset target function $G : \mathfrak{Z} \otimes \mathfrak{Y} \to \mathfrak{V}$, where $\mathfrak{V}$ is a multitude of quality characteristics of the system operation $\boldsymbol{S}$, to find solution $z$, minimizing function $G(z, P(z))$. If multitude $\mathfrak{U}$ is more powerful, the selection layer cannot find the solution, since both the output and, generally speaking, the target function become indefinite.

The task of the *training layer* is to narrow the multitude of uncertainties $\mathfrak{U}$ by using all the information available in the system. Here, semantic information is assumed on the points of the multitude of uncertainties, defined by the methods of mathematic informatics [50] as a filter of multitude $\mathfrak{U}$, comprising a family of sub-multitudes $\mathfrak{U}$, containing this point, and also of intersections and super-multitudes of them. An absolute setting of a point of the set, that is, a complete removal of uncertainty, corresponds to complete information. Pieces of information that refer to the given point of the set and have equal reliability are comparable with each other and form a distributive lattice. Therefore, the task of the training layer is to build maximum information about the point of the set $\mathfrak{U}$, sufficient for decision-making in the selection layer.

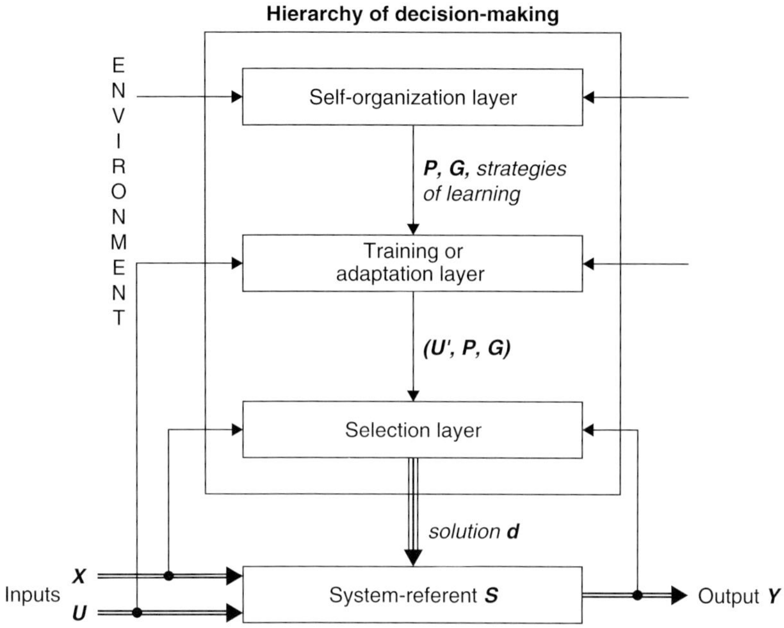


Fig. 1.1. Multi-layer decision-making system.

The ***self-organization layer*** is intended to define the structure, functions and strategies used in lower layers in such a way as to ensure getting closer to the global aim of the system $\boldsymbol{S}$, as far as possible.

***Multi-echelon*** hierarchies of decision-making systems represent models of complex system, which have competing sub-systems of decision-making, each with its own aim. Levels in these systems are usually treated as ***echelons***. While stratified systems and hierarchies of decision-making layers have formally one element in every layer, each echelon comprises several of them. The elements of the upper echelon possess a priority to interfere into functioning of the lower echelon elements, usually into definition of objectives for them.

Altogether, every hierarchy is set as a family of systems $\Im = \mathfrak{S}_l, l \in \mathfrak{L}$, where $\mathfrak{L}$ is a multitude of indices, on which a partial strict order relation $\succ$ is set. Then, couple $(\Im, \succ)$ is called a ***hierarchy of systems***. If $\Im$ is a hierarchy of decision-making systems, and relation $\succ$ is such that $i \succ j$, if $\mathfrak{S}_i$ has action priority in relation to $\mathfrak{S}_j$, then $(\Im, \succ)$ is called a *hierarchy of decision-making*. ***A multi-echelon hierarchy*** is such a hierarchy of decision-making $(\Im, \succ)$, if for any $i, j \in \mathfrak{L}$ there exists no more than one $k \in \mathfrak{L}$, such that for any $l \in \mathfrak{L}$ out of $l \succ i$, $l \succ j$ it follows that $l \succ k$.

The mechanism of the MHST allows describing systems also in the cases when interference of the upper level changes the structure of the lower level in accordance with its aims, which implement also the aim of the whole system. In this case, a ***coordinator*** acts in the system, which manages the inter-influence of the layers in accordance with this or that selected coordination strategy.

### *1.1.2 Hierarchical space*

Presently, the MHST has reached its ultimate development in the concept of hierarchical space created by S. I. Novikova [258, 259], which unites evolvement and development of all natural hierarchies into a united process of changing hierarchical coordinates. In this approach, the system itself and its interactions with the environment are presented as a set of coordinates, which are hierarchical systems themselves and are included into the global process of level growth, the basic moments of which are reflected in the law in inter-level interaction.

**Definition 1.1.1. (Law of inter-level interaction)** The units of the given level are combined, in the process of their interaction, into the units of a new level, which multiply and alter the designs of lower levels.

This fundamental law of hierarchical space reflects the processes which are specific both for stratified and for hierarchical decision-making systems, and also for multi-echelon hierarchies. Indeed, if we neglect the formation of new level units, we obtain a set of non-interacting levels of a stratified system, each with its own laws of motion and interaction of units.

The authors of this concept called the process of presentation of a system by means coordinates in the hierarchical space to be the AED-presentation[2].

By using the notation system of the AED-presentation, a hierarchical structure may be described by a set of *universal variables* in all structural levels. A complete set of universal variables includes the following: $\Lambda$, $\lambda$—level (time); $\Gamma$, $\gamma$—interaction (links); P, $\rho$—action (process); $\Omega$, $\omega$—unit object (state) $\Sigma$; $\sigma$—construction (contents); B, $\beta$—new time (evolving level); A, $\alpha$—control (coordinator). AED-construction $A^\lambda$ in a current structural level $\lambda$ is described by a symbolic expression ${}^\times\alpha^\lambda$ in the following set of linked expressions:

$$
\begin{aligned}
&\mathrm{A}^\lambda \underset{\rho}{\overset{\gamma}{\leftrightarrow}} \left\{ {}^{\beta}_{\omega} \overset{\gamma}{\underset{\rho}{\mathrm{A}}} {}^{\lambda}_{\sigma} \right\}^{\lambda \overset{\mathrm{A}^\lambda}{\underset{\rho}{\rightarrow}} \beta}, \quad
\Lambda^\lambda \underset{\rho}{\overset{\gamma}{\leftrightarrow}} \left\{ {}^{\beta}_{\omega} \overset{\gamma}{\underset{\rho}{\Lambda}} {}^{\lambda}_{\sigma} \right\}^{\lambda \overset{\Lambda^\lambda}{\underset{\rho}{\rightarrow}} \beta}, \\
&\Gamma^\lambda \underset{\rho}{\overset{\gamma}{\leftrightarrow}} \left\{ {}^{\beta}_{\omega} \overset{\gamma}{\underset{\rho}{\Gamma}} {}^{\lambda}_{\sigma} \right\}^{\lambda \overset{\Gamma^\lambda}{\underset{\rho}{\rightarrow}} \beta}, \quad
\mathrm{P}^\lambda \underset{\rho}{\overset{\gamma}{\leftrightarrow}} \left\{ {}^{\beta}_{\omega} \overset{\gamma}{\underset{\rho}{\mathrm{P}}} {}^{\lambda}_{\sigma} \right\}^{\lambda \overset{\mathrm{P}^\lambda}{\underset{\rho}{\rightarrow}} \beta}, \\
&\Omega^\lambda \underset{\rho}{\overset{\gamma}{\leftrightarrow}} \left\{ {}^{\beta}_{\omega} \overset{\gamma}{\underset{\rho}{\Omega}} {}^{\lambda}_{\sigma} \right\}^{\lambda \overset{\Omega^\lambda}{\underset{\rho}{\rightarrow}} \beta}, \quad
\Sigma^\lambda \underset{\rho}{\overset{\gamma}{\leftrightarrow}} \left\{ {}^{\beta}_{\omega} \overset{\gamma}{\underset{\rho}{\Sigma}} {}^{\lambda}_{\sigma} \right\}^{\lambda \overset{\Sigma^\lambda}{\underset{\rho}{\rightarrow}} \beta}, \\
&\mathrm{B}^\lambda \underset{\rho}{\overset{\gamma}{\leftrightarrow}} \left\{ {}^{\beta}_{\omega} \overset{\gamma}{\underset{\rho}{\mathrm{B}}} {}^{\lambda}_{\sigma} \right\}^{\lambda \overset{\mathrm{B}^\lambda}{\underset{\rho}{\rightarrow}} \beta}, \quad
\mathrm{A}^\beta \underset{\rho}{\overset{\gamma}{\leftrightarrow}} \left\{ {}^{?}_{\omega} \overset{\gamma}{\underset{\rho}{\mathrm{A}}} {}^{\beta}_{\sigma} \right\}^{\beta \overset{\mathrm{A}^\beta}{\underset{\rho}{\rightarrow}} ?}.
\end{aligned}
\tag{1.1.3}
$$

---

[2] In classical Greece, *aeds* were wandering singers who performed, among other works, epic songs, that is, delivered information.

Thus, all the hierarchical levels can generate their own unit states $A^{\lambda}$, with all their properties; $\leftrightarrow$ means relations of hierarchical objects. A hierarchical structure of the respective level $\Lambda$, $\Gamma$, B, P, $\Sigma$, $\Omega$ is directly connected both with its unit state $A^{\lambda}$ and with its own construction (with new interactions). Because of this, all the AED-levels can be resumed (duplicate states) in case of changing any random level. The effect of the hierarchical adding operator on the current state of the system leads to designing of the unit element of the upper level $A^{\beta}$:

$$A^{\beta} \underset{\rho}{\overset{\gamma}{\leftrightarrow}} \left\{ {}^{?}_{\omega} \underset{\rho}{\overset{\gamma}{A}} {}^{\beta}_{\sigma} \right\}^{\beta \underset{\rho}{\overset{A^{\beta}}{\rightarrow}} ?} . \tag{1.1.4}$$

Thus, the original state of AED $A^{\lambda}$ is defined so far as the image of its usual state is defined ($\Lambda$, $\Gamma$, B, P, $\Sigma$, $\Omega$), but the state of the following (in the sense of belonging to the upper hierarchical level) unit $A^{\beta}$ comprises the symbol of uncertainty '?' in its construction [263]. It is equivalent to the fact that a new state (a new position in a hierarchical space) $\beta$ cannot be precisely defined in the current state $\lambda$.

Such 'indefinite' behavior of a macroscopic mechanical system at the level of defining correlations is reflected in the variational formulation of mechanics, when for a system there is a principal possibility to choose this or other trajectory of development. Realization of a particular way is defined by imposition of additional conditions (laws in the MHST terminology).

## 1.2 Peculiarities of parameter space structure associated with fracture

For some time past, the theory of plastic deformation and fracture has been successful in explaining real processes [384]. Along with that, a number of effects observed in the process of deformation within classic presentations either lack a comprehensive explanation, or are explained insufficiently well [57, 141, 184]. Among these problems we see, for example, a correct description of plastic deformation of media with structure defects, with internal residual stresses, or problems of dynamical deformation.

For a medium with structural macroscopic and microscopic defects, description of plastic flow is complicated by the fact that the problem has a bound character [291]. It means that the plastic flow is distorted by the field of defects, and at the same time defects grow in the process of accumulation of plastic deformations. Here, we confront different dynamics for macroscopic and microscopic defects. Among microscopic defects we shall classify dislocations, disclinations, point defects of crystalline structure, which are realized in the microscopic structural level. Macroscopic defects—primordial cracks, damage cracks and main cracks—are realized in the mesolevel and instrument (device) level.

In accordance with the definitions of §1.1.1, this self-coordinated behavior indicates that a mathematical presentation of a deformed body should be an incompletely stratified system, in which a structure of layers has been realized. Since the deformed body achieves the set aim, in accordance with pre-set external conditions (for example, preset deformation) in conditions of uncertainty (stochastic dispersion of component properties, random microstructure), it is a hierarchy of decision-making layers and requires, as a minimum, three levels for its functional description.

Then, the problem of building combined plasticity equations with account of macroscopic damages is a problem of comparison of the set of universal hierarchical variables with the fracture parameters, and recording of equations (1.1.3) in the space of fracture parameters. It causes a necessity to study the structure of the given space of parameters. But first, let us deal with certain additional aspects of the geometrical structure of the values connected with the fracture.

### *1.2.1 Continual approximation in damage description*

The description sources of a macroscopic damage lie in an implicitly adopted hypothesis that definition of macroscopic characteristics and explanation of macroscopic processes should be held on the basis of consideration of the processes of the microscopic structural level. This approach was awarded the name of *continual fracture mechanics* and was formulated long before the principle of the hierarchical nature of fraction. Depending on the required detail level of macroscopic consideration, we may choose different microscopic levels. It is easily understood on the basis of the main provisions of the MHST. Since the uncertainty of a system description grows proportionally to the total number of operator actions $\rho^+$, the most efficient is the description of the upper level activity from the viewpoint of the preceding hierarchical level. Therefore, depending on the nature of the problems to be solved, as a dependent level (level of microscopic consideration), we may take either defects of an ideal crystal (imperfections, dislocations, point defects) or microscopic and primordial cracks.

#### *1.2.1.1 Microcracks*

When considering macroscopic fracture of a sample, Vakulenko and Kachanov [357] took a unit crack as the microscopic level. A break of dimension $S$, which can be correlated with a crack was considered as a value jump of the displacements field of the points of the body

$$\mathbf{b} = \mathbf{b}(M) = \mathbf{u}_+(M) - \mathbf{u}_-(M)$$

in point $M \in S$. In this expression $\mathbf{u}$ is the field of displacement vector of the body points at transition from the initial ideal state to the state with a crack. In this case, in the small neighborhood of any point $M \in S$ for smooth surface $S$ the break may be characterized by tensor $\mathbf{bn}ds$, where $ds$ is an element of the break surface, $\mathbf{n}$ is a unit normal vector to the break surface, and $\mathbf{bn}$ are the dyads taken in this point. In this case, the field describing the 'spatial' distribution of cracks in the body, containing one crack, should go to zero in any area except for the crack itself. Then:

$$T'_\alpha = \mathbf{bn}\boldsymbol{\delta}(S), \tag{1.2.1}$$

where $\boldsymbol{\delta}(S)$ is a delta-function concentrated in the surface $S$, and $T_\alpha$ has the sense of fissuring tensor (for multiple cracks) or a crack tensor.

If a body counts $N$ cracks, the tensor field will be a superposition of $N$ tensor fields of individual cracks:

$$T'_\alpha = \sum_{k=1}^{N} \mathbf{bn}\delta(S). \tag{1.2.2}$$

Field (1.2.2) will be singular at any integer $N > 0$ because of the presence of $N$ delta-functions. We can get rid of this sort of singularity, if we manage to respectively 'smear out' the delta-function. To do this, we may use one approximation of the delta-function as a limit of a sequence of smooth functions. The decomposition may preserve the required number of first terms.

Within the usual assumptions of the continuum mechanics, the local form of the second law of thermodynamics may be expressed as follows:

$$\rho\frac{df}{dt} = \sigma^{ij}\frac{d\varepsilon_{ij}}{dt} - \rho s\frac{d\tau}{dt} - \tau\theta. \tag{1.2.3}$$

Here $\rho$, $f$, $s$ are the densities of the matter, free energy and entropy, $\sigma^{ij}$, $\varepsilon_{ij}$ are components of the stress tensor $T_\sigma$ and deformation tensor $T_\varepsilon$, $\tau$ is the absolute temperature, and $\theta$ is the power of internal entropy sources. On the basis of the second law of thermodynamics $\theta \geq 0$, and $\theta = 0$ only when the process is reversible.

At small deformations, it is assumed that the deformation and fissuring tensors are uniquely presented in the form of the sum of the elastic (reversible) and plastic (residual) parts:

$$T_\varepsilon = T_e + T_p, \quad T_\alpha = T_\alpha^e + T_\alpha^p, \tag{1.2.4}$$

where $T_e$, $T_\alpha^e$ represent the reversible part of the respective tensors at a reversible isothermal transition of a medium element from the current state into state $T_\sigma = 0$. Parts $T_e$, $T_\alpha^e$ may be formally presented as $T_p = T_\varepsilon - T_e$, $T_\alpha^p = T_\alpha - T_\alpha^e$.

**Remark 1.2.1.** The adopted assumption (1.2.4) is justified from statistical mechanics. It is known that, in the general case, the distribution function of a quantity is a sum of an equilibrium and non-equilibrium parts [250], and in this connection decomposition (1.2.4) is justified. While developing the idea of [357], we may similarly introduce the general tensor field that characterizes the material. This tensor field of the continuum should comprise both components (1.2.4), and additional terms, which correspond to microdefects. The introduction of the general tensor field causes a necessity to expand the continual theory of deformation and to build additional gauge fields [115, 116, 210].

### *1.2.1.2 Medium thermodynamics with account of fissuring*

The medium described by models (1.2.2–1.2.4) differs from the ideally elastic one by the presence of fissuring, and the fissuring tensor [357] should enter the list of arguments of the state function. That is why the basis of the state space is:

$$\{\mathfrak{e}'_i\} = \{T_e, T_\alpha, \tau, f = f(T_e, T_\alpha, \tau)\}. \tag{1.2.5}$$

Since basic functions (1.2.5) in statistical approximation are continuous, we can switch over, if necessary, by using the known procedure of coordinate replacement, to the basis functions of coordinate presentations. In other words, we can write down any relations, for example, the second law of thermodynamics (1.2.3) in coordinates $x^1, x^2, x^3$.

With account of the effect of fissuring on thermomechanical properties of the continuum, the second law of thermodynamics (1.2.3) takes the form of:

$$\rho\left(\frac{\partial f}{\partial e_{ij}}\frac{de_{ij}}{dt}+\frac{\partial f}{\partial \alpha_{ij}}\frac{d\alpha_{ij}}{dt}+\frac{\partial f}{\partial \tau}\frac{d\tau}{dt}\right)=\sigma^{ij}\frac{d\varepsilon_{ij}}{dt}-\rho s\frac{d\tau}{dt}-\tau\theta, \tag{1.2.6}$$

where $e_{ij}, \alpha_{ij}, \varepsilon_{ij}$ are components of the respective tensors in the basis, which does not depend on $t$.

**Remark 1.2.2.** In the local consideration, basis (1.2.5) has discontinuous character. Operations with discontinuous functions need certain care. Since the correct formulation of the second law of thermodynamics (1.2.3) is related to partial moving surface derivation, then, for covariance of obtained transformations it is necessary to use the covariant $\delta$-derivative [113, 348]. In this context, the complete time derivative requires an additional definition.

Since at reversible processes $\theta = 0$ and, besides, $dT_\varepsilon/dt = dT_e/dt, dT_\alpha/dt = dT_\alpha^e/dt$, equation (1.2.6) may be rewritten as:

$$\rho\left(\frac{\partial f}{\partial e_{ij}}\frac{de_{ij}}{dt}+\frac{\partial f}{\partial \alpha_{ij}}\frac{d\alpha_{ij}}{dt}+\frac{\partial f}{\partial \tau}\frac{d\tau}{dt}\right)=\sigma^{ij}\frac{de_{ij}}{dt}-\rho s\frac{d\tau}{dt}. \tag{1.2.7}$$

By transferring the right-hand part of (1.2.7), we may have:

$$\left(\rho\frac{\partial f}{\partial e_{ij}}-\sigma^{ij}\right)\frac{de_{ij}}{dt}+\left(\rho\frac{\partial f}{\partial \tau}-\rho s\right)\frac{d\tau}{dt}+\frac{\partial f}{\partial \alpha_{ij}}\frac{d\alpha_{ij}^e}{dt}=0. \tag{1.2.8}$$

If we assume that the change speeds of all the values included in expression (1.2.8) are independent, coefficients of the speeds may be equated to zero. Condition (1.2.8) can be fulfilled in two cases:

$$\frac{df}{d\alpha_{ij}}=0, \tag{1.2.9a}$$

$$\frac{d\alpha_{ij}^e}{dt}=H_{ij}^{kl}\frac{de_{kl}}{dt}+P_{ij}\frac{d\tau}{dt}, \tag{1.2.9b}$$

where $H_{ij}^{kl}$, $P_{ij}$ are components of some random tensors, which are functions $T_\alpha$, $T_e$, and of other parameters.

For the case of a medium with a random distribution of cracks, fulfilment of condition (1.2.9a) is deemed problematic, and condition (1.2.9b) is fulfilled. It is important for us that the fissuring tensor depends on the chosen state space basis and, naturally, the tensor will change with changing the basis.

### 1.2.2 Deformed continuum fibering

In most cases, the mathematical structure of the space related to elastic or plastic deformation is taken to be a trivial Euclidean continuum. However, a deeper analysis [9] indicates that deformation involves more complicated mathematical objects. Let us consider a plastically deformed body without accounting for its microstructure. We assume that the Treska plasticity condition is fulfilled. It is known that the choice of the Treska plasticity condition has deeper physical justifications [137, 138, 383]. Let the stressed state be executed in a plastico-rigid body that responds to the edge of Treska prism. The presentation of the stress tensor in the main axes will have the form [291]:

$$\sigma_{ij} = \sigma_1 l_i l_j + \sigma_2 m_i m_j + \sigma_3 n_i n_j, \tag{1.2.10}$$

where $\sigma_{(1,2,3,)}$ are the eigenvalues (main stresses) of the stress tensor; $l_i,\ m_i,\ n_i$ is the orthonormalize basis from own vectors of the stress tensor.

For the given stressed state, we can always choose the main axes of the stress tensor in such a way that $\sigma_3$ be either minimum or maximum, that is:

$$\sigma_1 = \sigma_2 = \sigma_3 \pm 2k, \tag{1.2.11}$$

where $k$ is the yield point at a pure shift.

Since the basis is orthonormalized, the following is valid:

$$\delta_{ij} = l_i l_j + m_i m_j + n_i n_j. \tag{1.2.12}$$

With account of relations (1.2.10)–(1.2.12), we obtain:

$$\sigma_{ij} = (\sigma_3 \pm 2k)\delta_{ij} \mp 2k n_i n_j. \tag{1.2.13}$$

It follows from equation (1.2.13) that stresses in a non-damaged plastico-rigid body are defined by scalar field $\sigma_3$ and unit vector field $n_i$. With account of expression (1.2.13), the equilibrium equation may be written down in the invariant form:

$$\operatorname{grad} \sigma_3 \mp 2k \operatorname{div}(\mathbf{n} \otimes \mathbf{n}) = 0. \tag{1.2.14}$$

In expression (1.2.14), we are able to switch over to random curvilinear coordinates $\xi^1, \xi^2, \xi^3$. Let us assume that $\Sigma = \sigma_3 / \mp 2k$. Equation (1.2.14) takes the form:

$$\operatorname{grad} \Sigma + \operatorname{div}(\mathbf{n} \otimes \mathbf{n}) = 0. \tag{1.2.15}$$

With the help of the Foss–Weil formula [113] we may pass over to divergence in random curvilinear coordinates:

$$(\operatorname{div}(\mathbf{n} \otimes \mathbf{n}))_l = \frac{1}{\sqrt{g}} g_{kl} \frac{\partial}{\partial \xi^m} \left( \sqrt{g} n^k n^m \right) + n^r n^s \gamma_{rs,l},$$

where $g_{kl}$ are components of the metric tensor; $g = \det(g_{kl})$, $\gamma_{rs,l}$ are Christoffel symbols of first kind.

Then we may write down the equilibrium equation (1.2.15) in the covariant form:

$$\frac{\partial \Sigma}{\partial \xi^l} + \frac{1}{\sqrt{g}} g_{kl} \frac{\partial}{\partial \xi^m} \left( \sqrt{g} n^k n^m \right) + n^r n^s \Gamma_{rs,l} = 0. \tag{1.2.16}$$

*1.2.2.1 1-Forms and fibering of vector field* **n**

From the mathematical viewpoint, equation (1.2.16) has a very interesting structure of components. Since the first term of the expression is a 1-form [9, 43], the remaining terms should also be 1-forms. Out of all possible 1-forms, the most interesting for us are fibered vector fields **n**. In the simplest case [291], the following definition is correct.

**Definition 1.2.1.** A vector field defined in a certain space domain is called fibered, if there exists a family of planes filling this domain, such that the vector field of unit normals to the family surfaces coincides with field **n**.

*1.2.2.2 Account of macroscopic damage*

Hereinafter, we shall be interested in the effect of damage on changes of the equilibrium law (1.2.16). For this purpose, we need to write down the metric tensor and Christoffel symbols in the expanded basis of the space of states with account of the macroscopic damage. In more detail, this problem will be considered in the chapters to follow.

## 1.3 Hierarchy in continuum models of a deformed solid

Mechanical behavior of materials is highly sensitive to changes in their micro-structure. Thus, changing it by adding certain alloy elements, machining, thermal treatment or recrystallization, etc. can essentially improve not only mechanical characteristics of the material but also its reliability. To describe the mechanical behavior, we need to have a complete and precise description of the material's intrinsic structure and a complete description of the influence of the intrinsic structure on the macroscopic (structural) characteristics. In a general sense, this influence is known and understandable, but a lot of issues important in critical conditions should be stated more accurately.

As a rule, analysis of a material's intrinsic structure is achieved by means of its imaginary decomposition into a number of homogeneously filled regions. Therefore, the whole methodological problem of describing the properties of some material in the hierarchical multilevel system can be presented in accordance with equation (1.1.1) as a Cartesian product of a number of subproblems: to define the mechanical properties of an ideal continuum within the boundaries of preset regions, and the interaction of them. The imaginary decomposition is restricted by the fact that at a certain level of decomposition we are confronted with a qualitative change of physical properties of the continuum. Moreover, this change is not necessarily clear; there are certain materials that change gradually and fuzzily along the transition path. Additionally, according to the theoretical provisions of the hierarchical multilevel system (see p. 3): the properties of any upper layers never make a simple sum of those of lower layers. Therefore, reducing the continuum mechanics methods applicable at the macrolevel down to the microlevel, just as an inverse problem, is a non-trivial task. It can be accomplished by means of the *continuous-approximation method* [270] within the limits of micromechanics.

### *1.3.1 Mechanical properties of an ideal continuum*

A detailed mathematical analysis of the space structure in the classical mechanics was made back in the 1950s [9, 256, 257, 351]. It was shown [368] that the intrinsic metric of the perfect body[3] did not correspond to the metric of a real material.

As long as each body can be viewed as a differentiable manifold, it generates two related dissections that have an internal constraint. This internal constraint must be at least a Riemannian one. The constraint of the real material does not necessarily coincide with the Riemannian one. Only in isotropic bodies or symmetrical materials may the constraint of the Riemannian intrinsic metric stand for the material metric [368, p. 37]. In developing this idea further and taking into consideration the intrinsic structure, it was suggested to consider a real physical body as a certain object endowed with an additional kinematical structure and implanted into the Euclidean continuum [110–112][4]. This provision is equivalent to the statement that any body is endowed with the structure of the Cosserat continuum. Intrinsic structure is necessary, for example, for correct introduction of the thermodynamic properties of the continuum. In Cosserat media, the intrinsic structure is described by an additional parameter, a vector called director. Depending on the group of problems, we can consider Cosserat media with one or several directors. Director movement causes the appearance of additional 'kinetic fields', for example, volume forces. In the theory of liquid crystals [250], the director field is connected with small mean turnings of molecules in liquid crystals. It is also noted that for transient processes it is not sufficient to consider one director to describe the state of the crystal, there is a need to introduce the parameters of tensor nature, for example, the tensor order parameter.

The above structure in real bodies restricts the use of traditional operations applied in the continuum mechanics: single representative volume element operation, averaging operation [270]. Thus, it is possible to analyze defects in crystal lattice and imperfections in crystal state down to the particle level. But it is impossible to compare this state with the ideal one, and to show the geometrical constraint between the ideal and non-ideal crystals. According to L. I. Sedov's terminology [316, p. 67], a real transition from the 'initial' state into the given real one may not exist at all. Therefore, in geometrical considerations the main emphasis is made on the objects which do not change their value when the coordinates are transformed (i.e., on invariant objects).

The recently discovered carbon nanotubes [83] are real objects which help to limit the choice of representative volume. As regard nanotubes, the unit volume corresponds to the nanotube itself and the ideal continuum, which is presented as a set of interacting

[3] '... smooth materially uniform simple body' [256].

[4] 'We suppose here that each material point of the body manifold in $\mathcal{E}^3$ is endowed with an additional structure represented by an independent vector field called a director' [110].

carbon rings, prevents evident entrance of defects. Deviation from ideality (chirality of nanotubes) is one of the basic characteristics of the material. One may get an impression that a nanotube in itself is one single macroscopic defect. In this connection, definition of metric properties of a nanotube's 'initial' state and investigation of a possible smooth transition to the metric of the nanotubes under study are of undoubted interest.

#### *1.3.1.1 Generalized mechanics of continua*

Consideration of invariant values in different structural levels makes the basic subject of the continuum mechanics. The steps of its development may be linked with the set of geometries used to describe a microstructure [270]. According to E. Kröner [170], the split goes along three branches (Fig. 1.2). These branches reflect the genealogy of geometries:

1. Micropolarity (number of microscopic functional degrees of freedom).
2. Configuration complexity (number of functional degrees of freedom).
3. Nonlocality.

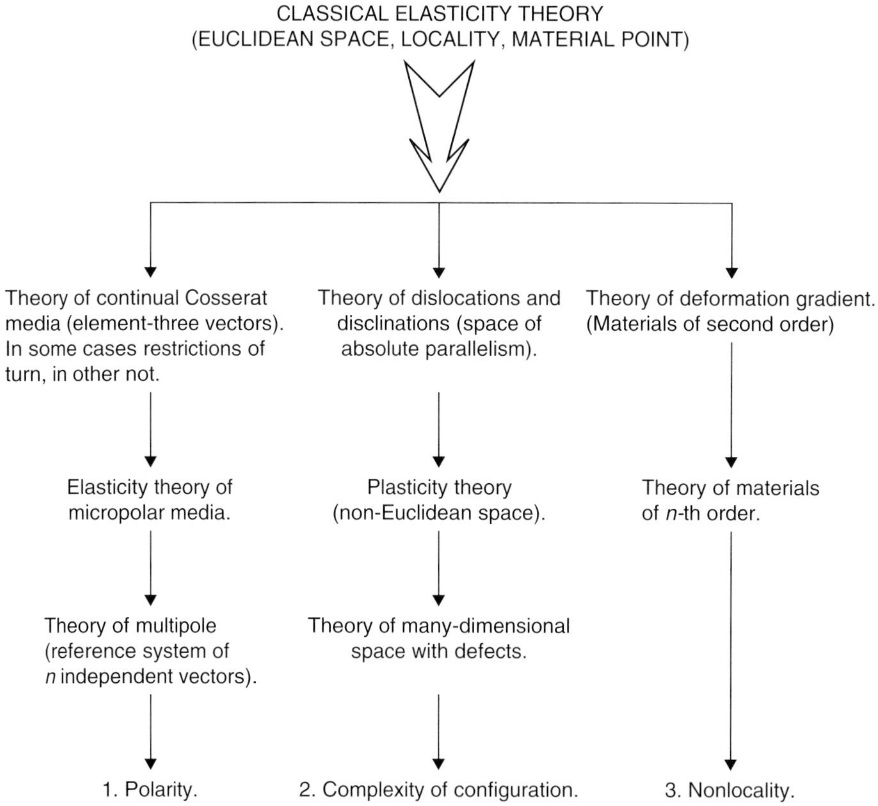


Fig. 1.2. Complexity growth of continuum models.

The first branch is the elasticity theory of micropolar media, which are a generalization of the continuous Cosserat medium. The second branch comprises the theories which consider the material space with defects being a generalization of an ideal continuum. The third branch accounts for the task to choose the measure of finite distances, to take into consideration the effects of displacement of the particles from the neighborhood of the point [270].

All these theories—the elasticity theory of micropolar materials, dislocations, plasticity and nonlocality of the continuum—are interrelated. For example, the Cosserat medium with rotation restrictions corresponds to second-order materials in the elasticity theory of micropolar materials. For the second branch, an expression in the space of stresses of the torsion tensor is nothing more but a description of couple stresses; therefore, respectively, they are linked with the elasticity theory of micropolar materials (the first branch) and the continuum nonlocality theory (the third branch of Fig. 1.2).

The statistical description (method of correlation functions) of structure and properties of solids is standing somewhat apart [58]. With the help of a body's correlation function we may also obtain a presentation about a 'configuration complexity' and 'nonlocality'. Hereinafter, we shall not consider any statistical description.

The generalized continuum mechanics, which unites all three branches, is very complicated and cannot be used in practice so far in its full scope. In the following chapters we consider in detail the second branch of micromechanics, which can be understood as *geometromechanics*, or an application of geometry to mechanics problems.

## 1.4 Hierarchy of systems and structures in fracture mechanics

In the mechanics of a deformed solid, the presence of a hierarchical deformation structure was progressively understood [187, 219, 277, 279]. One of the first works on the hierarchy phenomenon in fracture processes of spatial scale in structures and articles was published by Ya. B. Fridman in 1956 [102]. A schematic presentation of the hierarchical fracture process is given in Fig. 1.3 up to machine level. The depicted levels have a definite meaning and sense. As a rule, a transition to the upper hierarchical level is accompanied by a change of the leading mechanism of energy absorption [215, 321]. Each level is dominated by a single mechanism of energy absorption. The atomic level is connected with realization of unit atomic shifts, destruction of interatomic bonds and defect aggregation. The grain level is connected with the existence of the internal structure of deformed metal and the presence of structural borders. The mesoscopic level describes the change of the damage, growth of microcracks and transition to the stage of the mainline crack. Being the control level for the lower levels, when describing processes realized therein, it includes the descriptions and effects which are specific for dependent levels. The upper levels—the device level (or the testing instrument) and the human level—are rather transparent by their functions. Each cell of the presented diagram (see Fig. 1.3) represents a certain system (a set of elements), has its own structure and organization and realizes a state of a hierarchical system. The cells have different structures and organizations. These distinctions cause a hierarchy of mathematical structures, required

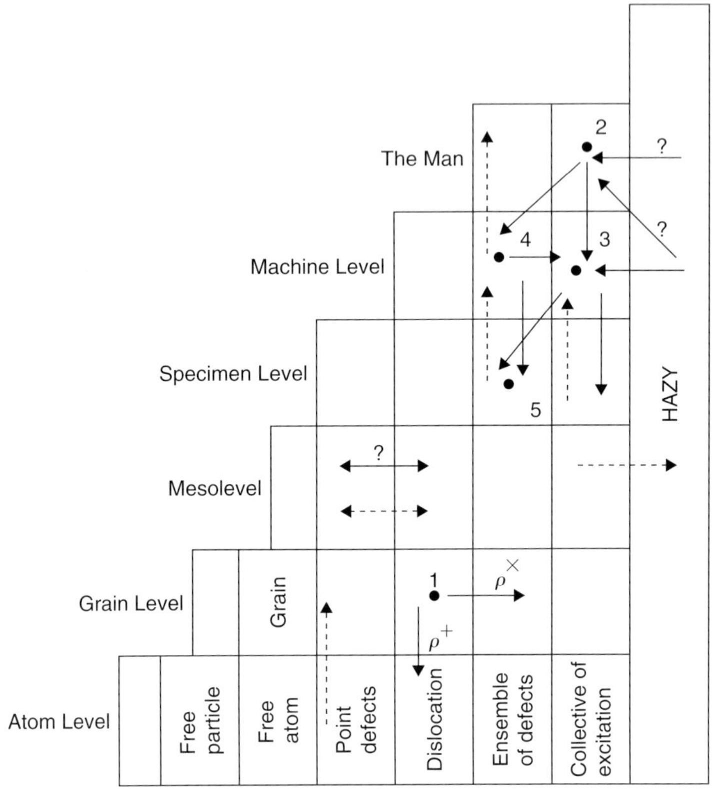


Fig. 1.3. Hierarchical presentation of the fracture process. Value $\rho^+$ is the summation operator (transition to a new level), $\rho^x$ is the multiplication operator; solid lines indicate control transfer, and dotted lines indicate signal transfer.

to describe the system. This provision was adopted back at the early stages of studying hierarchical systems [102]. Thus, in [75] a united formation hierarchy of solid structures was built (sequence molecules ⇒ grain ⇒ …⇒ body) based on mathematical objects classification group ⇒ ring ⇒ field. Whereby, the process of 'multiplication' of the initial structure takes place in the space in accordance with certain coefficients. Distinctions of cell organization cause distinctions in interaction laws, which is displayed in the necessity for different degrees of detailing when considering fracture processes (see Fig. 1.4). The diagram corresponds to one of the cells of hierarchical presentations at the grain level (see Fig. 1.3). In this case, the scale factor is not universal but depends on the considered level and requires that a respective machinery be applied. Thus, for computer simulation of plastic deformation (that is, evolution of dislocation structure), depending on the considered dimension range [41], one should use different simulation methodologies. At the microscopic scale ($10^{-10}$m) it is the simulation of method of molecular dynamics; the long-range elastic interaction of dislocations at the mesoscopic level ($10^{-6}$m) is described by the methods of dislocation dynamics.

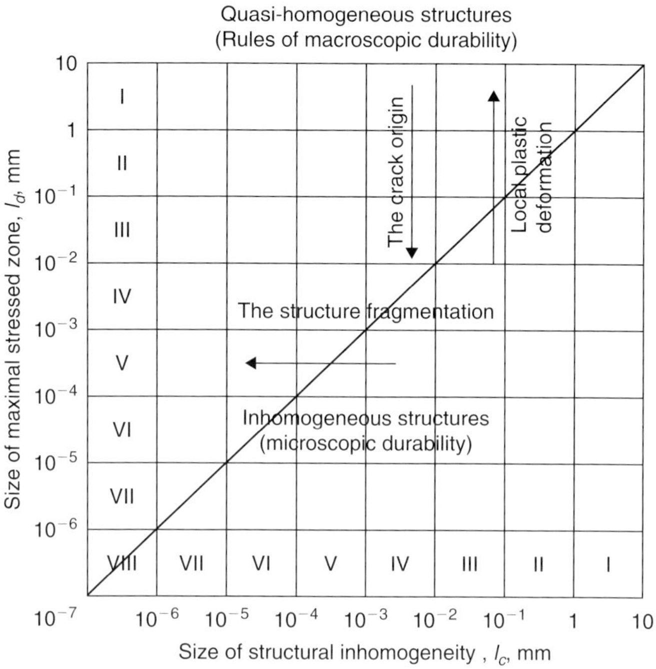


Fig. 1.4. Diagram of relative structural inhomogeneity. $l_d$ are linear dimensions of the maximum stressed zone, and $l_c$ are linear dimensions of structural inhomogeneities.

Possible objects, connected with various geometric deformation scales for structures and structural elements, are presented in Table 1.2. Each hierarchical system is interacting with different efficiency both with the systems in the same hierarchical level, and with the systems located in the neighboring levels. For the systems located above the diagonal line, the conditions of macrostrength balance are observed (structural inhomogeneity is displayed statistically). Below the line, there are systems where the structure inhomogeneity is essential.

**Example: Geometrical invariance for rocks.** For such an actual application of mechanics of deformed solids like mining mechanics, the hierarchy organization has its peculiarities [175, 309], connected, in the first place, with a clear scale gradation. In this context, a *geometric invariant* of mountain rock fracture is introduced:

$$\mu_\Delta(\delta) = \frac{\delta_i}{\Delta_i} = \theta_0 \quad \forall\, i, \tag{1.4.1}$$

where $\delta_i$ is the average crack opening (the distance between the edges), $\Delta_i$ is the diameter of the blocks of the $i$-th hierarchical level, $\theta = 0,5 \div 2$ is the coefficient.

As a rule, the crack opening for mountain rocks is understood as a zone of rock crushing around a tectonic fracture.

By itself, the representation about the hierarchical structure and behavior of geological and planetary structures is currently broadly acknowledged [309]. Thus, distribution

Table 1.2
Physical realization of unit objects of different structural levels of fracture process for structures.

| Area | Characteristic dimensions, mm | Typical objects realizing a stressed zone, $l_d$ | Objects of structural inhomogeneity $l_c$ |
|---|---|---|---|
| I | $10^8$ | Planet systems. | Geological blocks. |
| II | $10^6$ | Geological systems. | Blocks of Earth's crust. |
| III | $10^3$ | Construction systems. | *In situ* rock. |
| IV | 10 | Normal testing sample. | Reinforced concrete. |
| V | 1 | Microsample. Zone at impact test. Zone at coarse cutting. Zone of start of fatigue fracture of big parts. Technological explosion. | Big graphite in cast iron. Particles of soil. Particles by the explosion fracture. |
| VI | $10^{-1}$ | Zone at fine cutting. Zone of start of fatigue fracture of samples. | Big grains in steels. |
| VII | $10^{-2}$ | Zone at cavitation. Macrocrack. | Small graphite in cast irons. Small grains in steels. |
| VIII | $10^{-3}$ | Zone of affection of concentrated energy sources (laser). Particle beams. Mechanical activation by processing. | Thickness of martensite plate. Globule of quartz sand. Particles of ground peat. |
| IX | $10^{-4}$ | Zone near the crack tip. | Big liberations in duraluminum. |
| X | $10^{-5}$ | Zone at irradiation. | Small liberations in duraluminum, carbides in steel. |
| XI | $10^{-6}$ | Defects in crystals. Nanotubes and nanoparticles. | Atomic lattice deformations. |

of blocks by dimensions in a broad range of scales—from immense geoblocks down to micron particles obtained when crushing rock by mills—an unexpected property of dimensional distribution was obtained. There exist *preferential* dimensions, which form a geometrical progression with the index:

$$K = \frac{\mathfrak{length}_{i+1}}{\mathfrak{length}_i}. \tag{1.4.2}$$

Within a broad range of scales (15 orders!), $K = 2\div 5$ with an average value of $K = 3,5$. The hierarchical structure of geological massifs is extremely important at earthquakes. The total share of energy emitted in the form of seismic waves makes about 5% of the overall energy connected with the block deformation. The remaining 95% of energy is transformed between various structural levels (rock degradation, friction, lift of mountain masses, plastic deformation).

The real physical existence of structural hierarchy in a rock is theoretically described by means of the geometrical invariant $\mu_\Delta(\delta)$ (1.4.1). The availability of the geometrical invariant causes a range of nontrivial consequences.

The stability criterion of the defective continuum in rock mechanics is closely connected with the existence of block structure. According to [390], stability is defined by the ratio

$$Q = \frac{L}{l},$$

where $l$ is the linear dimension of the defect, $L$ is the distance between defects. For $Q < 3$ the system is stable, in case of $Q \geq 3 \div 5$, an energetic interaction of defects arises (crack growth), which results in fracture or formation of a defective structure of a higher hierarchical level.

It is worth noting here that permanent oscillation activity of mountain rocks is also closely related with the block structure. The range of oscillations spreads from acoustic and ultra-acoustic noises in the Earth's crust to oscillations of the hertz range and long-period own oscillations of the Earth. This difference in frequencies is directly connected with different modes of own oscillations of blocks of different sizes.

### *1.4.1 Hierarchical pattern of fracture process: applications of general systems theory*

Upon recognition of the reality of existence of a definite hierarchy in fracture processes, intensive studies were started of hierarchical structures in fracture mechanics, and a new scientific discipline, or more correctly, a concept—physical mesomechanics has formed. This concept unites different, sometimes non-traditional approaches to explain a complicated complex of issues connected with a real process of deformation and fracture. Summing up of the results obtained within this concept was made in the collective monograph edited by V. E. Panin [278]. After *'putting into practice'* of the idea of joint action of mechanisms from different levels for correct explanation of the plasticity mechanism, the concept of mesomechanics has acquired an essential popularity due to its heuristic productivity. At the microlevel, these are the issues of quantum calculations of connection strength and definition of mechanical characteristics of materials *ab initio* [55, 56, 219]. In this case, to describe hierarchical process $\rho$, it is necessary to consider energy dissipation of oscillations of molecular and atomic systems on elementary states of higher levels (for example, grains), or on elementary states of a more complex structure (for example, phonons, ensembles of defects). A relative complexity of elementary states may be assessed by the number of translation acts required to transit from one state into another. A relative complexity of an elementary state also enables defining a relative ranking of states. It is necessary to account for interaction with a higher level, since the elementary state of a higher level $\Omega$ is realized by means of the unit state $\omega$, generated by a set of universal hierarchical variables dependent level. Thus, in computer simulation of dislocation formation [358] it has been shown that an essential role is played by different scale bars. Whereby, shifts of unitary atoms at formation of dislocation core are causing a long-range elastic interaction, which forms a dislocation aggregate (an object of mesoscopic level).

Unification of models of various scale levels (characteristic range of $10^{-10}$–$10^{-6}$m) for explanation of the mechanism of elementary plasticity of crystals and motion of dislocation loops is offered in [41]. Practically in the same scale range (from $10^{-9}$m, atomic dimensions, up to $10^{-4}$m, grain dimensions), the crack development was studied [158].

Let us consider a possibility of changing the state of a hierarchical system [215]. A change of a system's state corresponds to a shift (displacement) from a cell to a cell under the action of hierarchical $\rho$ (process). For example, if the system is in state 3, $\lambda = 5$, the action of operator $\rho^+$ results in a transition into state 2 (see Fig. 1.3):

$$ {}^{\lambda}\rho^{+} :\mapsto \omega^{\lambda} = {}^{5}\rho^{+} :\mapsto \omega^{5} = \omega^{6}, \tag{1.4.3} $$

where $:\mapsto$ means the action of the hierarchical operator. To transfer the system from state 5 into state 2, we can suppose the existence of several trajectories in the hierarchical space, for example:

$$ {}^{\lambda}_{[5]}\rho^{+} :\mapsto {}^{\lambda}_{[4]}\rho^{+} :\mapsto {}^{\lambda}_{[4]}\rho^{\times} :\mapsto \omega^{\lambda}_{[5]} = \omega^{[6]}, \tag{1.4.4} $$

and a different trajectory:

$$ {}^{\lambda}_{[5]}\rho^{\times} :\mapsto {}^{\lambda}_{[6]}\rho^{+} :\mapsto {}^{\lambda}_{[3]}\rho^{+} :\mapsto \omega^{\lambda}_{[5]} = \omega^{[6]}. \tag{1.4.5} $$

Here, indices in square brackets indicate the chosen starting point. The identity of trajectories (1.4.4) and (1.4.5) is not obvious[5]. Moreover, the requirement of identity is a rather serious condition, not always realizable in reality.

The general description of the motion process in the hierarchical space with account of all universal hierarchical variables [263] at realization at particular processes requires a respective presentation of the hierarchical space. This is the main difficulty of consistent application of the MHST.

Sometimes, it becomes possible to build motion equations in the hierarchical space without using any additional information about the structure and topology of the space. Thus, equations (1.4.4), (1.4.5) have been derived making an assumption about absence of control (that is, the effect of the control level on the dependent one). In reality, this effect usually exists (solid arrows in Fig. 1.3). For structurally non-uniform bodies (for example, a medium with microstructure), the physical control mechanism is the energy relaxation through various mechanisms [134], while the control parameter may be the Grunaizen parameter for translational and rotational degrees of freedom. Whereby, as it should be, the number of degrees of freedom with the control is much higher than with the dependent levels (due to a possibility to realize a larger set of mechanisms of energy relaxation). A signal transfer from the dependent levels to the coordinator (dashed lines in Fig. 1.3) is interpreted, from the viewpoint of the systems theory, as a desire of the

[5] In the simplest case, a mismatch of trajectories corresponds to different results when changing the order of deformation.

system to obtain a mathematical expression of control and, by means of that, the status of a coordinator. In deformation and fracture processes, it corresponds to the existence of feedback in the synergetic system (sample). In deformation, a weak nonlinear feedback regulates engagement of relaxation modes of this or that order, and stabilizes unstable modes [134].

To obtain motion equations (of a system state), it may be sufficient to have an interaction law of the systems of different levels. Thus, work [328] shows the existence of three types of behavior (differing in their attitude to self-organization—stable, mode with self-organization, and chaotic mode) for a hierarchical system with a probable appearance of a defect in any hierarchical levels. The probability $P$ of defect appearance was taken proportional to the level number:

$$P(\lambda) = P_0 c^{\lambda},$$

where $0 \leq P_0 < 1$, $0 < c = \text{const} < 1$.

### 1.4.2 *Crack fractality, hierarchy and behavior*

One of the basic distinctive characteristics of a fractal structure is its self-similarity or self-repetition on different space scales [196]. This fact allows use of the fractal language for description of the fracture generality in the macroscopic level [4, 15, 16, 135, 319]. From the hierarchical point of view, it is necessary to consider the fractal objects as a *special limited case* of hierarchical structure, since only space scales in this object are self-similar [215]. Taking into account the properties of hierarchical objects, all universal parameters (except for the elementary state) are degenerated. In this case, as follows from (1.1.3)

$$\Omega^{\lambda} \overset{\gamma}{\underset{\rho}{\leftrightarrow}} \left\{ {}^{\beta}_{\omega} \overset{\gamma}{\underset{\rho}{\Omega}} {}^{\lambda}_{\sigma} \right\}^{\lambda \overset{\Omega^{\lambda}}{\underset{\rho}{\to}} \beta} \Rightarrow \left\{\Omega^{\lambda}\right\}^{\lambda \overset{\Omega^{\lambda}}{\underset{\rho}{\to}} \beta}. \qquad (1.4.6)$$

Let the elementary state be the dimension of the length $\mathfrak{lenght}$, $\lambda$ be the level, and $\beta$ be the index of level on the considered hierarchical scale (for example, $i$); and the originating level $\beta$ be the index of the level transition as a result of action of the process (for example, $j$). Then, expression (1.4.6) can be rewritten as

$$\left\{\Omega^{\lambda}\right\}^{\lambda \overset{\Omega^{\lambda}}{\underset{\rho}{\to}} \beta} = \{\mathfrak{lenght}\}^{\lambda} \Rightarrow \mathfrak{lenght}^{i \overset{\mathfrak{lenght}^{i}}{\underset{\rho}{\to}} j}. \qquad (1.4.7)$$

In accordance with the choice of the corresponding process (universal parameter $\rho$), we can obtain different types of the transformation law of the elementary state. Thus, not only the geometrical dimensions, but informational interaction between the elementary components (structures) of material can be considered as the defining parameters for modern materials.

### *1.4.2.1 Hierarchical representation of structural deformation level*

**The law of structural deformation levels.** According to Panin and co-workers [278, 279], the law of structural deformation levels for the description of a sequence of structural changes in a deformed solid can be defined as

$$\sum_{i=1}^{N} \operatorname{rot} J_i = 0, \tag{1.4.8}$$

where $i$ is the number of the structural deformation level, $N$ is the total number of structural levels, and $J$ is the angular momentum of the defects flux.

This equation shows that **the sum of rotors of all fluxes of deformation defects into the solid in the hierarchy of $N$ structural levels should be equal to zero**. Taking into account the explicit expression for the fluxes of deformation defects, the formula (1.4.8) for the $\mu$-component can be written down [278] as

$$\sum_{i=1}^{N} \left\{ \frac{\partial R^a_{\mu i}}{\partial t} - \frac{V_i}{C_t^2} \frac{\partial S_i^{a\mu}}{\partial t} - V_i f_i^{abc} \left[ A_i^b \times R_i^c \right]^{\mu} \right.$$
$$\left. - f_i^{abc} \left[ A_i^b \times S_i^c \right]^{\mu} - \frac{V_i}{l_i^{2D_{fi}}} J_i^{a\mu} \right\} = 0, \tag{1.4.9}$$

where $V$ is the velocity of accommodation fluxes of defects in the corresponding level; $R^a$ is the gradient of the component of the tensor of bending torsion; $C_t$ is the limit propagation velocity of gauge field in the media; $S^a$ is the variation of the gradient of component of distortion tensor; $f_i^{abc}$ are the structural constants of Lie-algebra of gauge field; $A^b$ is the gradient of component of distortion tensor reflecting the gauge field; $S^c$, $R^c$ are the components of tensor of gauge field strength; $l$ is the dimension parameter of structural level; $D$ is the fractal dimension; and $J^a$ are the fluxes, caused by modification of the frame in the space.

By analogy with (1.4.7), the length is chosen as an elementary state. Taking account of movement of the system within the range of one level, the law of transformation (1.4.7) can be written down as a system of equations under length and action:

$$\{\mathrm{A}\}^{\lambda \underset{\rho^+}{\overset{\Omega^\lambda}{\rightarrow}} \beta} = \{\mathrm{A}\}^{\lambda \underset{\rho^+}{\overset{\Omega^\lambda}{\rightarrow}} \lambda} = \{\mathrm{J}\}^{\lambda \underset{\rho^+}{\overset{\mathfrak{lenght}^\lambda}{\rightarrow}} \lambda};$$
$$\{\mathrm{A}\}^{\lambda \underset{\rho^\times}{\overset{\Omega^\lambda}{\rightarrow}} \beta} = \{\mathrm{A}\}^{\lambda \underset{\rho^\times}{\overset{\Omega^\lambda}{\rightarrow}} \lambda} = \{\mathrm{J}\}^{\lambda \underset{\rho^\times}{\overset{\mathfrak{lenght}^\lambda}{\rightarrow}} \lambda};$$
$$\{\Omega\}^{\lambda \underset{\rho}{\overset{\Omega^\lambda}{\rightarrow}} \beta} = \{\Omega\}^{\lambda \underset{\rho^\times}{\overset{\Omega^\lambda}{\rightarrow}} \lambda} = \{\mathfrak{lenght}\}^{\lambda \underset{\rho^\times}{\overset{\mathfrak{lenght}^\lambda}{\rightarrow}} \lambda}. \tag{1.4.10}$$

At the same time, it is necessary to note that the law has a specific contradiction (1.4.8). In the general description of the structure of deformation levels [278], a possibility of existence of interacting substructures in the given structural level is admitted. But in expression (1.4.8), a possibility of action of operators $\rho^\times$ is not admitted. In the general case, as is noted in §1.1.1, the movement of the system within the scope of one structural level is described by means of $\rho^{\times}$ operator, and the transfer between the levels by means of $\rho^{+}$. Therefore the law (1.4.8) should be written down as follows

$$\sum_{k=1}^{M}\sum_{i=1}^{N} \operatorname{rot} J_{ki} = 0, \tag{1.4.11}$$

where the order of summation is important, $k$ is the level number, and $i$ is the subsystem number in the level.

*1.4.2.2 Local and global fracture parameters*

A correct transition from the microscopic description to the macroscopic one is a complicated and not uniquely defined problem. On the one hand, it is necessary to take into account the properties of elementary structural units (for example, the existence of the grain boundaries in steel). On the other hand, it is important to obtain only integral (averaged) features of the medium. Besides, according to the macroscopic approach, it is necessary to have some parameter connected with the microscopic scale. As Nowozhilov has claimed [261, 262], some hidden parameter of the length dimension is present in fracture mechanics. Thus, Shaniavsky [319] showed that under the consideration of the crack growth on the basis of the Sih criterion, it is necessary to take into account not only the normal opening of the crack, but also the lengthwise and transversal displacement. This fact demonstrates the *local* character of predicting the crack trajectory. Moreover, it is well known that to describe the *global* crack trajectory, it is necessary to take into account the scale effects. In this context, Ivanov has noted [133] that the fracture process depends not only on the typical cracks length $A = 2\varepsilon$, but also on the size of the considered objects. This does not contradict the ASTM agreement [38]. According to ASTM the linear fracture mechanics take place in the case of

$$\varepsilon > 2.5(K_{Ic}/\sigma_y)^2,$$

where $K_{Ic}$ is the fracture toughness, and $\sigma_y$ is the yield stress. For a mild steel ($\sigma_y \approx 300\,\text{MPa}$, $K_{Ic} \approx 50\,\text{MPa} \cdot \text{m}^{1/2}$), $\varepsilon \approx 0.07\,\text{m}$; for the high-strength steel ($\sigma_y \approx 1200\,\text{MPa}$), $\varepsilon \approx 0.003\,\text{m}$. It means that modern micromechanical devices cannot be evaluated in linear approximation and difference between the linear fracture mechanics, and more detail description obtain 50% [341]. Furthermore, for small size objects ($1 \times 10^{-9} < L < 1 \times 10^{-8}$ m, nanoobjects) we need to use not classical elasticity but nonlocal continua models [286].

For short cracks, the criterion of the linear fracture mechanics should be modernized [261, 262]. This is connected with the outcome of linear theory, according to which the solid must fail, when it has a cut under any amount of small limit stress $\sigma$. Taking into account the fact that around the crack tip the stress and the gradient are large, the criterion of the fracture should be written down in integral form. This leads to disappearance of the singularity near the crack tip and to the appearance of dimensional parameter $a$, which is the critical opening of a crack. This coincides with the results of Ståhle [341]. According to [341], the length of the zone of weakened connections $\varepsilon - \varepsilon_0$ in front of the tip of the incipient crack reaches the maximum

$$\varepsilon - \varepsilon_0 = \pi E u^* / [2(1 - \nu^2)\sigma_D],$$

where $E$ is the Young modulus, $u^*$ is the critical opening of the crack, $\nu$ is the Poisson ratio, and $\sigma_D$ is the maximum stress in the zone of the beginning of the crack.

When this length is reached, the zone of weakened bonds collapses and the continuum loses stability [261, 341]. It means that a crack is formed and the crack tip begins its movement. Namely, in the area of the loss of stability by the continuum, the interaction of the beginning of the microcrack with the defect media structure is important, since this structure has some inner energy and can be either the source of redundant energy or the accumulator of elementary excitation energy of lattice. The maximum length of the zone of weakened bonds corresponds to critical opening of the initial crack $u^*$. Application of the linear fracture mechanics can take place only for cracks where the length (opening) exceeds this value. According to comparison of the precise model and the ASTM standard, the maximum of results divergence is around 5%.

The multiple factors of the process and the hierarchical character of the mechanism involved in the fracture realization lead to the fact that the front of the crack and the surface of the fracture have complicated character, which is so far from the classical Barenblatt–Dugdale model. Thus, in [346] the fracture as a result of application of tearing strain and the velocity of the crack growth and its profile are investigated. It is obtained that the profile of the crack begins sharply changing under a definite critical shift $u_{cr}$. Besides, the appearance of the state of chaos, when the crack is moved, was considered. Moreover, it was shown that if $\Delta = u/u_{cr}$ is small, the randomness is greater.

The profile distortion can be interpreted as nonstability of the transition process from the condition of the microdamage accumulation to the condition of the macrocrack propagation. The value of the crack propagation velocity has been obtained as 1.05–1.1 from the velocity of sound. The instability of the crack profile depends on the point of application of the tearing stress [267]. By means of the disturbance theory methods, the existence of the sector inside which the applied symmetric point forces cause the nonstability of crack trajectory was shown. From another side, in [246] it was illustrated that instability of crack propagation takes place as the lower bound of the velocity range so in the upper bound, under $V \approx 0.4V_R$, where $V_R$ is the Rayleigh wave speed in the material. This instability is connected with two points of bifurcation on the curve of the dependence of macroscopic tensor of microdefects density on the applied stress. These bifurcation points are defined by means of the similarity parameter, which is the function of two scales: the typical size of defect nuclei and the average distance between them. Drastic ordering of the ensemble of mesodefects under the process of transition towards the bifurcation points can be caused by means of turning on the next cascade of the energy transfer between the scale levels [107]. According to §1.1.1, a possibility of existence of that mechanism of the crack behaviour is connected with the fact that the crack forms the fractal surface of fracture, but the energy can be considered as a control, which is transmitted between the structural levels [213].

By the fracture investigation on the lower hierarchical level as the evolution of the defect ensemble, we require considering statistic singularity of the evolution [61, 246].

In addition, namely the energy fluctuation, which is caused under the process of the defect ensemble development, is one of the factors, defining the fracture development. The solution of the kinetic equation for the tensor of microdefects density has a wave character of different type depending on the stress area where the system is located.

By the crack study on the higher hierarchical level, the instability of the front growth can be investigated in the wave approximation [240, 241]. In this case, the wave equation, the parameters of which are defined from the elasticity and plasticity theory, is put in correspondence to the crack growth [293]. And as a result, the evolution of the crack surface does not depend on the crack front development [292] and it is realized in the self-similar way.

### *1.4.3 Self-organization processes at plastic deformation and fracture*

A macroscopic manifestation of self-organization in the deformation process is localization of deformation, which finalizes the stage of the uniform plastic flow [14]. Prior to start of localization in the shift bands, break neck and other macroscopic manifestations, at least two deformation modes are realized. Whereby, changes of the internal structure take place in various scale levels—from microns up to the sample dimensions [320–322]. The description of the complete deformation process in unified variables is rather complicated because of the multilevel character. From the general theory of hierarchical systems, it is possible in principle, provided we know the law and the state of the control system of the upper level [29, 258, 259]. However, the complete knowledge of the law from the viewpoint of the dependent system is impossible (the system always has some hierarchical haziness). This is confirmed by the fact that for a fracture crack, the microscopic $10^{-8}$m and mesoscopic $0,05 \div 5 \times 10^{-6}$ m structural levels are subject to deterministic analysis, while in the macroscopic $> 5 \times 10^{-6}$ m level uncertainties arise related to stochastization and appearance of determinate chaos. The way out of the situation can be an introduction of additional parameters, which characterize, for example, a collective change of the structure at all the deformation levels.

Such parameters may have a kinetic character [14]. Since mobile defects are physical carriers of plastic deformation (for example, dislocations and point defects), the appearance of the structure at the microscopic level should be viewed as a collective movement of the dislocation ensemble. The equations of dislocation dynamics allow one, in principle, to describe the basic regulations of self-organization processes [119].

### Certain outcomes

- The standard way to solve the problems of the mechanics of deformed solids, when the mathematically possible behaviour of the solution 'is limited' by imposition of starting and boundary conditions, from the viewpoint of the system theory, is a direct consequence of control from the upper hierarchical level.
- Evolution of any dynamical system (of a fractured solid, in particular) can be described in terms of phase trajectories. The basic problem lies in correct choice of the basis

of the states of the phase space and definition of mathematical properties of this state space (for example, the notions of norm, metric, etc.). The hierarchical space reflects, by virtue of its generality, only the most essential regularities of the fracture process. In each particular case, the variables of the hierarchical space require specification and detailing.

- The concept of universal 'elementary volume' for fracture is groundless. According to the MHS theory, it is possible to speak only about the 'elementary volume' in a given hierarchical level.

# 2 Space Geometry Fundamentals

## Contents



Practical experience indicates that a 3D vector space is a geometrical model of a continuous medium (a solid). The choice of this model is defined by the fact that under classical (Newtonian) mechanics points (material particles) of the continuum are *individualized*, i.e., different from each other, and each particle relates to a radius-vector $\boldsymbol{r}$, which corresponds to a certain axis set within a certain frame of reference. The material deformation process is accompanied by changing the mutual location of medium points. One of the fundamental kinematical assumptions is the hypothesis of continuity. It requires a continuous mapping of the initial state $\xi^i$ into the current state $x^i = x^i(\xi, t)$. In this case, a Jacobian of the coordinate transformation has the following form:

$$\frac{\partial x^i}{\partial \xi^i} \equiv \partial_{\xi^i} x^i \neq 0. \tag{2.0.1}$$

The mapping continuity condition is an additional postulate, the correctness of which demands an additional study in particular problems. It was shown by Sedov [316] that under a seemingly natural choice of the initial state, in which the structure of each continuum element is organized and no forces affect the element, the metrics of the resulting space may be non-Euclidean, and there may exist no transition between the initial state and the current state. L. I. Sedov called this ideal state to be 'initial' as different from the real initial state.

This is due to the fact that the conditions imposed by the equation (2.0.1) are not sufficient to define the geometric and thermodynamic structure of the space, associated with deformation and destruction. If we take the notions of point and radius-vector (vector) to be known and non-definable, we must set definite additional intuitive relations between these values and additional properties of the material. Thus, the deformation thermodynamics moves an additional hypothesis of equal presence [113]: ***all the defined fields are taken dependent from one and the same set of parameters***. Besides, consideration of the energy balance at displacement of a rigid body indicates that its internal energy undergoes no changes in the process, which makes the essence of the *material indifference hypothesis*.

As consequence of formulating these intuitive relations and various limitations imposed thereon, we arrive to the presentation of *a space with a different geometrical structure*. Let us briefly consider certain properties of possible spaces to be used by us below.

## 2.1 Construction principles of various space types

### *2.1.1 Affine space*

Most general intuitive relations include [294] four basic axioms.

**Definition 2.1.1.** At least one point exists.

According to this definition, we can always fix the point of origin.

**Definition 2.1.2.** Each pair of points $A$, $B$ defined in a certain order corresponds to one and single vector.

The vector may be denoted as $\overrightarrow{AB}$ or by a separate bold letter. For example, **a**.

**Definition 2.1.3.** For each point $A$ and each vector $\mathbf{x}$ exists one and single point $B$ such as:

$$\overrightarrow{AB} = \mathbf{x}.$$

This vector definition corresponds to a presentation about the vector not as about an ordered segment, but as about a parallel shift, which corresponds to mechanical presentation of the motion.

**Definition 2.1.4.** *Parallelogram axiom.* If

$$\overrightarrow{AB} = \overrightarrow{CD}, \Rightarrow \overrightarrow{AC} = \overrightarrow{BD}.$$

The vivid sense of axiom 2.1.4 lies in the fact that under equality and parallelism of one pair of opposite sides of a tetragon, the same is valid for the other pair.

An additional group includes the axioms related to vector-by-number multiplication. When we imply real numbers, we will get real affine spaces, and if we imply complex numbers we will get complex affine spaces. Of principle here is the fact that an affine space may be built over a certain algebraic field, a vector field or an operator field.

**Definition 2.1.5.** Each vector $\mathbf{x}$ and each number $\alpha$ has been assigned a definite vector. This vector is designated $\alpha\mathbf{x}$ and is called a vector-by-number product.

**Definition 2.1.6.**

$$1\mathbf{x} = \mathbf{x}.$$

The main axiom sense implies that by various products of vectors by numbers we may exhaust all the vectors of the given space.

**Definition 2.1.7.**

$$(\alpha + \beta)\mathbf{x} = \alpha\mathbf{x} + \beta\mathbf{x}.$$

**Definition 2.1.8.**

$$\alpha(\mathbf{x} + \mathbf{y}) = \alpha\mathbf{x} + \alpha\mathbf{y}.$$

The axioms 2.1.7 and 2.1.8 express distributive laws: one to multiply a vector by a sum of numbers, the other to multiply a sum of vectors by a number.

**Definition 2.1.9.**

$$\alpha(\beta\mathbf{x}) = (\alpha\beta)\mathbf{x}.$$

**Definition 2.1.10.** *Dimension axiom.* There exist $n$ linearly independent vectors, but any $n + 1$ vectors are mutually dependent.

**Definition 2.1.11.** *Space definition.* We shall call an $n$-dimension affine space to be a set of points and vectors, which satisfy the axioms 2.1.1–2.1.10.

We can demonstrate [315] that the thus introduced space is generating a coordinate transformation, which is a definite linear *affine group* $G_a$. In this case the following takes place.

**Definition 2.1.12.** *Geometry definition.* We shall call the affine geometry to be the theory of all invariant properties of figures in the affine space, relative to the affine group $G_a$.

### *2.1.2 Vectors, covectors, and 1-forms and tensors*

On the basis of introduced vectors, we can present objects of a different mathematical origin. Let us consider vector space[1] $V$. Different operators may be considered in this space. Of most interest is the vector, the effect of which on any vector out of $V$ produces a real number. Let $\boldsymbol{\omega}$ be the operator:

$$\boldsymbol{\omega} : V \mapsto \mathbb{R}; \quad \mathbf{a} \mapsto \boldsymbol{\omega} \cdot \mathbf{a}. \tag{2.1.1}$$

Operator $\omega$ is called linear if:

$$\boldsymbol{\omega}(k\mathbf{a} + \mathbf{b}) = k(\boldsymbol{\omega} \cdot \mathbf{a}) + \boldsymbol{\omega} \cdot \mathbf{b}. \tag{2.1.2}$$

---

[1] Definition of this space is provided in Appendix A.

We shall call such linear operators *covectors*. Relation between vectors and covectors is symmetrical. It means that we may view a vector as an operator in the space of covectors. A covector has a transparent geometrical sense [43].

Let us consider the set of all the vectors $\mathfrak{v} \in V$, such that:

$$\boldsymbol{\omega} \cdot \mathfrak{v} = 1. \tag{2.1.3}$$

Then, this set (Gothic vectors) represents covector $\boldsymbol{\omega}$. In the 2D vector space, set $\mathfrak{v}$ represents a straight line. Let us take two different vectors $\mathfrak{a}$, $\mathfrak{b}$, such as:

$$\boldsymbol{\omega} \cdot \mathfrak{a} = 1,$$

$$\boldsymbol{\omega} \cdot \mathfrak{b} = 1.$$

From the property of linearity (2.1.2):

$$\boldsymbol{\omega} \cdot (\mathfrak{a} - \mathfrak{b}) = 0, \tag{2.1.4}$$

therefore, for every $k$ we have:

$$\boldsymbol{\omega} \cdot [k(\mathfrak{a} - \mathfrak{b}) + \mathfrak{a}] = 0. \tag{2.1.5}$$

Thus, if vectors $\mathfrak{a}$ and $\mathfrak{b}$ belong to set $\omega$, the whole straight line defined by the vectors belongs to this set. The complete set of lines of the operator may be defined from the condition:

$$\boldsymbol{\omega} \cdot \mathfrak{v} = \ldots, -1, 0, 1, 2, \ldots \tag{2.1.6}$$

It is clear that the line set of a level is a set of lines, which are parallel to $\mathfrak{v}$ and equidistant from each other.

Thus, we obtain fibration of the covector space.

**Exterior forms.** One of the simplest implementations of the covector notion is the notion of exterior 1-form [9, 84].

**Definition 2.1.13.** The form of degree 1 (or shortly, 1-form) is a linear function of vector, $\omega$: $\mathbb{R}^n \mapsto \mathbb{R}$,

$$\omega(\lambda_1\vec{\xi}_1 + \lambda_2\vec{\xi}_2) = \lambda_1\omega(\vec{\xi}_1) + \lambda_2\omega(\vec{\xi}_2) \quad \forall\, \lambda_1, \lambda_2 \in \mathbb{R},\ \vec{\xi}_1, \vec{\xi}_2 \in \mathbb{R}^n,$$

where $\vec{\xi}_1, \vec{\xi}_2$ are derivative vectors[2].

[2] It is essential that the Euclidean structure is not specifically fixed on $\mathbb{R}$. It allows working with the form in non-Euclidean spaces without any modifications.

**Tensors.** An extension of definition (2.1.1) to linear vector spaces of arbitrary dimensionality brings us to the notion of tensors. A general type tensor is a random linear operator, which makes a mapping of the type $V \mapsto V^*$, $V \times V\mathbb{R}$, $V^* \mapsto V$, $V^* \times V^*\mathbb{R}$, $V \mapsto V$, $V^* \times V^*$, and also similar higher order operators. Here, $*$ defines a dual conjugation of spaces.

Below, we give another constructive definition of tensors to be used in the forthcoming considerations.

### *2.1.3 Euclidean space*

In the considered affine space, an important notion, essential from the mechanics viewpoint, has not been defined, namely, the distance between the points. To introduce the notion of distance, we need to define *a scalar, or interior product of vectors*. A scalar product transfers the affine space into metric affine space and induces all other metric space properties.

Let us set in $n$-dimensional affine space a certain bilinear scalar function $\varphi(x, y)$ of two vector arguments $\mathbf{x}$, $\mathbf{y}$. Function $\varphi$ meets the condition of nondegeneracy and symmetry:

$$\varphi(\mathbf{x}, \mathbf{y}) \neq 0, \tag{2.1.7}$$

$$\varphi(\mathbf{x}, \mathbf{y}) = \varphi(\mathbf{y}, \mathbf{x}). \tag{2.1.8}$$

**Definition 2.1.14.** *Definition of Euclidean space.* We shall define the $n$-dimensional Euclidean space as the $n$-dimensional affine space, in which a fixed bilinear form has been set, meeting the conditions of (2.1.7) and (2.1.8).

Function $\varphi(\mathbf{x}, \mathbf{y})$ will be understood as *a scalar product*. For simplicity, we define it as $\mathbf{xy}$. A scalar vector square $\mathbf{x}$ is defined by formula:

$$\mathbf{x}^2 = \mathbf{xx}. \tag{2.1.9}$$

We define the *length* of vector $\mathbf{x}$ as $\sqrt{\mathbf{x}^2}$ and define it as $|\mathbf{x}|$:

$$|\mathbf{x}| = \sqrt{\mathbf{x}^2}. \tag{2.1.10}$$

We define the *distance* between two points $A$, $B$ as the length of vector $\overrightarrow{AB} = \mathbf{x}$:

$$AB = \sqrt{\mathbf{x}^2}. \tag{2.1.11}$$

Euclidean spaces are segregated into two large classes: *real* and *complex spaces*—depending on the type of affine space used initially. For our further purposes we shall use, as a rule, the real space $\mathbb{R}^n$, the index $n$ defining the space dimension.

*Real Euclidean spaces*, in their turn, are separated into two classes: *proper Euclidean spaces*, in which for any $\mathbf{x} \neq 0$

$$AB = \sqrt{\mathbf{x}^2}, \tag{2.1.12}$$

and *pseudo-Euclidean spaces*, in which a vector square may be positive, zero and negative. This is equivalent to the fact that the vector length $|\mathbf{x}|$ may be real, imaginary and zero.

Assigning a bilinear scalar function is equivalent to assigning a twice covariant tensor $\varphi_{ij}$ of its components:

$$\varphi_{ij} = \varphi(\mathbf{e}_i, \mathbf{e}_j), \quad \varphi(x, y) = \varphi_{ij}\mathbf{x}^i\mathbf{x}^j.$$

In a special case of a scalar product, an individual name and definition are introduced for this tensor. Hereafter, we define this coefficient tensor as $g_{ij}$ and call *a metric tensor* of the Euclidean space. According to definition [294], we have:

$$g_{ij} = \mathbf{e}_i\mathbf{e}_j, \tag{2.1.13}$$

$$\mathbf{xy} = g_{ij} = g_{ij}\mathbf{x}^i\mathbf{y}^j. \tag{2.1.14}$$

The symmetry condition (2.1.8) is equivalent to symmetry of coefficient tensor (metric tensor)

$$g_{ij} = g_{ji}. \tag{2.1.15}$$

An essential fact here is that metric tensor (2.1.13) may be presented as an indicatrix[3] in the Euclidean space. An important peculiarity of the Euclidean space is the fact that any principal differences between covariant and contravariant values disappear in it, and any covariant value may be 'transformed' into a contravariant one and vice versa with the help of the respective metric tensor:

$$x_i = g_{ij}x^j, \quad x^i = g^{ij}x_j, \tag{2.1.16}$$

where value $g^{ij}$ is the inverse value of the metric tensor:

$$g^{ij}g_{jk} = \delta^i_k.$$

**Cartesian product of vector spaces.** The notion of Cartesian (direct) product is a certain generalization of a scalar product. Let the $E$ and $V$ be two vector spaces. Then, the following is correct.

**Definition 2.1.15.** *Definition of a Cartesian product.* We define a Cartesian product $E \times V$ to be a vector space, where elements are vector pairs, one vector originating from $E$, the other from $V$.

For this product we can define operations of multiplication by a number and addition as:

$$k(e, f) = (ke, kf), \quad (e_1, f_1) + (e_2, f_2) = (e_1 + e_2, f_1 + f_2),$$

where $e \in E,\ f \in F$ and $k \in \mathbb{R}$.

Thus, for instance, a plane (a set of all pairs of real numbers) is $\mathbb{R}^2 = \mathbb{R} \times \mathbb{R}$.

[3] Indicatrix is a sphere of a unit radius.

### *2.1.4 Generalization: affine connectivity spaces and Riemann space*

The introduced notions are, as a rule, sufficient for description of microscopic motions of a homogeneous isotropic ideal rigid body in the nonrelativistic case. However, studying more delicate effects related to deformation of nonhomogeneous rigid bodies and effects of self-organization requires introduction of additional geometrical objects, which are richer in their structure and properties.

*Curvilinear coordinates in affine space*

According to definition (2.1.12), a transition from one affine coordinate system to another is achieved by a linear transformation of coordinates:

$$x^{i'} = A_i^{i'} x^i + A^{i'}, \tag{2.1.17}$$

where the transformation $A^i$ coefficients are randomly chosen with the sole condition following from condition (2.0.1):

$$\mathrm{Det}\,|A|_i^{i'} \neq 0.$$

It is thought that the simplest generalization is a possibility to choose a coordinate transformation law in a more general way than the one set up by the equation (2.1.17).

**Definition 2.1.16.** Let in $n$-dimensional connected domain of the affine space $\Omega$ we have set $n$ continuously differentiable unambiguity functions of affine coordinates $f_k(x^1, \ldots, x^n), k = 1, \ldots, n$. Let us introduce new variables $x^{1'}, \ldots, x^{n'}$ by coupling equations:

$$x^{i'} = f_i(x^1, \ldots, x^n), \quad i = 1, \ldots, n. \tag{2.1.18}$$

The variation domain of new variables is $\Omega'$. We impose the requirement of inversibility on the functions defined by equation (2.1.18), i.e., a possibility to express from (2.1.18) coordinates $x^i$ as continuously differentiable functions from $x^{i'}$:

$$x^i = g_i(x^{1'}, \ldots, x^{n'}), \quad i = 1, \ldots, n. \tag{2.1.19}$$

In this case, we shall name the set of values $x^{i'}$ to be curvilinear coordinates in domain $\Omega$ of the affine space.

An important property follows from our definition: Jacobians of direct and inverse transformation are different from zero and mutually inverse in the whole domain of their existence.

In the mechanics of a deformable solid body, the values are practically always set or exist in every point of domain $\Omega$. Then we say that value *field* in $\Omega$ has been set.

*Parallel displacements. Connectivity object*

Local coordinate systems within the affine or Euclidean geometry are independent. Therefore, the values in the neighboring points are in principle not connected with each other. However, from the general geometry characteristics, it is possible to define a value displacement operation from the local affine space in point $\xi^i$ into the local affine space in point $\xi^i + d\xi^i$. In fact, we need to build a random curve passing through points $\xi^i$ and $\xi^i + d\xi^i$. In each point of the curve we shall lay off constant vector $\boldsymbol{\xi}_0$. The coordinates of this constant vector change due to changes from point to point of the reference field (they depend on the natural parameter of curve $s$): $\boldsymbol{\xi}^i = \boldsymbol{\xi}^i(s)$. Our task is to determine the law according to which the $\boldsymbol{\xi}^i(s)$ vary at least in the infinitely small travel section.

Since functions $x^i(s)$ are continuously differentiable, the change of vector $\boldsymbol{\xi}(t)$ is continuous along the displacement travel. According to the general rule of vector decomposition by basis set, we may write down:

$$\boldsymbol{\xi}_0 = \xi^i(s)\mathbf{x}_i(x^1, \ldots, x^n). \tag{2.1.20}$$

**Remark 2.1.1.** It is necessary to take into account that arguments $x^1, \ldots, x^n$ themselves depend on the natural parameter.

By differentiating (2.1.20) with account that $\boldsymbol{\xi}_0 = \text{const}$, we have:

$$0 = d\xi^i\mathbf{x}_i + \xi^i d\mathbf{x}_i. \tag{2.1.21}$$

According to the usual formula for total differential

$$d\mathbf{x}_i(x^1, \ldots, x^n) = \mathbf{x}_{ij}dx^j, \tag{2.1.22}$$

where

$$\mathbf{x}_{ij} = \frac{\partial^2\mathbf{x}(x^1, \ldots, x^n)}{\partial x^i \partial x^j}. \tag{2.1.23}$$

Since $\mathbf{x}_{ij}$ is a vector, it may be resolved into vectors of the local reference field $\mathbf{x}_i$ in an arbitrary point:

$$\mathbf{x}_{ij} = \Gamma_{ij}^k(x^1, \ldots, x^n)\mathbf{x}_k. \tag{2.1.24}$$

**Definition 2.1.17.** *Definition of connectivity parameters.* We shall call the values $\Gamma_{ij}^k$, defined according to (2.1.24) for this curvilinear coordinate system $x^i$ for any point $M$ of domain $\Omega$, to be called connectivity parameters.

With account of equation (2.1.21) and definition 2.1.17, equation (2.1.21) takes the form:

$$0 = d\xi^k\mathbf{x}_k + \xi^i\Gamma_{ij}^k\mathbf{x}_k dx^j. \tag{2.1.25}$$

Since vectors $\mathbf{x}_k$ are linearly independent, their linear combination may become zero only if all its coefficients also become zero. This condition yields:

$$d\xi^k\mathbf{x}_k + \Gamma_{ij}^k\xi^i dx^j = 0,$$

which is equivalent to:

$$d\xi^k \mathbf{x}_k + \Gamma^k_{ij}\xi^i dx^j = 0. \tag{2.1.26}$$

Equation (2.1.26) is the *law of parallel displacement of vector in infinitesimal.*

Let us consider a special case, when coordinates $x^i$ are affine ones. Then:

$$\mathbf{x}(x^1, \ldots, x^n) = x^i \mathbf{e}_i, \quad \mathbf{x}_i = \mathbf{e}_i,$$

and from the expression (2.1.23) it follows:

$$\mathbf{x}_{ij} = 0; \Rightarrow \ \Gamma^k_{ij} = 0. \tag{2.1.27}$$

*Properties of connectivity object*

Let us assume that we have passed from preset point $M$ of domain $\Omega$ into another curvilinear coordinate system $x^{i'}$, and calculated their values $\Gamma^{k'}_{i'j'}$. A natural question arises as to in what way the values $\Gamma^k_{ij}$ and $\Gamma^{k'}_{i'j'}$ are linked together. A double differentiation of the new vector coordinates by the old ones yields [294] the dependence:

$$\Gamma^{k'}_{i'j'} = \frac{\partial^2 x^k}{\partial x^{i'}\partial x^{j'}}\frac{\partial x^{k'}}{\partial x^k} + \frac{\partial x^i}{\partial x^{i'}}\frac{\partial x^j}{\partial x^{j'}}\frac{\partial x^k}{\partial x^{k'}}\Gamma^k_{ij}. \tag{2.1.28}$$

Analysis of equation (2.1.28) indicates that value $\Gamma^k_{ij}$ is transformed at displacement not as a value of tensor character, since the first member of the transformation law (2.1.28) comprises second-order derivatives.

For our aims, the paramount importance of the connectivity object of the affine space lies in the fact that it *defines the whole geometry of the affine space in the domain where the object is defined.* A respective theorem exists in this connection [294, p. 350].

**Remark 2.1.2.** It is necessary to note an important peculiarity of the connectivity object. If mathematic objects $\Gamma^k_{ij}(x^1, \ldots, x^n)$ of random structure are set, values $(x^1, \ldots, x^n)$ may not always be understood as curvilinear coordinates. It is possible only if we consider that object $\Gamma^k_{ij}(x^1, \ldots, x^n)$ sets a certain geometry, which is a generalization of the affine geometry (*sets the space of affine connectivity*).

To make the law of coordinate transformation (2.1.28) set affine transformations, we need to demand that $\Gamma^{k'}_{i'j'}(x^1, ..., x^n) = 0$ in accordance with condition (2.1.27). This requirement allows us to write down:

$$\Gamma^k_{ij} = \frac{\partial^2 x^{k'}}{\partial x^i \partial x^j}\frac{\partial x^k}{\partial x^{k'}}. \tag{2.1.29}$$

Multiplying term by term by $\partial x^{l'}/\partial x^k$ and summing by $k$, we obtain a second-order differential equation system:

$$\Gamma^k_{ij}\frac{\partial x^{l'}}{\partial x^k} = \frac{\partial^2 x^{k'}}{\partial x^i \partial x^j}\frac{\partial x^k}{\partial x^{k'}}\frac{\partial x^{l'}}{\partial x^k} = \frac{\partial^2 x^{k'}}{\partial x^i \partial x^j}\delta^{l'}_{k'} = \frac{\partial^2 x^{l'}}{\partial x^i \partial x^j}. \tag{2.1.30}$$

This indicates that a second-order differential equation system may be interpreted as a structure that sets a certain affine space [85].

Let us consider an important characteristic of the field of connectivity objects. Let us designate:

$$S_{ij}^{k} = \Gamma_{ij}^{k} - \Gamma_{ji}^{k}. \tag{2.1.31}$$

Let us now show the change character $S_{ij}^{k}$. Let us rewrite (2.1.29), having interchanged indices $i'$, $j'$ and interchanging the designations of summing indices $i$, $j$. We obtain:

$$\Gamma_{j'i'}^{k'} = \frac{\partial^2 x^k}{\partial x^{j'} \partial x^{i'}} \frac{\partial x^{k'}}{\partial x^k} + \frac{\partial x^j}{\partial x^{j'}} \frac{\partial x^i}{\partial x^{i'}} \frac{\partial x^{k'}}{\partial x^k} \Gamma_{ji}^{k}. \tag{2.1.32}$$

By subtracting term by term (2.1.32) from (2.1.29) and taking into account definition (2.1.31), we obtain:

$$S_{i'j'}^{k'} = \frac{\partial x^l}{\partial x^{i'}} \frac{\partial x^j}{\partial x^{j'}} \frac{\partial x^{k'}}{\partial x^k} S_{ij}^{k}. \tag{2.1.33}$$

Object $S_{ij}^{k}$ is called a torsion tensor of the said affinity connectivity space.

*Manifolds*

When defining affine coordinate systems, we need to satisfy the requirements of a bijective (one-to-one) analytical $N$ times continuously differentiable mapping of the coordinate system:

$$x^{i'} = f_i(x^1, \ldots, x^n); \quad x^i = f_i(x^{1'}, \ldots, x^{n'}). \tag{2.1.34}$$

An additional condition is the requirement of linearity of coordinate transformation. This particular condition decodes on the reference of a coordinate system to affine spaces. If we refuse from the linearity requirement (it is equivalent to recognition of parity of all the coordinate systems), then we cannot demand the affinity of the space, and have just a certain set of objects. Geometrical properties of this set are quite poor. This set represents *an elementary manifold*.

Sometimes, it is additionally required for the manifold to be differentiable. From the viewpoint of mathematics, we can state the following.

**Definition 2.1.18.** *Differentiable manifold.* Differentiable manifold $\mathcal{M}$ of dimensionality $n$ is called a Hausdorff space, which has the following properties [364, 371]:

- $\mathcal{M}$ is locally Euclidean, i.e., for any point $P \in \mathcal{M}$ there exists a neighborhood of this point $U \subset \mathcal{M}$ and homeomorphism $\phi$ on the open subset in $\mathbb{R}^n$. This homeomorphism is written down in a standard way $\phi : U \to \mathbb{R}^n$. That is, for $P \in U$:

  $$\phi(P) = x^1(P), x^2(P), \ldots, x^n(P) \in \mathbb{R}^n.$$

  The set (point, homeomorphism $(U, \phi)$) is called a map. The set of real numbers $(x^1(P), x^2(P), \ldots, x^n(P))$ is called coordinates of this point $P \in U \subset \mathbb{R}^n$ relative to the map.

- There exists a set of maps $\{(U_a, \phi_a\}$ such that:
  - $\underset{a}{\cup} U_a = \mathcal{M}$;
  - if the set $U_{ab} = U_a \cup U_b \neq \oslash$, then, the mapping $\phi_{ab} = \phi_b \circ \phi_a^{-1}$ from $\phi_a(U_{ab}) \subset \mathbb{R}^n$ in $\phi_b(U_{ab}) \subset \mathbb{R}^n$ is a differentiable mapping.

A set of maps with these properties is called an atlas.

*Elementary manifold*

**Definition 2.1.19.** *Elementary manifold.* The elementary manifold (of $n$ measures and class $N$) is called any set $\mathcal{M}$, for which a biunivocal (one-to-one) mapping is set onto a connected domain of changing of $n$ variables $x^1, \ldots, x^n$. This mapping is set only with the accuracy up to random transformation of these variables into new variables according to (2.1.34).

Briefly, the mapping may be written down as:

$$M \leftrightarrow (x^1, \ldots, x^n) \in \Omega, \tag{2.1.35}$$

where $\Omega$ is a *connected* variation domain of variables $x^1, \ldots, x^n$.

Under the whole poorness of geometrical properties induced by this general mapping, we may still define certain properties. The manifold elements $\mathcal{M}$ may be considered to be a sort of *points*, the preset mappings (2.1.35) to be *coordinate systems* in the preset manifold $\mathcal{M}$, and values $x^1, \ldots, x^n$, corresponding to point $M$ in mapping (2.1.35), to be *coordinates* in the respective coordinate system.

Basically, we may state that an elementary manifold (of degree $N$) reflects those properties of changing domain of variables $x^1, \ldots, x^n$, which are invariant at any bijective and continuously $N$-times differentiable transformation of these variables into new variables $x^{1'}, \ldots, x^{n'}$. The higher is $N$, the smaller the number of transformations possible, and the higher the number of properties the manifold has. *A manifold of a higher degree comprises all the properties, without exclusions, of lower degree manifolds and, besides, possesses certain additional properties.* Transformation of coordinates may be set by $n$ equations:

$$x^{i'} = x^{i'}(x^1, x^2, \ldots, x^n) \quad (i', j' = 1, \ldots, n). \tag{2.1.36}$$

Sometimes, to underline the number of measures of manifold $\mathcal{M}$, it is written in the form of $\mathcal{M}_n$.

*Geometrical objects and geometrical values on a manifold*

The presence of various transformation types of coordinate systems in $\mathcal{M}_n$ allows existence of mathematical structures which are not usually defined in the Euclidean geometry. Let in manifold $\mathcal{M}$ exist a set of permissible coordinate systems $(\varpi)$. Then the following definition [315] is correct.

**Definition 2.1.20.** If in a definite point $M$ of manifold $\mathcal{M}$ there is a conformity between ordered sets $K$ of numbers $\Phi_\Lambda$, $\Lambda = 1, \ldots, K$ and $(\varpi)$, near $M$ such that: (a) each $(\varpi)$ corresponds to one and only one set $\Phi_\Lambda$; (b) set $\Phi_{\Lambda'}$ may be expressed only through $\Phi_\Lambda$

and values $A^i_{i'}$, $\partial_j A^i_{i'}$, $\partial_j \partial_k A^i_{i'}, \dots$ in point $M$, then it is said that $\Phi_\Lambda$ are *components relative to* ($\varpi$) *geometrical object in point* $M$.

**Definition 2.1.21.** If additionally to the requirements of definition (2.1.20), the expression for $\Phi_{\Lambda'}$ is linear and uniform by $\Phi_{\Lambda'}$, algebraically uniform by $A^i_{i'}$, and comprises no derivatives $A^i_{i'}$, then $\Phi_\Lambda$ are *components relative* ($\varpi$) *geometrical value in point* $M$.

According to our definitions, the connectivity objects defined in accordance with expression (2.1.28) are not a geometrical value, but a geometrical object.

*Tensors in manifolds*

Since very little remains from the object geometrical properties in a manifold, it is necessary to define anew the known mathematical constructions in the manifold geometry. Usually, it is done by analogy with introduction of a respective value into the Euclidean space with account of the properties of considered manifolds. By reasoning in this way, we may introduce a formal definition of a tensor and a manifold.

**Definition 2.1.22.** We assume that in a given point $M$ of manifold $\mathcal{M}$ a tensor is set, $n$ times contravariant and $m$ times covariant, if in every coordinate system $x^1, \dots, x^n$ there exists a set of numbers $V^{j_1, j_2, \dots, j_n}_{k_1, k_2 \dots, k_m}$. This set of numbers varies, at transition to another coordinate system $x^{1'}, \dots, x^{n'}$, according to the law:

$$V^{j'_1, j'_2, \dots, j'_n}_{k'_1, k'_2 \dots, k'_m} = \frac{\partial x^{j'_1}}{\partial x^{j_1}}(M) \frac{\partial x^{j'_2}}{\partial x^{j_2}}(M) \dots \frac{\partial x^{j'_n}}{\partial x^{j_n}}(M) \frac{\partial x^{k_1}}{\partial x^{k'_1}}(M) \times \\ \times \frac{\partial x^{k_2}}{\partial x^{k'_2}}(M) \dots \frac{\partial x^{k_m}}{\partial x^{k'_m}}(M) V^{j_1, j_2, \dots, j_n}_{k_1, k_2 \dots, k_m}, \tag{2.1.37}$$

where derivatives are calculated in point $M$ of the manifold.

Calculation of the derivative in point $M$, which evidences setting of a tensor in this point, has a great importance for calculation of tensor values in generalized spaces. We may speak about a tensor on the manifold, which is set continuously as a function in each point of the manifold. In this case, one usually speaks about a tensor field:

$$V^{j_1, j_2, \dots, j_n}_{k_1, k_2 \dots, k_m} = V^{j_1, j_2, \dots, j_n}_{k_1, k_2 \dots, k_m}(M) = V^{j_1, j_2, \dots, j_n}_{k_1, k_2, \dots, k_m}(x^1, \dots, x^n). \tag{2.1.38}$$

Since we are considering an elementary manifold of degree $N$, function (2.1.38) should be differentiable $N - 1$ times. This is connected with the requirement that the expression of new coordinates as a function of old ones $x^{i'}(x^1, \dots, x^n)$ and vice versa should be differentiable $N$ times. Since the manifold has no local and global affine reference fields, we may and should perceive the coordinates themselves $(x^1, \dots, x^n)$ as local reference fields of the manifold.

**Remark 2.1.3.** Geometrical definition of tensors is also possible in the case of defining tensors on the manifold. For the problems of mechanics, it leads to definition of the *symplectic structure* on the manifold [9].

The law of component transformation (2.1.37) will be valid also at transition to another coordinate system $x''$. It may be demonstrated by a consequent application of the transformation law (2.1.37) [294]. The law allows to demand in the usual way to execute all the operations of tensor algebra for tensors on the manifolds, defined in one and the same field point.

At the same time, it is not possible to make joint operations on the tensors defined in different points of the manifold. This is caused by the fact that when comparing the values of any origin, they should be expressed in the unit of measure defined relative to one and the same basis (coordinate system). Therefore, when passing from point $M_1$ of the manifold to point $M_2$, we need to know the law of basis transformation, and this law is not defined in the general case. To explain the situation, let us consider [294] two tensors in points $M_1$ and $M_2$, which have the same coordinates in the given coordinate system $x^i$:

$$V^{j_1,j_2,\dots,j_n}_{k_1,k_2\dots,k_m}(M_1) = V^{j_1,j_2,\dots,j_n}_{k_1,k_2\dots,k_m}(M_2).$$

Nevertheless, these tensors cannot be considered equal, since when passing over to a new coordinate system $x^{i'}$, in accordance with the transformation law (2.1.37), the equality of the components is broken. This is related to the fact that when transforming $V^{j_1,j_2,\dots,j_n}_{k_1,k_2\dots,k_m}(M_1)$ in accordance with (2.1.37), members will appear of the type $\dfrac{\partial x^{x'}}{\partial x^i}(M_1)$, and for the tensor $V^{j_1,j_2,\dots,j_n}_{k_1,k_2\dots,k_m}(M_2)$, members will appear of the type $\dfrac{\partial x^{x'}}{\partial x^i}(M_2)$, which are in no case equal to each other.

However, operations of tensor algebra may be partially transferred to tensor fields in the manifold. In this case, operations with fields are defined as operations with tensors in local points. Thus, for tensor fields of the *same* geometry $V^j_{km}(M)$, $U^j_{km}(M)$ the addition may be defined:

$$W^j_{km}(M) = U^j_{km}(M) + V^j_{km}(M);$$

tensor multiplication $U^j_{km}(M)$ and $V^l_n(M)$ as:

$$W^{jl}_{kmn}(M) = U^j_{km}(M) + V^l_n(M);$$

contraction of indices for $U^{jlp}_{km}(M)$ by the first upper and the second lower indices yields the tensor field:

$$W^{lp}_k(M) = U^{jlp}_{kp}(M);$$

substitution of indices for the field (sometimes this operation is called *isomer formation*) $U^{jlp}_{km}(M)$ yields a tensor field of the same geometry $Z^{jlp}_{km}(M)$ according to the standard procedure:

$$Z^{jlp}_{km}(M) = U^{ljp}_{mk}(M).$$

**Remark 2.1.4.** All the operations of vector algebra may be defined, preserve their sense and have a usual character *only because they have been taken in one and the same point of the manifold.*

**Remark 2.1.5.** Operation of absolute differentiation of the tensor field has no sense in the manifold. The standard differentiation procedure of a random value assumes definition of the difference of the values in two infinitely near points. It was already mentioned that for the manifold this operation has no mathematic sense.

*Vectors in manifolds*

Let us consider a new system of local coordinates defined by equations (2.1.36). Let us assume that our manifold $\mathcal{M}$ is completely covered by this coordinate system. It means that any point $M$ from $\mathcal{M}$ is represented as a set $n$ of independent variables $x^i$, $i = 1, 2, \ldots, n$. Let us consider in manifold $\mathcal{M}$ a curve passing through point $M$.

**Definition 2.1.23.** A curve in manifold $\mathcal{M}$ is understood as a set of the points defined by parametric equations:

$$x^i = x^i(t), \tag{2.1.39}$$

where $t$ is an invariant parameter.

Since we consider a manifold of degree $N$, function $x^i(t)$ is $N$ times continuously differentiable. We assume also that values $dx^i/dt$ do not go to zero simultaneously.

We have a possibility in principle to perform a definite mathematical operation, which would relate manifold $\mathcal{M}$ with a certain affine space [294]. For this purpose, we choose in $\mathcal{M}$ point $M$. Out of all contravariant tensors $\xi^i$, let us take $n$ linearly independent:

$$\xi^i_{(1)}, \xi^i_{(2)}, \ldots, \xi^i_{(n)}.$$

Linear independence is equivalent to observance of the condition:

$$\mathrm{Det}|\xi^i_{(j)}| \neq 0.$$

Then, any tensor belonging to $\mathcal{M}$ may be presented as a linear combination of reference (basic) tensors:

$$\xi^i = \alpha^{(1)}\xi^i_{(1)} + \alpha^{(2)}\xi^i_{(2)} + \ldots + \alpha^{(n)}\xi^i_{(n)}, \quad (i = 1, 2, \ldots, n). \tag{2.1.40}$$

Expansion coefficients $\alpha^{(i)}$, $i = 1, \ldots, n$ are defined from $n$ equations for expansions of tensor components $\xi^i$.

Let us consider affine space $\Omega_n$, the dimensionality of which coincides with the dimensionality of the manifold. Let us randomly choose in it $n$ linearly independent vectors:

$$\boldsymbol{\xi}_{(1)}, \boldsymbol{\xi}_{(2)}, \ldots, \boldsymbol{\xi}_{(n)}.$$

Then, each tensor (2.1.40) may be uniquely referenced to a vector $\boldsymbol{\xi} \in \Omega$:

$$\boldsymbol{\xi} = \alpha^{(1)}\boldsymbol{\xi}_{(1)} + \alpha^{(2)}\boldsymbol{\xi}_{(2)} + \ldots + \alpha^{(n)}\boldsymbol{\xi}_{(n)}. \tag{2.1.41}$$

*Riemann space*

The Riemann space is a manifold provided with a metric. It is defined by setting a metric tensor in the manifold.

**Definition 2.1.24.** We shall call the Riemann space $V_n$ to be a manifold $\mathcal{M}_n$, in which a tensor field is defined

$$g_{ij}(M) = g_{ij}(x^1, \ldots, x^n) \tag{2.1.42}$$

which is twice covariant, symmetric and non-degenerate

$$\mathrm{Det}\left|g_{ij}\right| \neq 0, \quad g_{ij} = g_{ji}. \tag{2.1.43}$$

One and the same manifold may be overlapped by a metric in different ways.

With account of the definition of the vector (2.1.41), we can introduce a definition of a scalar product of any two vectors $\boldsymbol{\xi}$, $\boldsymbol{\eta}$ in the given point $M$ as

$$\boldsymbol{\xi}\boldsymbol{\eta} = g_{ij}(M)\xi^i\eta^j. \tag{2.1.44}$$

Following this definition, the vector length $\boldsymbol{\xi}$ is expressed by formula

$$|\boldsymbol{\xi}| = \sqrt{\boldsymbol{\xi}^2} = \sqrt{g_{ij}\xi^i\xi^j}, \tag{2.1.45}$$

and the length of the curve arc (more precisely, the length of the vector connecting the start and end points along the curve) is

$$|\mathbf{x}| = \sqrt{\mathbf{x}^2} = \sqrt{g_{ij}dx^idx^j}. \tag{2.1.46}$$

Thus, the square of the arc differential is expressed by the differential quadratic form.

It follows from the definition of the Euclidean space that it is also provided with a metric tensor, and this tensor defines the whole geometry of this space. Consequently, the Euclidean space may be considered as a special case of the Riemann space. The specific feature of this special case is connected with the fact that in the Euclidean space we may also pass over into such special coordinate systems, where coordinates of the metric tensor become constants. For the Riemann space it may be done locally, and not always, therefore, is the Euclidean space in a certain sense 'straighter' than the Riemann one.

Until now, we have considered the geometry set by connectivity and the geometry, initiated by metric tensor, without their mutual correlation. There exists a respective theorem [161, IV.3], asserting the topology coincidence for the Riemann metric space[4].

[4] We underline again that this is correct for metric spaces only. Hereafter, in case of media with microstructure and generalized spaces, this proposition acts as an additional postulate.

### 2.1.5 Tangent spaces

By the help of (2.1.41), a point of manifold $\mathcal{M}$ is mapped on a randomly marked point $O \in \Omega$ of the affine space. Thus, *for each manifold $\mathcal{M}$ we may build an affine space $\Omega$, having one common point with the manifold. In doing so, linear relations between tensors in the manifold are transiting into linear relations between vectors in the affine space.* Then, space $\Omega_n$ is called a *tangent affine space*, and its vectors $\boldsymbol{\xi}$ are tangent vectors in this point $P$ of manifold $\mathcal{M}$.

**Remark 2.1.6.** One should remember that tangent spaces $\Omega_n$ in different points of the manifold do not coincide with each other. That is why, strictly speaking, for a proper choice of a tangent space, we need to set both the coordinates of a manifold point (tangent point), and the set of tangent vectors.

The tangent vector is not uniquely defined. It is related with the fact that under continuously differentiated and reversed transformation of parameter $\tau = \tau(t),\ t = t(\tau)$, we have:

$$\frac{dx^i}{d\tau} = \frac{dx^i}{dt}\frac{dt}{d\tau}.$$

It allows building a tangent fibration to manifold $\mathcal{M}$ in point $M$.

**Definition 2.1.25.** A tangent fibration is called to be a unification of tangent spaces $\mathfrak{T}(\mathcal{M}) = \bigcup_{x\in\mathcal{M}} \Omega_n$.

Equation (2.1.40) is equivalent to mapping of:

$$\pi : \mathfrak{T}(\mathcal{M}) \to \mathcal{M}, \quad \pi(\xi) = x, \quad \forall\, \xi \in \Omega_n. \tag{2.1.47}$$

**Definition 2.1.26.** The set $(x^i, \xi^i)$ is called to be canonical coordinates of space $\Omega_n$, defined by coordinates $(x^i)$ of the mapping point (2.1.47) with base $M$. The mapping $\pi$ is the canonical projection.

We can demonstrate [231] that $\mathfrak{T}(\mathcal{M})$ possesses the structure of a differentiable manifold of dimensionality *2n*.

For our tangent space $\Omega_n$ we can build (see §2.1.2) a dual space $\Omega_n^*$, called *co-tangent space*, the elements of which are covectors.

*The tangent map of real Cartesian spaces*

Up to now we have been discussing certain abstract spaces. These will be important in the context of our objective of modelling physical phenomena by various mathematical spaces with mappings between them. For vector spaces, in general, and $\mathbb{R}^n$, in particular, we have linear mappings, and, more generally, nonlinear mappings. For vector spaces we have an important device for dealing with the more general nonlinear mappings. We can approximate them by linear mapping derivatives. In our applications we will need to use for our domains spaces more general than $\mathbb{R}^n$, including spaces which are not vector spaces (surfaces in space of state, for example). We will use spaces for which we can describe a device such as we have for vector spaces; a way of approximating mappings between such spaces by linear ones: 'derivatives'. These spaces will be our differentiable manifolds.

The structures, tangent and cotangent spaces, that we need for the concept of a 'derivative' of a mapping of differentiable manifolds can be described for the special cases of vector spaces, and, in particular, of Cartesian spaces [371]. It will turn out that we will be able to make a straightforward generalization of these concepts in the general setting of manifolds. Thus, we can temporarily avoid some of the abstract properties of manifolds, and see, in the special case of Cartesian spaces, some of the important concepts that can be defined on them.

**Definition 2.1.27.** Real Cartesian $n$-dimensional space, $\mathbb{R}^n$, is the set of all ordered $n$-tuples of real numbers.

A point in this space is an $n$-tuple $(a^1, a^2, \ldots, a^n)$, which we will frequently abbreviate by $a$. We will focus our attention on an arbitrary but fixed point $a \in \mathbb{R}^n$ and its neighborhoods $\mathfrak{U}_a$. These are defined by the *usual* topology of $\mathbb{R}^n$; that is, the topology given by the distance $d(a, b) = \left[\sum_{i=1}^{n}(a^i - b^i)^2\right]^{1/2}$.

We can, on the one hand, consider functions on $\mathfrak{U}_a$, $f : \mathfrak{U}_a \to \mathbb{R}$, from a neighborhood of $a$ to the reals, and, on the other hand, we can consider curves in $\mathbb{R}^n$ *through* $a$, $\gamma_a : I \to \mathbb{R}^n$ from an open interval of $\mathbb{R}$ containing 0, and such that $\gamma_a(0) = a$.

There are very important examples of functions where $\mathfrak{U}_a = \mathbb{R}^n$, namely, the $n$ *natural coordinate functions* (or natural projections), $\pi^i : \mathbb{R}^n \to \mathbb{R}$ given by $a \mapsto a^i$. There are $n$ corresponding special curves through $a$, $\theta_{ak} : \mathbb{R} \to \mathbb{R}^n$, given by $u \mapsto (a^1, a^2, \ldots, a^k + u, \ldots, a^n)$. They are the $n$ natural coordinate curves through $a$ (or natural injections). Note that the $\pi^i$ are linear, but the $\theta_{ak}$ are not.

With a given curve through $a$ we can form its component functions, $\gamma_a^i = \pi^i \circ \gamma_a$. With a given function on $\mathfrak{U}_a$ we can form the partial functions $f \circ \theta_{ak}$. The value of $f \circ \theta_{ak}$ at $u$ is $f(a^1, a^2, \ldots, a^k + u, \ldots, a^n)$.

Since $\mathbb{R}$ and $\mathbb{R}^n$ are topological spaces, and we have the concept of continuity for mappings between any two topological spaces, we have the concept of continuity for functions and curves through $a$. (In particular, it is clear that $\pi^i$ and $\theta_{ak}$ are continuous on their domains.)

Now, since Cartesian spaces are normalized vector spaces, we can go further and make the following definitions for maps between $\mathbb{R}^n$ and $\mathbb{R}^p$ (and, more generally, for maps between any two normalized vector spaces $V$ and $W$).

**Definition 2.1.28.** Two maps $\psi$ and $\phi$ (continuous at $a$) are *tangent* (to each other) at $a$ if

$$\lim_{v \to a} \frac{\psi(v) - \phi(v)}{v - a} = 0.$$

A map $\psi : \mathcal{U}_a \to \mathbb{R}^p$ ($\mathcal{U}_a \subset \mathbb{R}^n$) is *differentiable* at $a$ if it is tangent at $a$ to an affine map, $A$, that is, a map $A(v) = \psi(a) + D_0\psi(v - a)$, where $D_0\psi$ is a linear map from $\mathbb{R}^n$ to $\mathbb{R}^p$. $D_0\psi$ is called the derivative of $\psi$ at $a$.

### *2.1.6 Main fibration**

The notion of fiber (bundle) space plays an important role in physical applications of differential geometry. The fact is important for us that when considering a deformed solid, the fibration appears in the natural way from the structure of defining equations without involvement of additional postulates. Generally speaking, fibration appears when a structure of layers (bundles) is introduced on the manifold $\mathcal{M}$. It may be done in various ways [364]. In the simplest way, we split the $\mathcal{M}$ into certain subsets (layers) and isomorphically correlate these layers with a certain set $F$. The layers are numbered by points of a certain set $M$, which is called *a base*. In other words, it means that the canonical projection $\pi : \mathcal{M} \mapsto M$ is set, mapping all the points of one layer (and only them) onto point $x \in M$. Besides, locally, in a certain vicinity $U \subset M$ fibration $P$ is organized in the same way as a direct product $U \times F$. However, the global structure of fibration may be non-trivial. In this context, it is important in what way various parts of the fibration are glued with each other.

The notion of fibration appears also in the case when, with each point of the manifold $\mathcal{M}$, a set of objects is related characterizing the $\mathcal{M}$ in this point. In the previous abstract, such set was a set of tangent vectors, which formed a tangent fibration $\mathfrak{T}(\mathcal{M})$.

**Main fibered space.** An important special case of fibration (bundled) spaces is the main fibration space (main fibration) [231, 364]. The main fibration connects two manifolds $\mathcal{M}, \mathcal{P}$ by means of a Lie group $G$. This fibration is defined as $\mathcal{P}(\mathcal{M}, G, \pi)$.

Usually, $\mathcal{P}$ is called a space of fibration: $\mathcal{M}$—base, $G$—structural group, $\pi$—projection. Additional requirements and details of organization of the main fibration may be found in the literature [206, 364].

*Fibration in the Euclidean space*

Since the Euclidean space should comprise the properties of the manifold, which are implemented thereon, let us consider a possibility to exist for the fibration of the Euclidean space. Let us assume that a vector field exists in a certain domain of the Euclidean space. A vector field is called fibered if there exists a family of surfaces, which fills in this domain, such that the vector field of unit normal to the surfaces of the family coincides with $\mathbf{n}$.

In order to have a vector field $\mathbf{n}$ to be fibered in a certain domain of the space, it is necessary and sufficient [291] that everywhere in this domain the following correlation be observed:

$$\mathbf{n} \cdot \operatorname{rot} \mathbf{n} = 0. \tag{2.1.48}$$

The layers of the vector field $\mathbf{n}$ are compiled from vector lines of field $\operatorname{rot} \mathbf{n}$ according to the following procedure. First, a surface is chosen in such a way that vectors $\mathbf{n}$ touch it in every point. Then, a family of trajectories is built on this surface, orthogonal to **n**, and from each point of an orthogonal trajectory, vector lines are emitted rot **n** to form a field layer **n**. In the particular case the set of equipotential surfaces generated the layers vector field.

If we introduce direction cosines of the normal:

$$\mathbf{n} = \sin\psi \sin\vartheta\,\mathbf{i} - \cos\psi \sin\vartheta\,\mathbf{j} + \cos\vartheta\,\mathbf{k},$$

then the fibration condition (2.1.48) takes the form:

$$\cos\psi \frac{\partial\vartheta}{\partial x_1} + \sin\psi \frac{\partial\vartheta}{\partial x_2} - \frac{1}{2}\sin 2\vartheta \sin\psi \frac{\partial\psi}{\partial x_1} + \\ + \frac{1}{2}\sin 2\vartheta \cos\psi \frac{\partial\psi}{\partial x_2} + \sin^2\vartheta \frac{\partial\psi}{\partial x_3} = 0. \tag{2.1.49}$$

According to condition (2.1.49), all the flat and axially symmetric vector fields are fibered.

*Tangent vector and directional derivative*

The notion of the tangent vector is connected with the directional derivative. Let us consider a smooth curve $\alpha(t)$, which passes through point $P_0 \in \mathcal{M}$, $\alpha(0) = P_0$. Parameter $t$ adopts a value from interval $(a, b)$. By definition, a directional derivative of function $f$, set by curve $\alpha(t)$, is:

$$\frac{d}{dt} f(\alpha(t)) \,|_{t=0} = \frac{d}{dt} \tilde{f}\left(\alpha^1(t), \ldots, \alpha^n(t)\right)|_{t=0} = \frac{d\alpha^i(t)}{dt}|_{t=0} \frac{\partial \tilde{f}(x^1, \ldots, x^n)}{\partial x^i}|_{P_0}.$$

If we define $\kappa^i = \frac{d\alpha^i(t)}{dt}|_{t=0}$, the directional derivative may be written down as:

$$\kappa^i \frac{\partial f}{\partial x^i}\,|_{P_0} = \left(\kappa^i \frac{\partial}{\partial x^i}\right)|_{P_0} f = X_{P_0}(f), \tag{2.1.50}$$

where $X_{P_0}$ is a linear mapping from the algebra of differentiable functions $f$ on $\mathcal{M}$ in $\mathbb{R}$. We shall define this algebra as $\mathfrak{F}(\mathcal{M})$. Then $\mathbf{X}_{P_0}$ is a vector, tangent to curve $\alpha(t)$ in point $P_0$. Mapping $\mathbf{X}_{P_0}$ satisfies the Leibniz rule:

$$\mathbf{X}_p(fg) = \left(\mathbf{X}_p(f)\right) g(p) + f(p)\mathbf{X}_p(g).$$

We can demonstrate [364] that compliance of linear mappings $\mathbf{X}_p : \mathfrak{F}(\mathcal{M}) \to \mathbb{R}$ to the Leibniz rule is a sufficient condition for $\mathbf{X}_p$ being a tangent vector.

**Definition 2.1.29.** If for any differentiated function $f \in \mathfrak{F}(\mathcal{M})$ function $Xf$ is also differentiable, then field **X** is called a differentiable vector field. Later, we shall not specifically mark the vector nature of field **X**.

Coordinates $\kappa^i(p)$ of such field in the local basis are differentiable functions. The set of all differentiable vector fields on $\mathcal{M}$ is $\mathfrak{X}(\mathcal{M})$. Field $X \in (\mathcal{M})$ is a linear mapping of algebra on itself $X : (\mathcal{M}) \to (\mathcal{M})$. On the set $\mathfrak{X}(\mathcal{M})$ a multiplexer (bracket operation) is defined:

$$[X, Y](f) = X(Y(f)) - Y(X(f)).$$

In relation to bracket operation, $\mathfrak{X}(\mathcal{M})$ is the Lie algebra. It means that the following relations are observed:

$$[X, Y] = -[Y, X] \tag{2.1.51}$$

$$[X, [Y, Z]] + [Y, [Z, X]] + [Z, [X, Y]] \tag{2.1.52}$$

for $X, Y, Z \in \mathfrak{X}(\mathcal{M})$. Expression (2.1.52) is called Jacobi identity.

*Nonlinear connectivity*

Let the main fibered space $\mathcal{P}(\mathcal{M}, G)$ be set. Let us consider tangent space $\mathfrak{T}(\mathcal{M})$ to $\mathcal{M}$ in point $M \in \mathcal{M}$. This tangent space may be split into two subspaces. One of them comprises the vectors tangent to the layer, and is defined uniquely (*vertical subspace*). This vertical direction (direction along the layer) is fixed by the action of the structural group. The vector along the base needs an additional definition. For this purpose, a nonlinear connectivity is introduced.

**Definition 2.1.30.** Nonlinear connectivity $N$ of a tangent fibration $\mathfrak{T}(\mathcal{M})$ is a regular distribution $N : y \in \mathfrak{T}(\mathcal{M}) \to N_y \subset \mathfrak{T}(\mathcal{M})_y$, such as:

$$\mathfrak{T}(\mathcal{M})_y = N_y \oplus \mathfrak{T}(\mathcal{M})_y^v. \tag{2.1.53}$$

In other words, a nonlinear connectivity is a rule of mapping a point of the manifold on a certain subspace. Then, the local derivative has the basis:

$$\frac{\delta}{\delta x^i} = \frac{\partial}{\partial x^i} - N_i^k(x, y)\frac{\partial}{\partial y^k}. \tag{2.1.54}$$

**Definition 2.1.31.** Values $N_i^k(x, y)$ are called coefficients of nonlinear connectivity $N$.

We can show [231] that $N_i^k(x, y)$ when transforming coordinates changes according to the law as follows:

$$N_{i'}^{k'}(x, y) = \frac{\partial x^{h'}}{\partial x^h} N_k^h + \frac{\partial^{h'}}{\partial x^l}\frac{\partial^2 x^l}{\partial x^{k'}\partial x^{m'}} y^{m'}. \tag{2.1.55}$$

Similarly to the Riemann geometry, the nonlinear connectivity may be obtained as a solution for the Pfaffian system [84, 85]:

$$\delta y^i = dy^i + N_k^i dx^k = 0.$$

### 2.1.7 Vertical and horizontal lift

A complete fibration of a differentiable manifold may be split into certain submanifolds, which have special properties.

**Vertical lift.** Let us consider a differentiable manifold with canonical coordinates $(x^i, y^i)$. Let $\mathfrak{T}(\mathcal{M})_y$ be a tangent space in point $y$ to the differentiable manifold $\mathfrak{T}(\mathcal{M})$. The set $\left(\frac{\partial}{\partial x^i}, \frac{\partial}{\partial y^i}\right)$, $i = 1, 2, \ldots, n$ is a natural basis $\mathfrak{T}(\mathcal{M})_y$. We can consider submanifold $\mathfrak{T}(\mathcal{M})_y^v$, generated only by basis $\partial/\partial y^i$. According to the general rule (2.1.37), the basis is transformed as:

$$\frac{\partial}{\partial y^{i'}} = \frac{\partial x^i}{\partial x^{i'}} \frac{\partial}{\partial y^i}.$$

Then, the mapping $\mathfrak{T}(\mathcal{M})^v : y \in \mathfrak{T}(\mathcal{M}) \to \mathfrak{T}(\mathcal{M})_y^v$ is called a vertical distribution of the tangent fibration. The vertical fibration has dimensionality $n$. We can introduce isomorphic mapping:

$$l_v : \Omega_n \to \mathfrak{T}(\mathcal{M})_y^v, \quad (\pi(y) = x), \quad \forall\, x \in \mathcal{M}. \tag{2.1.56}$$

This mapping transforms the tangent vector space into the vertical one and is called *a vertical lift.*

**Horizontal lift.** Similar to the vertical lift, we can introduce the horizontal lift. Let us have subspace $\mathfrak{T}(\mathcal{M})_y^h$, generated by basis $\partial/\partial x^i$. According to definition (2.1.54) for local derivative, we have:

$$\frac{\delta}{\delta x^i} = \left(\frac{\partial}{\partial x^i} - N_k^i \frac{\partial}{\partial y^k}\right). \tag{2.1.57}$$

**Definition 2.1.32.** The horizontal lift of the vector field $X$ on manifold $\mathcal{M}$ is a unique vector field $X^*$ on the main fibration, which is projected on $X$:

$$l_y : M_x \to N_y\,(x = \pi(y)). \tag{2.1.58}$$

A split into horizontal and vertical lifts is equivalent to presentation of a random vector $\mathbf{X} \in \mathfrak{T}(\mathcal{M})$ as:

$$\mathbf{X} = \mathbf{Y} + \mathbf{Z}, \quad \mathbf{Y} \in \mathfrak{T}(\mathcal{M})_y^v, \quad \mathbf{Z} \in N_y.$$

## 2.2 Minimum paths

Mechanics—the subject of our main consideration—may be built on the basis of variation of minimum principles. Therefore, in connection with the considered geometrical presentations, of interest are the problems related with detection of the minimum (maximum) of certain values in space. In the Euclidian geometry, the minimization problem is connected with detection of a derivative (differential) of a particular function. Since, according to remark 5, the operation of absolute differentiation for manifolds has no sense, it is necessary to find its analogue.

### 2.2.1 Covariant differentiation

In order to introduce the differentiation notion in fibrations, similar to affine spaces, we need to define parallel displacements. This issue has been studied in differential geometry, where it was shown that a parallel displacement is connected with the horizontal lift [364]. In the general case, covariant differentiation may be defined in accordance with the standard procedure, which is similar to the procedure in §2.1.4. Let a covariant vector field $v^{\varkappa}$ be set. Then, the field value in point $\xi^{\varkappa}+d\xi^{\varkappa}$ is $v^{\varkappa}+dv^{\varkappa}=v^{\varkappa}+d\xi^{\mu}\partial_{\mu}v^{\varkappa}$. Let us define by means of $v^{\varkappa}+\overset{*}{d}\,v^{\varkappa}$ a vector transferred into $\xi^{\varkappa}+d\xi^{\varkappa}$. Then, from the basic regularities of parallel displacements [315], a combination of values

$$\delta v^{\varkappa}\overset{def}{=}dv^{\varkappa}+\overset{*}{d}\,v^{\varkappa} \tag{2.2.1}$$

is transformed as a vector, although the summands are not vectors. Then, $\delta v^{\varkappa}$ is *a covariant differential of field* $v^{\varkappa}$ *relative to a preset displacement.* Covariant differentiation is an operation acting in the tangent space to the manifold. Since $\delta v^{\varkappa}$ is a differential in relation to the displaced coordinate system, then the $\delta v^{\varkappa}$ is *a field differential from the viewpoint of the observer, whose local reference system is subject to displacement.* Actually, in a certain sense, it is equivalent to the Euler coordinate system.

The $h$- (horizontal) and $v$- (vertical) covariant derivative of an arbitrary tensor field are distinguished by '$_|$' and '$|$' respectively.

*Geodesic*

In what way can we introduce elementary notions of the Euclidian geometry in the case of affine geometry and manifolds? One of the simplest notions of the Euclidean geometry is a right line. This notion may be related with the notion *of the shortest distance* as the curve-right line length for the Euclidean space between the preset points.

**Definition 2.2.1.** *Definition of geodesic line.* A geodesic line $z^i=z^i(t)$ of the given connectivity $N$ is called such a curve, along which the covariant derivative of vector field $T^i=dz^i/dt$ is zero.

According to G. Weil, this definition is equivalent to the one that the direction of the tangent to the line when moving along the geodesic line remains constant. The constancy of direction is understood in the sense of the internal space metric. For the space of affine connectivity, according to results of §2.1.4, the parallel displacement of vectors is set by the law (2.1.26):

$$d\xi^k\mathbf{x}_k+\Gamma^k_{ij}\xi^i d\mathbf{x}^j=0.$$

Let the geodesic line be set in the form of $x^i = x^i(t)$, $a \le t \le b$. Let the parallel displaced tangent vector be $\boldsymbol{\xi}^i$. From the fact that colinear vectors are parallel, we may pass over to the canonical parameter along curve $\tau$. For this parameter:

$$\frac{dx^i}{d\tau} = -\xi^i. \tag{2.2.2}$$

Making use of the definition of the geodesic line and the definition of parallel displacement (2.1.26), we have:

$$d\frac{dx^k}{d\tau} = -\Gamma^k_{ij}\frac{dx^j}{d\tau}dx^i.$$

By dividing the right- and left-hand parts by $d\tau$, we have a geodesic line equation in the form referred to the canonical parameter:

$$\frac{d^2x^k}{d\tau^2} = -\Gamma^k_{ij}\frac{dx^i}{d\tau}\frac{dx^j}{d\tau}. \tag{2.2.3}$$

Formula (2.2.3) may be considered as a system of second-order nonlinear differential equations, the integral lines of which are geodesics.

The geodesics have an important feature: *since the equations are of the second order, in the affine connectivity space via every point with coordinates $x^k_0$ in each direction, defined by derivatives $\left(\frac{dx^k}{ds}\right)_0$, one and only one geodesic line may pass.* There is an important theorem for geodesics [294].

**Theorem 2.2.1.** *Connectivity object $\Gamma^k_{ij}$ defines in the given manifold the same geodesics as the connectivity object without torsion $\widetilde{\Gamma}^k_{ij}$, obtained by its symmetrizing:*

$$\widetilde{\Gamma}^k_{ij} = \frac{1}{2}\left(\Gamma^k_{ij} + \Gamma^k_{ji}\right).$$

□ To prove the theorem, let us write down the differential equation of geodesic line for connectivity $\widetilde{\Gamma}^k_{ij}$:

$$\frac{d^2x^k}{d\tau} = -\widetilde{\Gamma}^k_{ij}\frac{dx^i}{\tau}\frac{dx^j}{\tau} = -\frac{1}{2}\Gamma^k_{ij} - \frac{1}{2}\Gamma^k_{ji}. \tag{2.2.4}$$

By changing designations of summing indices in the second member of the right-hand part ($i$ into $j$ and vice versa), we make sure that both members are equal. As a result, the doubled first member remains in the right-hand part, and we obtain:

$$\frac{d^2x^k}{d\tau^2} = -\Gamma^k_{ij}\frac{dx^i}{\tau}\frac{dx^j}{\tau}.$$

It coincides with geodesic line equations for connectivity $\Gamma^k_{ij}$ (2.2.3). ■

We shall actively use this theorem in the future, for example, when calculating the crack–dislocation interactions.

## 2.3 Effect of microscopic defects on continuum

The progress in modern theoretical and experimental mechanics of solids [2, 20, 41, 110, 112, 119, 285, 300, 367] opens new opportunities in understanding complex processes of spatial-temporal organization of internal structure of a deformed body. In particular, these opportunities lie in establishing the dependence of macroscopic material properties on microscopic characteristics. In accordance with the hierarchical presentation about the character of deformation and fracture, the processes in the underlying (microscopic) level can influence the control level by transfer of control and signals. Physical mechanisms for realizing control may be different. What is important is the fact that locally realized values and objects, for example, internal stresses or structure defects, cause an effect on the 'non-local' volume of the material. In this case, the processes, realized in the macroscopic level (for example plastic deformation or crack propagation), should comply with the microscopic conditions not yet realized (the variation principle 'foresees' the future). Thus, the residual stresses transform a homogeneous body into an anisotropic and nonhomogeneous one even in the absence of an externally applied stress [57, 141]. The defining equations should at the same time comprise both the material properties of non-deformed continuum, and the internal inhomogeneity and anisotropy of the field of residual stresses. It brings us to an assumption on incompatibility of deformations, which is clearly connected with the assumption that the curvature tensor in the material space differs from zero [57].

For imperfect crystals we can build a ***local mapping*** of the imperfect crystalline medium into an ideal crystal without defects. If we mentally isolate all the defect-free elements of the medium, we shall have a manifold of small separated perfect crystal elements, which already have no phase links and positions characteristic for them in the Euclidean space. A similar situation is observed in the continuum of nanotubes, when individual nanotubes represent an ideal object, and interaction of nanotubes is weak. However, even this weak interaction essentially influences the properties of nanotubes (for example, intersection of nanotubes leads to their bending [12, 105, 124]).

These separated areas exercise no effect from the surrounding objects, internal stresses in them are completely relaxed, and therefore the elements take the dimensions and form which are most stable in the present given state. This state is termed by K. Kondo as ***natural***, which coincides with L. I. Sedov's definition of the ***initial*** state.

**Compatibility conditions.** Let us consider deformation of a solid (continuous) body. Let in each point of the body displacement vector be defined $\mathbf{u}$[5]. Then, the *distortion* field is defined by the following relation:

$$\beta = \text{grad } \mathbf{u} \equiv \nabla \mathbf{u}. \tag{2.3.1}$$

[5] Depending on the frame of reference, in which the deformation is considered, we may get different deformation tensors. If the Lagrange frame of reference is considered, we have the Green–Lagrange tensor, if the Euler one, we have the Almansi–Euler tensor. Besides, in curvilinear coordinates one should differentiate tensor stress densities and stress tensors.

It follows from well-known mathematic relations that field $\beta$ should correspond to the following condition:

$$\beta \equiv \nabla \times \beta = 0. \tag{2.3.2}$$

Fulfilment of equation (2.3.2) ensures the existence of solution **u** in equation (2.3.1). Hence, expression (2.3.2) is the compatibility condition for $\beta$.

*Deformation (small deformation tensor)* is defined as the symmetrical part of $\beta$:

$$\varepsilon \equiv {}^1\!/_2 \left( \nabla \mathbf{u} + \mathbf{u} \nabla \right), \tag{2.3.3}$$

while the *rotation tensor* as the asymmetrical part of $\beta$:

$$\omega \equiv {}^1\!/_2 \left( \nabla \mathbf{u} - \mathbf{u} \nabla \right). \tag{2.3.4}$$

According to definition (B.1), instead of the antisymmetric tensor, we may use vector:

$$\vec{\omega} \equiv {}^1\!/_2 \langle \omega \rangle = {}^1\!/_2 \nabla \times \mathbf{u}. \tag{2.3.5}$$

Expression (2.3.5) arises from substituting of definition (B.1) into expression (2.3.4). For deformation and rotation tensors, conditions of compatibility [70] can be obtained in the following form:

$$\operatorname{inc} \varepsilon \equiv \nabla \times \mathbf{u} \times \nabla = 0, \tag{2.3.6}$$

$$\nabla \cdot \vec{\omega} = 0. \tag{2.3.7}$$

Formally, these conditions of compatibility are a particular case of the general condition of compatibility for distortion (2.3.2). However, expressions (2.3.6) and (2.3.7) do not define uniquely the displacement field **u**, but only with precision down to terms $\vec{\omega}_0$ for the rotation tensor $\omega$ and terms of the type $\nabla\nabla\phi$ for tensor **e**. This situation is far from being unique. Thus, for example, in electrodynamics, an additional random condition (Lorentz gauge) is imposed on the electromagnetic field. Thus, the observance of conditions (2.3.6), (2.3.7) as such does not guarantee the existence of field **u**.

*Internal incompatibility*

When a body experiences a *random* plastic deformation, the latter does not meet the conditions of compatibility. Unlike perfect crystals, the space, formed after relaxation of an imperfect crystal into the natural state represents a manifold of Euclidean microspaces. In this case, the transformation matrix $A_i^{i'}$ (see p. 33) need not be linear, and transformations of coordinate systems become non-holonomic.

**Definition 2.3.1.** At a preset plastic deformation $\mathbf{e}^p$, the incompatibility tensor is defined by the relation:

$$\eta \equiv -\nabla \times \varepsilon^p \times \nabla. \tag{2.3.8}$$

Since as for a joint plastic deformation $\eta = 0$, tensor (2.3.8) characterizes a deviation from compatibility. Expression (2.3.8) leads to the continuity condition:

$$\nabla \cdot \eta = 0. \tag{2.3.9}$$

One of the basic problems in the analysis of a plastically deformed body on the basis of equation (2.3.8) lies in the fact that tensor $\boldsymbol{\varepsilon}^p$ is not a state parameter. It results in the fact that this tensor is not defined uniquely, but is a functional of the preceding history of loading [184]. But elastic deformation is quite a definite value. Therefore, while defining elastic deformation as:

$$\varepsilon = \varepsilon^T - \varepsilon^p, \tag{2.3.10}$$

where $\varepsilon$ is elastic deformation and $\varepsilon^T$ is the total deformation, we have to demand fulfilment for it of the condition of the type (2.3.6):

$$\mathrm{inc}\ \varepsilon^T = 0.$$

With account of expression for incompatibility tensor (2.3.8), we obtain the equation for field $\eta$ [70]:

$$\nabla \times \varepsilon \times \nabla = \boldsymbol{\eta}. \tag{2.3.11}$$

**Remark 2.3.1.** Equation (2.3.11) expresses, *from the geometry viewpoint*, the fact that in case a body has defects, to ensure its continuity, we need to have elastic deformation field $\boldsymbol{\varepsilon}$.

In other words, ***incompatibility is a source of elastic deformation***. This existing incompatibility of the state causes essential mechanical problems when describing [31, 118] the behavior of materials. These difficulties are connected both with the structural (engineering) level (for example, in real constructions there are internal self-balanced stresses [51, 245]), and with the level of grain, or mesolevel—theoretical description of structure formation at deforming, self-organization of dislocation fields, and material aging processes. Unfortunately, the mathematical and conceptual means of classic mechanics were formed as applied to small deformations, the presence of essential restrictions (for example, the provision of compatibility of deformations) and fails to essentially satisfactorily describe the observed phenomena and processes.

In most cases, it is assumed that incompatibility is caused by the field of *geometrically necessary dislocations* [35] or field of disclinations [70]. In any case, the tensors of disclination density $\theta$ and dislocations $\alpha$ may be introduced as measures of incompatibility:

$$\theta \equiv -\nabla \times \varkappa^p, \quad \alpha \equiv -\nabla \times \boldsymbol{\beta}^p, \tag{2.3.12}$$

where $\varkappa$ is the tensor of bend-twist

$$\varkappa = \nabla \overrightarrow{\omega},$$

and index $p$ refers the value to the description of plastic deformation.

Apparently, it is possible to regard any given inhomogeneity as resulting from a suitable combination of chosen basic defects. In real bodies the same inhomogeneity may result from different combinations of screw and edge dislocational defects when judged on the basis of components [48]. And good theory stems from a criterion of mechanical consistency, namely from the remark that a change in the uniform reference must not affect the inhomogeneity, and leaves no ambiguity about the separate contribution of screw and edge dislocation defects.

### *2.3.1 Basics of continuous approximation for imperfect crystals*

We discussed earlier (see §1.2.1) a principal possibility of continual consideration of macrodefects, and microcracks among them. The use of this continual consideration is justified by the necessity in each particular case to define the notion of 'macroscopicity'. Thus, when considering an interaction of a crack with a dislocation, a fracture area having the length of $A \sim 10^2 \dots 10^4$ Burgers vectors may be considered to be macroscopic. In this case, a singular discontinuous field (1.2.2) is replaced by a continuous one with the help of a respective 'smearing' of $\delta$-function or application of the sinusoidal Frenkel potential. In this case, a continuous quantity—the fissuring tensor—enters the space state basis. Thus, from the formal viewpoint, a transition to the continuous description has caused an appearance of additional degrees of freedom. Similar problems also arise at the continual description of deformation at the microlevel, that is, at the level of structure defects.

A fundamental problem in mechanics of bodies with microstructure lies in calculation of fields of internal stresses and deformations, which arise from distributed defects of the structure. The theory of continuously distributed dislocations was initially developed by Kondo [162] and independently by Bilby with co-authors [27] on the basis of the work of Nye [264]. Later, the approach was developed in the works of E. Kröner, A. M. Kosevich, and L. I. Sedov. In all these geometrical theories, the deformation tensor played the role of the fundamental metric space tensor.

In a paper [257] W. Noll exploited the remark that material properties, more specifically material uniformity, may endow simple bodies with the structure of general manifolds. Thus, in contrast with K. Kondo, B. Bulby and E. Kröner the geometric properties are interpreted by Noll as a consequence of constitutive prescriptions.

The body is imagined as a collage fragment of the same material, patched together after having been differently strained. Thus the inhomogeneity of strain is trapped into the body and no smooth deformation can restore, in general, the original uniformity [57, 316]. The collection of uniform local placements (uniform reference) characterizes the *inhomogeneity* of the body[6].

Besides, already K. Kondo [164] established that the Cartan torsion (the first Cartan curvature tensor) is describing, when simulating the behavior of solids, movement of dislocations in the crystal. It leads to identifying the curvature with the Burgers vector of

[6] Maybe the best way is to unite both representations the first better for describing such kind of defects as dislocation and disclination, and the second one for describing boundaries of ideal grain, crystallites and others.

Table 2.1
Interconnection defects field—space model

| Curvature tensors | Space model | Real objects realization |
|---|---|---|
| $R = S = T = 0$ | Flat space | Ideal crystal |
| $R = T = 0,\ S \neq 0$ | State of absolute parallelism | Deformed body (dislocations in continuum) |
| $T = S = 0,\ R \neq 0$ | Riemann space | System with disclinations. Cosserat media |
| $R \neq 0,\ S \neq 0,\ T = 0$ | Riemann–Cartan space | Theory of dislocations and disclinations |
| $R \neq 0,\ S \neq 0,\ T \neq 0$ | Affine—metric (Finsler) space. | Real inhomogeneous body. Point, linear and surface defects |

the dislocation. Plastic deformation (start of plastic flow) may be considered as destabilization of a 3D body in the Euclidean space of a greater number of measures. Thus, for the presence of all types of defects the crystal requires the existence of three non-zero Cartan tensors [109, 143], Table 2.1.

From the mathematical viewpoint, any curved space should have some definite connectivity. It opens a natural way for introduction of parallelism. Since connectivity is non-symmetrical, the resulting space is non-Riemann in the general case [317]. It may be interpreted as lattice curvature in the presence of dislocation, while the evolving curvature is treated as the Nye curvature [264]. Ultimately, we can consider the deformation process in the following way. At the initial moment, prior to deformation, the body contains no defects and no deformations of the ideal lattice and is embedded into 3D Euclidean continuum. In this state, the 3D Euclidean geometry with Lagrange coordinates is sufficient to describe the body configuration. The state of the body after deformation is described not only by a displacement of the body after deformation from the Euclidean space but also by translation of its structure (actual configuration) into the non-Euclidean one.

For imperfect crystals, we can probably build a ***local mapping*** of the imperfect crystalline medium into an ideal crystal without defects. However, since for crystals with defects there exist the incompatibility tensor (2.3.8) and bend-twist tensor (2.3.12), this transition is impossible in the whole volume of the crystal. In physical terms, the respective incompatibility of deformations in microlevel, related to non-elastic behavior, is described in terms of Riemannian or non-Riemannian geometry [164]. Thus, for a non-ideal crystal we need to define non-holonomic coordinates [315]. Along with that, we may consider a closed-loop tracking, comprised of sections of four geodesics[7]. In this case, the nullity vector appears (the vector connecting the initial and transferred vectors) and a turn related to the non-holonomic object. Depending on the medium imperfection, we obtain various components of nullity and non-holonomity object. It is important to obtain the whole geometrical structure of the continuum with defects without a prior

[7] Naturally, this transition is defined in the tangent space.

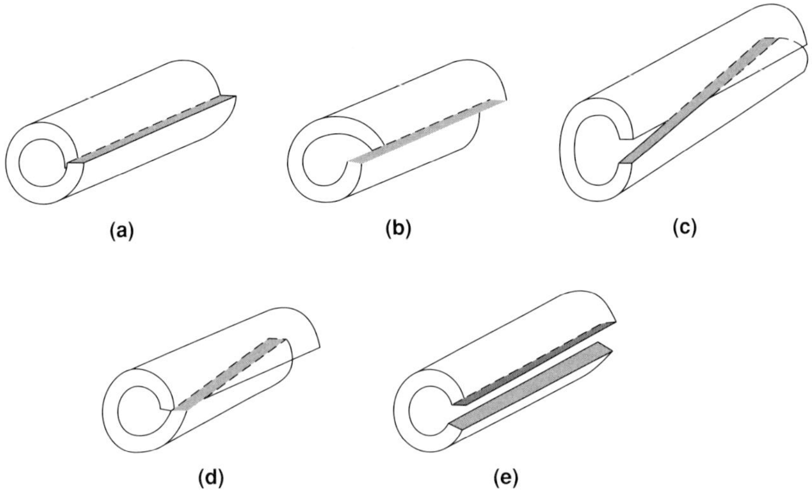


Fig. 2.1. The Riemann–Christoffel tensor component interpretation: (a) $R_{11}{}^{1}{}_{2}$, $R_{21}{}^{2}{}_{2}$, (b) $R_{32}{}^{3}{}_{1}$, (c) $R_{31}{}^{1}{}_{2}$, $R_{31}{}^{2}{}_{2}$, (d) $R_{11}{}^{3}{}_{2}$, $R_{11}{}^{2}{}_{2}$, (e) $R_{21}{}^{1}{}_{2}$, $R_{21}{}^{3}{}_{2}$.

introduction of the continuum metric. This corresponds to the fact that the defects are the ***reason*** for metric appearance in areas with defects. The presence of different defects required different geometrical space structure [109, 143]. Naturally, various types of defects are formed in the real material independently and deform the continuum in different ways [270]. In this case, as it was shown by Berdichevskiy and Sedov [317], we can introduce non-holonomic geometry, in which the curvature tensor of Riemann–Christoffel is equal to zero, and the affine connectivity has a more general character than the one taken by Bilby [27]. Should the Riemann–Christoffel tensor have components different from zero, we may obtain their geometrical and mechanical interpretation, Fig. 2.1.

**Remark 2.3.2.** Since defects in the body are formed and developed independently, in the ideal case we need an opportunity to independently define the tensors of curvature, torsion, and Riemann–Christoffel. It is ***principally impossible within the Riemann geometry, where there is a single-valued interconnection among all the tensors.***

## 2.4 Finsler geometry and its applications to mechanics of a deformed body

It was noted in §2.3 that to describe media with microstructure, it is necessary to use the space with some geometry, different from the Euclidean one. One of these affine spaces, rather broadly used in physics and mechanics of deformed body, is the Finsler space.

### 2.4.1 Finsler space

The notion of Finsler space was introduced into mathematics by Paul Finsler, who presented in 1918 to the university of Göttingen his dissertation 'Über Kurven und Flächen in allgemeinen Raumen'[8]. The general theory of Finsler space can be developed without introduction of metric [231]. In this case, the fundamental notion of the theory is nonlinear connectivity, defined in the tangential fibering of the differentiable manifold.

*Finsler geometrical objects on a manifold*

Let us assume that within this paragraph, capital Greek letters take the values $0, \ldots, N$. Then, the following is valid.

**Definition 2.4.1.** The field of Finsler geometrical objects of order $k$ in manifold $\mathcal{M}$ is called the set $N$ of functions $\mathfrak{O}^{\Lambda}(x, y)$, defined in each coordinate vicinity of atlas $\pi^{-1}(U)$ in the tangential fibering $\mathfrak{T}(\mathcal{M})$. When changing coordinates, the objects change according to the law:

$$\mathfrak{O}^{\Lambda}(x', y') = F^{\Lambda}\left(\mathfrak{O}^{\Sigma}(x, y), x^{i}, x^{i'}, \partial_j x^{i'}, \ldots, \partial_{jk\ldots jl} x^{i'}\right). \tag{2.4.1}$$

In this case, function $F^{\Lambda}$ meets the additional requirements:

$$\begin{cases} \mathfrak{O}^{\Lambda}(x, y) = F^{\Lambda}\left(\mathfrak{O}^{\Sigma}(x, y), x^{i}, x^{i}, \delta^{i}_{j}, 0, \ldots, 0\right); \\ F^{\Lambda}\left(F^{\Sigma}\left(\mathfrak{O}^{\Pi}(x, y), x^{i}, x^{i'}, \partial_j x^{i'}, \ldots, \partial_{jk\ldots jl} x^{i'}\right) x^{i'}, x^{i"}, \partial_{j'} x^{i''}, \ldots, \right. \\ \left. \partial_{j_{k'},\ldots j_{l'} x^{i''}}\right) = F^{\Lambda}\left(\mathfrak{O}^{\Sigma}(x, y), x^{i}, x^{i''}, \partial_j x^{i''}, \ldots, \partial_{jk\ldots jl} x^{i''}\right). \end{cases} \tag{2.4.2}$$

If we compare definitions 2.1.20 and 2.4.1, they are in fact different in the definition area of the objects. Finsler objects are only the function of coordinates and first-order derivative of coordinates. If we have function of high order derivatives the objects formed Lagrangian space.

Local map $\pi^{-1}(U), \phi$ in manifold $\mathfrak{T}(\mathcal{M})$ does not have $\left(\mathfrak{O}^{\Lambda}(x, y)\right)$ as function coefficients.

### 2.4.2 h- and v-connectivities

Field fibering tangents of Finsler geometrical objects enable introducing, according to general geometrical presentations, various derivatives depending on the leading variable. According to the general procedure [306], we have:

$$M_x = \Gamma^h_u \oplus \Gamma^v_u \oplus \mathfrak{T}(\mathcal{M})^v_x, \tag{2.4.3}$$

where $\Gamma^h_u$ is the horizontal connectivity, $\Gamma^v_u$ is the horizontal connectivity in point $u \in \mathfrak{T}(\mathcal{M})$, and $M^v_x$ is the vertical subspace of the tangent space $M_x$.

[8] 'On Curves and Surfaces in Generalized Spaces'.

Let us designate the coefficients of h-connectivity as $F^i_{jk}(x, y)$ and coefficients of v-connectivity as $C^i_{jk}(x, y)$. Then, there exists the single Finsler connectivity $(N, F, C)$. Coefficients of this connectivity are $N^i_j$, $F^i_{jk}(x, y)$ and $C^i_{jk}(x, y)$. We can show that $F^i_{jk}(x, y)$ is a Finsler linear differential object with the transformation law according to (2.1.55). Coefficients $C^i_{jk}(x, y)$ are transformed according to the law:

$$C^{i'}_{j'k'}(x', y') = \frac{\partial x^{i'}}{\partial x^i} \frac{\partial x^j}{\partial x^{j'}} \frac{\partial x^k}{\partial x^{k'}} C^i_{jk}(x, y). \tag{2.4.4}$$

We can use three coefficients $N^i_j$, $F^i_{jk}(x, y)$ and $C^i_{jk}(x, y)$ to build three curvature tensor fields and five torsion tensor fields [231, 306]. Thus, for example, for the curvature tensors we have:

$$R^i_{jkl} = \frac{\delta F^i_{jk}}{\delta x^l} - \frac{\delta F^i_{jl}}{\delta x^k} + F^m_{jk} F^i_{ml} - F^m_{jl} F^i_{mk} + C^i_{jm} F^m_{kl}, \tag{2.4.5}$$

$$P^i_{jkl} = \frac{\partial F^i_{jk}}{\partial y^l} - C^i_{jl|k} + C^i_{jm} P^m_{kl}, \tag{2.4.6}$$

$$S^i_{jkl} = \frac{\partial C^i_{jk}}{\partial y^l} - \frac{\partial C^i_{jl}}{\partial y^k} + C^m_{jk} C^i_{ml} - C^m_{jl} C^i_{mk}, \tag{2.4.7}$$

where

$$P^i_{jk} = \frac{\partial N^i_j}{\partial y^k} - F^i_{kj}, \quad R^i_{jk} = \frac{\delta N^i_j}{\delta x^k} - \frac{\delta N^i_k}{\delta x^j}.$$

**Remark 2.4.1.** We remark here that all the introduced tensors were built without using the metric. Taking into account the mechanical nature of our consideration, the metrical properties, according to our presentation, can be connected with the natural state of the continuum, while the properties of the tangent space are defined by the defective structure of the continuum.

**Remark 2.4.2.** The curvature and torsion tensors, obtained from the Finsler connectivity $N, F, C$, are not absolutely free. They are linked by the eleven Bianchi identities [231].

### 2.4.3 *Fracture geometry of solids*

For a long time, the fracture and deformation processes have been viewed from two weakly interconnected positions [276]: as a continuum mechanics subject, when a material is characterized by integral characteristics; and as a subject of the dislocation theory, when local characteristics of the medium and ensembles of defects are considered. Since the phenomena, which are taking place near the peak of the propagating crack, represent a complex combination of processes of local and global character—elastic and plastic material deformation, generation and accumulation of microscopic cracks and microdefects—a correct description of crack propagation requires building some sort of synthetic theory.

*Building principles of synthetic theory*

The general provisions of the multilevel hierarchical system theory (MHST) indicate that such synthetic theory should be built as a microscopic theory with imposition of the laws and control of the macroscopic level. The microscopic level can be described in the continual dislocation theory. The basic idea underlying the continual dislocation theory, starting from the fundamental works of B. Bilby, E. Kröner, A. Kosevich and I. Kunin, lies in establishing the links existing between the defects presenting in a real solid and the geometry (metric properties) of the medium. These metrical properties make the control, i.e., they perform the role of coordinator. The geometrical approach allows making a correct and logical description of crystal imperfections. However, the connections of metrical medium properties with the processes of plastic deformation and fracture (crack motion) have been poorly studied so far. We shall consider the effect of the medium metrical properties (of the metric tensor) on the crack trajectory. We shall link the fractality of the fracture process with local geometrical properties of the space, which is associated with destruction.

*Geometrical account of microstructure*

As a rule, we consider a material body (continuum) $\mathcal{B}$ with some microstructure at the equilibrium configuration $C_0$, in which the density $\rho_0$ and the temperature $T$ have uniform values, the stress state is not uniform, and the heat flux is everywhere equal to zero.

The existence of internal degrees of freedom for media with microstructure requires a respective description tool. Back in 1955, K. Kondo suggested to use tangent fiberings for describing non-uniformly deformed crystals [163]; these fiberings being linked with six parameters. These parameters can be correlated with three coordinates of the centre of deformed grain and three non-holonomic coordinates, which describe the initial grain (or crystal) orientation. The more general Finsler geometry for describing the non-uniform plastic deformed state was introduced some time later [165, 171]. The Finsler geometry, as a metric generalization of the Riemannian, apart from the deformed solid mechanics [306, 307], has been widely used also in physical generalizations of gravitation theories and gauge field theory [10, 303]. One of the basic advantages of the Finsler geometry in the fracture theory is a possibility to describe, without introducing any additional hypotheses on the properties of the continuum, the defect medium structure and the history of deformation, since the metric and metric tensor of Finsler space are functions not only of coordinates, but also of the change rate of the coordinates along the parameter.

Since the geometrical structure is related with the symmetry group of the given space, the use of the generalized space also allows one to build a non-contradictory picture of introduction of the gauge invariance procedure and to find the Lagrangian of gauge fields from first principles [210]. However, the full picture of Finsler presentation of intrinsic state of deformed body is still far from completion.

The simplest effect arising in the presence of defects in a solid is the appearance of dedicated directions (for example, dislocation or disclination axis, direction of crack propagation, or a grain axis). This demands introduction into the structure of the ideal continuum model of additional free parameters. Thus, J. Saczuk [308] supposed that for a defected body, each material point of $\mathcal{B}$ located in the position $\mathbf{x}$ is endowed with an internal structure described by an additional independent microvector (director) $\mathbf{y}$. All such points are embedded into the continuum which contains heterogeneously distributed dislocations and voids (crack and dislocation arrays). The state space $B$ for the body without microstructure is a three-dimensional physical Euclidean space. The state space $\mathcal{B}$ for the body with microstructure is the fiber space

$$B \times \times M \subset \mathbb{E}^3 \times \mathbb{E}^3, \tag{2.4.8}$$

where $B$ is the body and $M$ is the microstructure in the reference configuration $C_0$, $E$ is the Euclidean space and $\times\times$ denotes the local Cartesian product (a local trivialization).

Then, in mathematical terms, deformation of a body is a mapping process from the initial configuration $C_0$ into the current one $C_t$,

$$C_0 \mapsto C_t, \Leftrightarrow \exists \quad \mathbf{X} = \xi(\mathbf{x}, \mathbf{y}), \quad \xi : \mathcal{B} \mapsto \mathbb{E}^3 \times \mathbb{E}^3. \tag{2.4.9}$$

A special case of mapping (2.4.9) is the mapping of the Cosserat continuum [308]. In the case of the Cosserat continuum, orientation of the space is achieved by introduction of the director vector.

When considering defects in the most general case, a continuum is related to a certain manifold $\mathcal{M}_n$, where $n$ is the dimensionality of the manifold. Depending on the selected interpretation of defect characteristics, the manifold has this or that geometrical character—from limited theories (in which the curvature tensor $R^l_{ij,k} = 0$) to metric affine-bound manifolds of the most general type.

Hereinafter, everywhere Latin indices will number space unitary vectors; for dummy indices Greek letters may also be used.

For random manifold $\mathcal{M}_n$, one of the simplest generalized spaces, the metric of which admits the existence of all the three non-zero curvature tensors and which is used to describe the media with microstructure, is the Finsler space [306, 307]. As we have already noted in §2.4.1, the coefficients of the local map $\mathfrak{D}^\wedge(x', y')$ for the Finsler space depend only from $x^i, \dot{x}^i$. Among other factors, connected with a possibility of correct description of plastic deformation of media with microstructure, the selection of these spaces is caused by the fact that for the Finsler space the Hamilton function of system $H(x, y)$ and metric function $F(x, \dot{x})$ are connected by usual canonical equations [303]:

$$\frac{\partial F(x, \dot{x})}{\partial x^i} = -\frac{\partial H(x, y)}{\partial x^i}, \quad \dot{x} = \frac{dx^i}{dt}. \tag{2.4.10}$$

In (2.4.10) values $x^i, \dot{x}^i, y_i$ are considered as independent variables of respective right- and left-hand parts of the equation. To reduce the notation, we also adopt an agreement on function arguments:

$$f(x^i, x^j \ldots) = f(x).$$

Coordinates, as a rule, are designated with Latin letters $x, y$.

### 2.4.4 Metrical Finsler spaces

The metrical Finsler space is a special case of Finsler space. Finsler connectivity $\Gamma^{\mu}_{\sigma\lambda}$ is the Finsler metric connectivity, only if $h$- and $v$-covariant derivative

$$g_{ij|k} = 0, \quad g_{ij|k} = 0, \tag{2.4.11}$$

where $g_{ij}$ is a symmetrical non-degenerate Finsler field of type (0,2).

Let in domain $R$ of random $n$-dimensional space $X_n$ the metric function $F(x^i, \dot{x}^i)$ be set. Transformation of coordinates is set by $n$ equations:

$$x^{j'} = x^{j'}\left(x^1, ..., x^n\right), \quad \left(j', i' = 1, ..., n\right). \tag{2.4.12}$$

We suppose that functions $x^i$ are at least class $C^2$, i.e., continuous in all their variables and possess a continuous derivative up to the second order inclusively. We believe also that the transformation Jacobian (2.4.12) does not turn identically into zero, that is $\det\left(\partial x^{i'}/\partial x^i\right) \neq 0$. Then, in space $X_n$, the distance may be introduced between points:

$$ds = F\left(x^i, \dot{x}^i\right). \tag{2.4.13}$$

The following obvious conditions are imposed on function $F\left(x^i, \dot{x}^i\right)$:

A. Function $F\left(x^i, \dot{x}^i\right)$ is positive, if not all $\dot{x}^i$ are equal to zero simultaneously, that is, $F\left(x^i, \dot{x}^i\right) > 0$ at $\sum_i \left(\dot{x}^i\right)^2 \neq 0$.

B. Quadratic form

$$F^2_{\dot{x}^i\dot{x}^j}\left(x^i, \dot{x}^i\right)\xi^i\xi^j \equiv \frac{\partial^2 F^2\left(x^i, \dot{x}^i\right)}{\partial \dot{x}^i \partial \dot{x}^j}\xi^i\xi^j > 0,$$

at all $\xi^i$, not equal to zero simultaneously.

In this case, space $X_n$ is a positively defined Finsler space [294].

The Riemannian space is tangent to the Finsler one in the point (in other words, the Finsler space is local Riemannian one). From the usual definition of distance for the Riemannian and Euclidean spaces, $ds^2 = g_{ij}x^ix^j$, condition B defines the metric tensor of tangent space $T_n$:

$$g_{ij} = g_{ij}\left(x, \dot{x}\right) = \frac{1}{2}\frac{\partial^2 F^2\left(x^i, \dot{x}^i\right)}{\partial \dot{x}^i \partial \dot{x}^j}, \tag{2.4.14}$$

which is, as the metric tensor should be, the tensor of second rank [294]. Value $\dot{x}^i$ is a random vector of tangent space in contact point $T_n(P)$; $\dot{x} = dx/dt$, where $t$ is the invariant parameter along the curve, $x^{i'} = x^{i'}(x^i(t))$. In this case, each random counter-variant vector $\dot{x}^i \in T_n$ may be correlated with covariant vector $y_i$ of dual tangency space:

$$y_i = g_{ij}(x, \dot{x})\dot{x}^j. \tag{2.4.15}$$

By the geometrical sense, values $\dot{x}^i$ define linear vector space $T_n$ of dimension $n$, tangent to the basic manifold $\mathcal{M}_n$, and are contravariant vectors of this $T_n$. The tangency point is manifold point $P(x^i)$. The Riemannian space is a special case of the Finsler space, for which $g_{ij} \neq g_{ij}(\dot{x}^i)$, that is, the metric tensor does not depend on the direction. For the Riemannian space, the tangent space is the Euclidean one, and the Riemannian space may be considered as locally Euclidean space. Similarly, the Finsler space is locally the Minkowski space [10].

We can show that the connectivity coefficients of the metric Finsler space have the form [303]:

$$P_{h\ j}^{*i}(x,\dot{x}) = g^{ik}(x,y)P_{hkj}^{*}(x,\dot{x}); \quad y \equiv g_{ij}(x,\dot{x})\dot{x}^j, \tag{2.4.16}$$

where $P_{h\ j}^{*i}(x,\dot{x})$ are symmetrical connectivity coefficients;

$$\begin{aligned} P_{ikj}^{*}(x,\dot{x}) = \gamma_{ikj}(x,\dot{x}) - \Big[ & C_{jkh}(x,\dot{x})P_{i\ l}^{\ h}(x,\dot{x}) \\ & + C_{kih}(x,\dot{x})P_{j\ l}^{\ h}(x,\dot{x}) - C_{ijh}(x,\dot{x})P_{k\ l}^{\ h}(x,\dot{x})\Big]\dot{x}^l; \end{aligned} \tag{2.4.17}$$

$$P_{k\ j}^{\ i}(x,\dot{x}) = \gamma_{k\ j}^{\ i}(x,\dot{x}) - C_{k\ l}^{\ i}(x,\dot{x})\gamma_{p\ j}^{\ l}(x,\dot{x})\dot{x}^p = \Gamma_{k\ j}^{\ i}; \tag{2.4.18}$$

where $\gamma_{k\ j}^{\ i}(x,\dot{x})$ are Christoffel symbols of the second kind,

$$\gamma_{k\ j}^{\ i}(x,\dot{x}) = g^{ip}(x,\dot{x})\gamma_{kpj}(x,\dot{x}).$$

Christoffel symbols of the first kind $\gamma_{kji}(x,\dot{x})$ are defined, like in the Riemannian geometry, by equations:

$$\gamma_{k\,j\,i}(x,\dot{x}) = \frac{1}{2}\left(\frac{\partial g_{ki}(x,\dot{x})}{\partial x^j} + \frac{\partial g_{ij}(x,\dot{x})}{\partial x^k} - \frac{\partial g_{jk}(x,\dot{x})}{\partial x^i}\right). \tag{2.4.19}$$

Despite the fact that the Finslerian Christoffel symbols are defined by the same rule (2.4.19) as in the Riemannian case, their transformation law under coordinate transformations $x^{j'} = x^{j'}(x^j)$ differs essentially from the transformation law of the Riemannian Christoffel symbols. This fact may be inferred from the dependence of Finslerian Christoffel symbols $\gamma_{jk}^i(x,y)$ not only on the point $x^i$ but also on tangent vector $y^i$. Since hereinafter, as a rule, we consider metric Finsler spaces, in equation (2.4.18) we have passed over to the usual designation of the affine connectivity, $\Gamma_{k\ j}^{\ i}$. The procedure, described in detail in [303, Chaper 2], leads to the following identity [10]

$$\Gamma_{i\ j}^{\ k}(x,y) = \gamma_{i\ j}^{\ k} - C_{i\ n}^{\ k}\frac{\partial G^n}{\partial y^j} - C_{j\ n}^{\ k}\frac{\partial G^n}{\partial y^i} + C_{ijn}\frac{\partial G^k}{\partial y^l}g^{ln}. \tag{2.4.20}$$

Coefficients (2.4.20) are called *the Cartan connections coefficients*. Value $C_{ijk}$ is the characteristic tensor of the Finsler geometry (Cartan torsion), which arises from the dependence of the metric from derived coordinates by parameter:

$$C_{ijk}(x,\dot{x}) = \frac{1}{2}\frac{\partial g_{ij}(x,\dot{x})}{\partial \dot{x}^k} = \frac{1}{2}\frac{\partial^3 F^2(x,\dot{x})}{\partial \dot{x}^i \partial \dot{x}^j \partial \dot{x}^k}, \tag{2.4.21}$$

where $g_{ij}(x,\dot{x})$ is the metric tensor, defined by the fundamental metric function $F(x,\dot{x})$.

Analogously to (2.1.54) we can build covariant partial derivatives of the Finsler space. The covariant partial derivative of any tensor which depends on $x^i$ alone, for example of the tensor field of the form $X_i^k(x)$ can be constructed by the comparison of the transformation laws of $\partial X_i^k/\partial x^j$ and (2.4.19) [303, Chapter 2]. This procedure leads to the definition

$$X_i^k(x)_{|j} = \frac{\partial X_i^k(x)}{\partial x^j} + \Gamma^{\ k}_{n\ j}(x, y)X_i^n(x) - \Gamma^{\ n}_{i\ j}(x, y)X_n^k(x). \tag{2.4.22}$$

As noted above, in the case if $g = g_{ij}(x)$, we have the Riemannian geometry, which coincides with the Finsler one in the local point.

According to the definition of three Cartan curvature tensors, for the Finsler space we have:

$$\tilde{K}_{j\ hk}^{\ i} = \left(\frac{\partial \Gamma^{*i}_{j\ h}}{\partial x^k} + \frac{\partial \Gamma^{*i}_{j\ h}}{\partial \dot{x}^l}\frac{\partial \xi^l}{\partial x^k}\right) - \left(\frac{\partial \Gamma^{*i}_{j\ k}}{\partial x^h} + \frac{\partial \Gamma^{*i}_{j\ k}}{\partial \dot{x}^l}\frac{\partial \xi^l}{\partial x^h}\right)$$
$$+ \Gamma^{*i}_{m\ k}\Gamma^{*m}_{j\ h} - \Gamma^{*i}_{m\ h}\Gamma^{*m}_{i\ k}. \tag{2.4.23}$$

Tensor (2.4.23) is called *the relative curvature tensor* in view of the fact that it depends on arbitrary vector fields $\xi$ together with its derivative $\xi^l_{\ ,k}$.

### 2.4.5 *Indicatrix and orthogonality condition in Finsler space*

The notions of angles and direction volumes in the Finsler space have certain peculiarities connected with selection of the supported (reference) element [303]. Let us consider certain notions, important for further applications.

*Indicatrix*

**Definition 2.4.2.** In each tangent space $T_n$ the Finsler metric function $F(x, y)$ defines a $(N-1)$-dimensional hypersurface

$$F(x, y) = 1, \tag{2.4.24}$$

where $x^i$ is assumed to be fixed, any $y^i$ is random, called *the indicatrix.* It can be seen from this definition that the indicatrix is formed by the end points of unit vectors $l^i(x, y) = y^i/F(x, y)$ supported by point $x^i$.

The indicatrix may be assigned parametrically by means of a set $N-1$ of scalar values

$$u^a = u^a(x, y), \quad a = 1, 2, \ldots, N-1, \tag{2.4.25}$$

on which certain additional conditions are imposed [10]:

1. The positive zero-degree homogeneity of $u^a$ with respect to $y^i$, i.e. $u^a(x, ky) = u^a(x, y),\ k > 0$.
2. The matrix having entries $\partial u^a/\partial y^i$ is of the highest rank; $\text{rank}(\partial u^a/\partial y^i) = N-1$.

3. The functions (2.4.25) obey the same differentiability condition as those imposed on the Finsler metric function. It means the smoothness of at least class $C^3$ and $C^5$ with respect to $x^i$ and $y^i$ respectively.

If we add function set $u^a$ by the null indices $a$ by function $\ln F(x, y)$ in such a way that:

$$z^a(x, y) = u^a(x, y), \quad z^0(x, y) = \ln F(x, y), \tag{2.4.26}$$

then, we can obtain the reverse transformation, and it means that there exist functions $w^i(x, z)$ of at least class $C^3$ and $C^5$ with respect to $x^i$ and $y^i$, respectively, such that:

$$y^i = w^i(x, x) \tag{2.4.27}$$

and

$$z^p(x, w(x, z)) = z^p. \tag{2.4.28}$$

As a result of choice $z^0 = \ln F$, equation (2.4.28) gives the unit $y^i$ in the case when function $w^i$ in (2.4.27) is evaluated for $z^0 = 0$. We can denote

$$t^i(x, y) = w^i(x^n, 0, u^a)$$

and obtain the expression for the projection factor

$$t^i_a = \frac{\partial t^i}{\partial u^a}, \quad u^i_a = \frac{\partial u^i}{\partial y^a}. \tag{2.4.29}$$

Values $t^i_a$ are connected with deformation of the Finsler space metric and define *the induced Riemannian metric tensor of the indicatrix* [10]

$$g^*_{ab} = g_{ij}(x, t(x, u))\, t^i_a(x, u) t^j_b(x, u). \tag{2.4.30}$$

Now, having the metric tensor at our disposal, we can build all the curvature tensors, associated with the indicatrix, in the usual way.

We can show [303] that since the Hamiltonian has the character of the reference surface for the indicatrix, functions $H$, $F$ are dual functions, and metric function $F(x, \dot{x})$ is the Lagrangian of the system [306, 307].

*Orthogonality in Finsler space*

The notion of orthogonality is important in the consideration of propagation of wave fronts and equienergy surfaces. Since the notion of angle for the Finsler space is defined non-uniquely, the orthogonality demands an additional consideration. Orthogonality is closely related with the notion of indicatrix.

Let us consider a contact of a hyperplane with the indicatrix in a given point $\dot{x}^i_{(0)}$ of tangential space $T_n(P)$. Similarly to the Riemannian geometry, this condition is written in the form of:

$$F_{\dot{x}^i}\left(x^k, \dot{x}^k_{(0)}\right)\left(\dot{x}^i - \dot{x}^i_{(0)}\right) = 0. \tag{2.4.31}$$

With account of expression (2.4.14) equation (2.4.31) takes the form:

$$g_{ij}\left(x^k, \dot{x}^k_{(0)}\right)\dot{x}^i_{(0)}\dot{x}^j = 1. \tag{2.4.32}$$

For a sphere of random radius $\xi$, the equation of the tangent hyperplane (2.4.31) is written as follows:

$$g_{ij}\left(x^k, \xi^k\right)\xi^i\dot{x}^j = |\xi|^2, \tag{2.4.33}$$

where $\xi^i$ is a random vector of length $|\xi| \equiv F(x, \xi)$, from centre O of space $T_n(P)$. Then any vector $\eta^j$, belonging to the hyperplane or parallel to it, is orthogonal (normal) to vector $\xi^i$. Since we can always represent $\xi^i = \dot{x}^i_{(2)} - \dot{x}^i_{(1)}$ [4], out of equation (2.4.33) we have:

$$g_{ij}\left(x^k, \xi^k\right)\xi^i\eta^j = 0. \tag{2.4.34}$$

It follows from equation (2.4.34) that due to the dependence of the metric tensor from arguments $x^i$, $\xi^i$, ***the notion of orthogonality in the Finsler space is asymmetric relative to*** $\xi^i$, $\eta^j$.

## 2.5 Description of plastic deformation in generalized space*

In the plasticity theory of ideal bodies, it is usual to represent the deformation process as the process of changing the yield surface in the nine-dimension space of stress tensor $\sigma^{ij}$. The dynamics equation of the yield surface is taken in the form of

$$f\left(\sigma^{ij}, g_{ij}, T, \mu_i\right) = 0, \tag{2.5.1}$$

where $f$ is the yield function or loading function, $T$ is the temperature, and $\mu_i$ are certain physicochemical parameters of the material. If the medium is isotropic, $\mu_i$ are scalars; otherwise, they can be tensor or vector values. The material models differ in the choice of the law defining the behavior of $\epsilon^p_{ij}$ and $\mu_i$. As a rule, no parameters of the internal structure are included in the material models. In this manner, the existing plasticity models are restricted by macroscopic volumes, for which it is possible to correctly isolate some representative volume and make averaging over this volume. Let us consider certain possibilities to build the material models, which are more adequate to the real physical origin of the material.

*Direct account of fissuring*

Equation (1.2.16) has been written down without account of microstructure and fissuring. Since coordinates $\xi^i$ in the equation are random curvilinear coordinates, we can compare them with extended basis (1.2.5). It was remarked in §2.1.4 that in order for random values to set a system of coordinates, they should be subjected to essential restrictions. Since we have to demand at least the affinity, the condition (2.1.29) should be observed:

$$\Gamma^k_{ij} = \frac{\partial^2 x^{k'}}{\partial x^i \partial x^j}\frac{\partial x^k}{\partial x^{k'}}.$$

Let us notice also that the metric space tensor may be written as (2.1.13):

$$g_{ij} = \mathbf{e}_i \mathbf{e}_j.$$

If we assume that the system of coordinates is orthogonal, then we can switch over from known basic unit vectors of a random curvilinear affine system of coordinates to new basic unit vectors with the use of Lamé coefficients:

$$\mathbf{e}_i = \frac{1}{H_i}\frac{\partial \mathbf{r}}{\partial q^i}, \quad \frac{\partial \mathbf{r}}{\partial q^i} = \frac{\partial x^k}{\partial q_i}\mathbf{e}_k$$

(Lamé coefficients are scalars, by index at summation coefficients are not!!)

$$H_i = \left|\frac{\partial \mathbf{r}}{\partial q^i}\right| = \sqrt{\left(\frac{\partial x^i}{\partial q^i}\right)}. \tag{2.5.2}$$

In the general sense, orthogonality should be understood as a 'possibility of decomposition by independent variables', similar to the well-known polar decomposition. With account of the expression for the Lamé coefficients (2.5.2), we can write down the basic unit vectors in the state space for a fissured body as:

$$\mathbf{e}_{T_e} = \frac{1}{H_{T_e}}\frac{\partial \mathbf{r}}{\partial T_e}; \ \mathbf{e}_{T^e_\alpha} = \frac{1}{H_{T^e_\alpha}}\frac{\partial \mathbf{r}}{\partial T^e_\alpha}; \ \mathbf{e}_\alpha = \frac{1}{H_\alpha}\frac{\partial \mathbf{r}}{\partial \alpha}; \tag{2.5.3}$$

where coefficients $H_{T_e} = \sqrt{\partial x^i/\partial T^e} = \sqrt{1/(\partial T^e/\partial x^i)}$; $H_{T^e_\alpha} = \sqrt{\partial x^i/\partial T^e_\alpha} = \sqrt{1/(\partial T^e_\alpha/\partial x^i)}$; $H_\alpha = \sqrt{\partial x^i/\partial \alpha} = \sqrt{1/(\partial \alpha/\partial x^i)}$. And further, we can write down equation (1.2.16) with account of fissuring by calculating the metric tensor of the medium and Christoffel symbols for the basic vectors, set by expression (2.5.3).

*Account of microstructure*

We now turn to the procedure of building the deformation tensor in the Euclidean space and the connection of this tensor with the metric tensor [200, 316]. Let a plastically deformed body fill a restricted domain $F_n$. The, the square of the distance measured along the curve in $F_n$ between two infinitely close points $P$ and $P + dP$ in Euclidean space is defined by the quadratic form:

$$ds^2 = g_{ij}(x, \dot{x}, t)dx^i dx^j, \tag{2.5.4}$$

where $g_{ij}(x, \dot{x}, t)$ is the extended basis of tensors found in $F_n$ in relation to $x^i, \dot{x}^i, t$.

**Definition 2.5.1.** $g_{ij}(x, \dot{x}, t)$ is called the deformation tensor.

Similar to (2.5.4), with account of definition (2.4.13) a length element can be defined in the reference configuration of $\mathcal{B}_F$ as

$$ds^2 = g_{ij}(x, y)dx^i dx^j + g_{ij}(x, y)Dy^i Dy^j,$$

and a length element in the actual configuration

$$d\tilde{s}^2 = \tilde{g}_{ij}(x, y)d\tilde{x}^i d\tilde{x}^j + \tilde{g}_{ij}(x, y)D\tilde{y}^i D\tilde{y}^j.$$

In this case, the total body deformation can be written down as

$$d\tilde{s}^2 - ds^2 = \mathbf{F}^T \tilde{g}\mathbf{F} - \mathbf{g}, \tag{2.5.5}$$

where $\mathbf{F}$ is the deformation gradient. Since the motion of the body $\mathcal{B}_F$ is realized in the non-Euclidean space, and taking into account the properties of our fibered space, its position vector $\mathbf{X}$ in the actual configuration is composed of the horizontal part $\mathbf{X}^h = \mathbf{X}(\mathbf{x}, \mathbf{y}(\mathbf{x}))$ and the vertical one $\mathbf{X}^v = \mathbf{X}(\mathbf{x}_p, \mathbf{y}(\mathbf{x}))$ where $\mathbf{x}_p$ denotes the coordinates of the path along which $\mathbf{y}$ is transported in parallel [306]. It means

$$\mathbf{X} = \mathbf{X}^h + \mathbf{X}^v. \tag{2.5.6}$$

Formally, $\mathbf{X}^h$ has a sense of classical position vector of a material point in the configuration space. In its turn, $\mathbf{X}^v$ is part of $\mathbf{X}$, which strictly depends on the topology of the internal space. Part $\mathbf{X}^v$ may be a function of the defects fields because the internal topology depends on the defects distribution.

Usually, a displacement (motion) is represented as a homeomorphism, continuously differentiable for any $t$, of radius-vector $\mathbf{X}$ of the initial position of the body by radius-vector $\mathbf{x}$ of the final position

$$\mathbf{x} = \mathcal{X}(\mathbf{X}, t)$$

with Jacobian

$$J = \det \mathbf{F},$$

where $\mathbf{F}$ with components

$$F = x_{k,K} = \frac{\partial \mathcal{X}_k}{\partial x_K}, \quad k, \quad K = 1, 2, 3 \tag{2.5.7}$$

is the direct deformation gradient.

With account of the classic definition of the deformation gradient (2.5.7) and decomposition (2.5.6), for the Finsler space we have decomposition of the gradient [306]

$$F = (\nabla^h + \nabla^v)X = F^h + F^v. \tag{2.5.8}$$

In this case:

$$F^h = \nabla^h \cdot X = {}_hF^{\beta}_{\alpha}\underline{\partial}_{\beta} \otimes dx^{\alpha}, \quad F^v = \nabla^v \cdot X = {}_vF^{\alpha}_{\beta}\partial_{\alpha} \otimes Dl^{\beta}. \tag{2.5.9}$$

Here, $\underline{\partial}_{\beta}$ is a unit vector of the current configuration, $\otimes$ the tensor product, and $v$-derivative and $h$-derivative are defined according to:

$${}_vF^{\beta}_{\alpha} = F^{v.\beta}_{\alpha} = L\partial_{\alpha|}X^{\beta} + A^{\beta}_{\delta\alpha}X^{\delta}, \tag{2.5.10}$$

$$ {}_hF_\alpha^\beta = (F^h)_\alpha^\beta = \partial_\alpha X^\beta - \partial_{\delta|} X^\delta \partial_{\alpha|} G^l + \Gamma_{\delta\alpha}^{*\beta} X^\alpha. \qquad (2.5.11) $$

In this case $\partial_\alpha$ is the partial derivative by $\alpha$-component of parameter $x$, $\partial_{\alpha|}$ is the partial derivative by $\alpha$-component of parameter $y$, $L$ is the metric function, $\Gamma_{\delta\alpha}^{*\beta}$, $\Gamma_{\delta\alpha}^{\beta}$ are connectivity coefficients of the Finsler and tangential Riemannian spaces, accordingly, $Dl = dl^\beta + \Gamma_{\delta\alpha}^\beta l^\delta dx^\alpha$ is the covariant derivative of the unit vector $l^\beta$, $A_{\alpha\beta}^\varepsilon = LT_{\alpha\beta}^\varepsilon$,

$$ G^l(\mathbf{x}, \mathbf{y}) = \frac{1}{2}\gamma_{jk}^l y^j y^k. $$

Eventually, the deformation gradient is only one of the summands of the complete gradient in conformity with decomposition (2.5.8).

Since the deformation gradient is defined, we can formulate all the defining relations of the classic mechanics for $g_{ij}$ according to the theorem of S. Minagava [66, 230]: 'Equilibrium equations for a stressed body for a material manifold are Bianchi identity relations or Codazzi equations, accordingly, for 3- and 2-D problems'. The form and expression for the deformation tensor depend on the deformation conditions, nature of the material and the shape of the body. Since actual systems are in equilibrium, there exists the homeomorphism of connectivity coefficients at transition of the system from the 'initial' into the real state [66].

For example, if the imperfection of the material is realized by means of dislocations, the $g_{ij}$ has the form [306]:

$$ g_{ij} = 2\int_{t_0}^{t} \Upsilon_{kij} v^k dt + \int_{x_0^k}^{x^k} (\Theta_{kij} + \Theta_{kji}) dx^k, \qquad (2.5.12) $$

where $\Upsilon_{kij}$ and $\Theta_{kji}$ are dislocation distributions (connectivity coefficients of the body material with dislocations [368]), defined accordingly by the density of moving dislocations and dislocations forest. The first term in expression (2.5.12) includes the distribution of mobile dislocations at the average speed $v^k$ in the time interval $t - t_0$, and the second term is connected with the forest dislocations in the average distance $x^k - x_0^k$, passed by the dislocation.

Once we have found the deformation tensor, we can build several different tensors of deformation and stress [311, 316] and choose the tensors, which are most adequately describing the problem.

It has been noted earlier that the Finsler space allows building three curvature tensor fields and five torsion fields. Thus, four tensor fields remain 'vacant' in principle, which may be used, for example, to account for other types of defects, thermal effects, hardening, etc. according to the way offered in [110, 111].

## 2.6 Geometry of nanotube continuum

Let us consider a possibility to describe carbon nanotubes by using the developed formalism. It is known [12, 83, 124] that chirality indices of nanotubes $i_1, i_2$ define

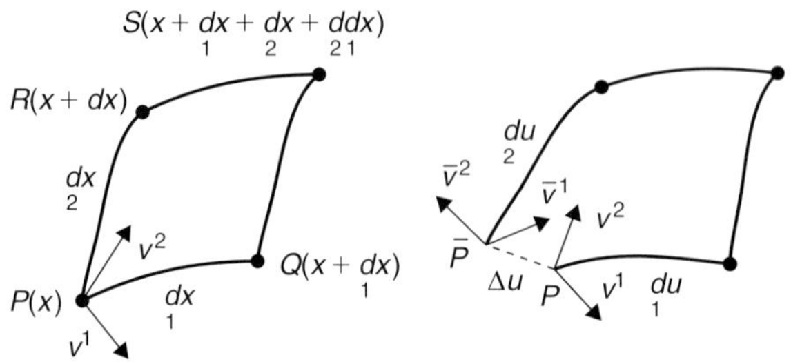


Fig. 2.2. Macroscopic contour and relaxation down to the natural state.

the vector $\mathbf{C} = i_1\mathbf{T}_1 + i_2\mathbf{T}_2$, where $\mathbf{T}_1, \mathbf{T}_2$ are vectors of elementary translations. These indices define two screw translations with angles and steps:

$$\delta\phi_i = 2\pi\frac{\mathbf{T}_i \cdot \mathbf{C}}{\mathbf{C}^2}; \qquad \Delta z_i = \mathbf{T}_i \cdot \mathbf{e}_i; \qquad i = 1, 2,$$

where $\mathbf{e}_i$ is the unit vector, directed along the band generator.

However, according to the Halphen theorem, the composite of two screw translations is the screw translation, the parameters of which $\phi, z$ can be defined in an elementary way [377]. Let $\varsigma$ be an elementary vector, parallel to the side of the carbon ring in the initial state.

According to the traditional procedure of encirclement in the space of imperfect (non-ideal) crystals (which is geometrically equivalent to the space of absolute parallelism [270]), the nullity vector of the end and start of a looped curve is equal to:

$$\Delta u^k = -2\Gamma^k_{[\mu\nu]} \underset{2}{d} x^{[\mu} \underset{1}{d} x^{\nu]} = 2S^k_{\mu\nu} f^{\mu\nu} \tag{2.6.1}$$

where $f^{\mu\nu}$ is the bivector of the 2-D area related to the infinitesimal loop and $S^k_{\mu\nu}$ is the torsion tensor. The vector rotation angle at encirclement is defined by the expression:

$$\Delta\mathbf{u}_\Lambda = \underset{2}{d}\,\underset{1}{d}\,\mathbf{v}_\Lambda - \underset{1}{d}\,\underset{2}{d}\,\mathbf{v}_\Lambda = R_{\lambda\mu\nu}^{\;\;k} f^{\mu\nu} \tag{2.6.2}$$

where $R_{\lambda\mu\nu}^{\;\;k}$ is the first curvature tensor (2.4.5).

Then, the deformation, connected with stretching and turn of the continuum (carbon monolayer), are presented as:

$$\delta\varsigma^{\varkappa} = -\frac{1}{2}\left(R_{\nu\mu\lambda}^{\cdot\cdot\cdot\varkappa}\varsigma^{\Lambda} + 2_{\mu\nu}^{\cdot\cdot\varkappa}\right) df^{\nu\mu}, \tag{2.6.3}$$

where $R_{\nu\mu\lambda}^{\cdot\cdot\cdot\varkappa}$ is the curvature tensor, $df_{\mu\nu}^{\cdot\cdot\varkappa}$ is the torsion tensor, and $df^{\nu\mu}$ is the bivector of the 2-D area related to the infinitesimal loop.

The torsion tensor can be easily related with the rotation vector, and the elastic deformation of the continuum is the deviation measure of the internal metric away from the external one. In this case, the compatibility conditions of Saint-Venant pass over into the condition of equality to zero of the Riemann–Christoffel tensor. From the elastic energy, we can define all the mechanical properties of nanotubes.

Coiled carbon nanotubes are a subject of additional interest [179]. A nanotube coiling is possible by two principally different methods: firstly, by coiling of a tube onto imaginary (virtual) cylinder having some definite radius $R$. In this case, the deformation is defined according to (2.6.3).

Secondly, by twisting a rectilinear tube around its own axis until appearance of instability, and deformation of the rectilinear generatix of the cylindrical shell into a spiral structure. In this case, the general deformation of the pipe continuum with account of the expression of relative curvature (2.4.23) will be defined as:

$$\delta \varsigma^{\varkappa} = -\frac{1}{2}\left(R_{\nu\mu\lambda}^{\cdots\varkappa}\varsigma^{\lambda} + \tilde{K}_{\nu\mu\lambda}^{\cdots\varkappa}\varsigma^{\lambda} + 2S_{\mu\nu}^{\cdots\varkappa}\right) df^{\nu\mu}. \tag{2.6.4}$$

Since in the second case the total deformation is much higher, the elasticity modulus of coiled nanotubes of the second type should have a compression–stretching asymmetry, and by stretching it should essentially exceed the elasticity modulus of coiled nanotubes of the first type.

## Certain outcomes

- Representation of the Euclidean space as of a model of a solid body space with defects is an *assumption*, which should be additionally justified. In the elasticity theory, this assumption is equivalent to the requirement of (2.3.6). For real bodies this postulate should be replaced with physical restrictions on incompatibility.
- The effect of microstructure on the intrinsic geometrical properties of the continuum can be accounted for by means of the non-Euclidean deformation tensor (p. 65). A particular tensor type (mathematical space classification) depends on the defects present in the material (Table 2.1).
- Mapping of non-Euclidean metric on real bodies makes it possible to obtain a series of *hidden parameters*, which cannot be described in non-Euclidean space. Thus, if a variable has dynamics along the fibering layer, its projection can be a point, parameters of which in the Euclidean space are constants. At the same time, a full behavior of the system can be sensitive to the dynamics on the layer.
- From the mathematical viewpoint, connectivity coefficients and curvature tensors can be obtained without any preliminary definition of the metric. In other words, curvature and twisting (second curvature) are introduced into space irrespective of the distance. Since the structure of curvature tensors is connected with the field of defects, it makes it possible to consider, in an independent way, the dynamics of defects and the dynamics of deformation. While building the damage tensor in this manner, we have obtained basis vectors (2.5.3).

# 3 Microscopic Crack in Defect Medium

## Contents



According to the basic principles of the theory of multilevel hierarchical systems, the behavior of the crack, as an object involved in a multilevel deformation and fracture process, is essentially different scale levels. It is logical to start the consideration of the fracture hierarchy from the level of quantum mechanics and unit fracture acts—recombination of interatomic bonds.

The consideration of unit fractures brings us to the thermoactivation fracture theory [295]. Along with that, the regulations of transition from the quantum level to the level of continual consideration are far from completely clear [28]. The basic difficulty lies in the fact that in the expression for stored energy of an elastic crystal,

$$\mathcal{E}(u) = \int_{\partial\Omega} E(\nabla u(x))dx, \tag{3.0.1}$$

an exact expression or even an approximation for the stored energy density $E$ is rarely available. The simplest model of calculations is based on a direct non-quantum summing up of contributions of interatomic potentials and derivation of certain averaged values [28, 106] for mechanical values.

With development of computing equipment, the basic regularities of formation of fracture surface may be found on the basis of 'point' approximation with the use of direct calculation methods and classical concepts of the elasticity and plasticity theory. The ideology of all these methods is connected with the well-known ideology of molecular dynamics, when researchers just monitor the behavior of individual particles in the whole ensemble by using the calculation capacities of a modern computer. Thus, reference [103] considers a sustainability model of a crack front using a model of weights on springs. In this case, the motion equations are written down in the simplest form:

$$m\ddot{U}_i = k_1(U_n - U_i) + k_2 U_i \theta(U_{fi} - U_i) + \sum_{j\in n.n} k_3(U_j - U_i) - b\dot{U}_i,$$

where $U_i$ is the shift of the $i$-th spring; $k_1$, $k_2$, $k_3$ are resilient constants of the springs, related accordingly to the $i$-th mass (located on the surface), mass in the other layer, and mass in the neighboring cell of the same layer; $U_N$ is the external offset applied to the plate; $\theta(x)$ is the Heaviside function; $U_{fi}$ is the threshold shift for the fracture of the $i$-th spring into two parts; and $b$ is the small dissipative parameter.

It follows from the equation that the interacting forces between the neighboring masses are taken to be linear. An interesting result is the presence of two bifurcation points in the curve: the propagation speed and applied edge offset. The first bifurcation arises at $\Delta \approx 1.0$, the second one at $\Delta \approx 1.05$, where:

$$\Delta \equiv \frac{U_n}{U_n^c}; \quad U_n^c \equiv \frac{U_n}{\sqrt{1 + k_2/k_1 U_f}}.$$

$U_n^c$ is the critical crack opening, at which the elastic energy is equivalent to the energy of the border break. The first bifurcation is obviously related with accumulation of internal energy in the material without crack growth (in this area, the crack propagation speed is equal to zero). Hereafter, the unsteady crack growth follows, similar to the results of §4.4 The next bifurcation is related to the mode of stable propagation. For the roughness indicator of the crack surface, $\alpha$, there also is bifurcation in the area $\Delta \approx 1.05$ as is stated in the next chapter. It has been established that instability of the crack front (physical curvature) arises in the whole area of changing parameters even for uniform media.

Analytically a similar result of bifurcation in simple atomic chain with linear interaction under external load was found fifty years ago [271].

Although in general utilization of direct methods allows one to obtain interesting results, it will be inevitably restricted by capacities of computers and will not allow one to obtain ***new*** regularities, since it is ***based*** on usage of known interaction laws[1].

With account of presentations of the MHST, a certain sequence of transition from a free atom to non-uniform continuum may be presented in conformity with Fig. 3.1.

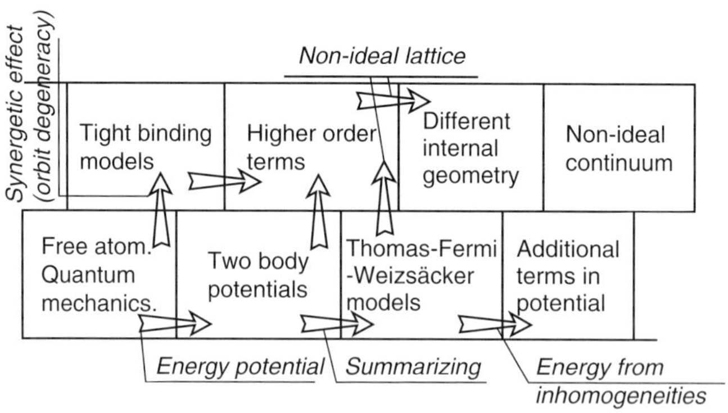


Fig. 3.1. Hierarchical representation of the elastoplasticity basis. Operators of hierarchical growth are in italics.

[1] In the well-known classification of all sciences: there is physics and there is stamp collecting; these methods should be referred to the latter area.

In conformity with this sequence, let us consider certain issues of the quantum fracture theory.

## 3.1 Fundamentals of quantum fracture theory

It is known that direct calculations cannot explain the physical mechanisms of an onset of the initial crack. For example, in the model considered in reference [103], a break of the link between the springs is considered upon fulfilment of a definite fracture criterion, which in its turn is postulated to be predefined in advance. Thus, we obtain here an example of logically looped chain of mutual references that is not satisfactory from the consistency viewpoint.

It appears that a sequential-consistent description of the fracture process should start from the quantum level and, after the respective continualization, pass over to the level of linear fracture mechanics [195]. A similar approach is developed in the works of Cherepanov [53, 54]. The basic methodological problem connected with development of quantum concepts in fracture is similar to the collapse paradox of the wave function of the classical quantum mechanics: how does the material volume (the wave function of the continuum of undamaged material) *know* about the start of fracture and in what manner and what is the 'quantum criterion' of fracture. Similar to one of the possible approaches to description of the maximum theoretical strength [195], this quantum criterion can be built as a function of the number of links[2]. While following hierarchical presentations, one should understand that the level of 'quantization' can be understood differently, depending on the hierarchical level of consideration. Thus, Novozhilov [258, 259], when considering the processes at the mesolevel, had introduced a concept of a *fracture cell* as a certain elementary volume, which is destructed as a whole integral volume, all simultaneously. These presentations have obtained their further development in the works of N. Morozov and Y. Petrov [238, 239, 288].

In what way can we adopt the quantum condition of indeterminacy

$$\Delta p_x \Delta x \geq \hbar, \qquad \Delta E \Delta t \geq \hbar$$

to the elementary cells? What processes are generated by this indeterminacy in classical fracture mechanics?

Another fundamental question lies behind the development of all these treatments of the boundary between two structural or hierarchical regions: What is the effect of the boundary on the result obtained using the chosen model? Since the sharp boundary in real bodies is only a fiction, it is generally hoped that its effect is negligible. In this discussion, we are interested in boundary problem and fracture description at the level of atoms. Naturally, it is possible to expect that the effect of any type of boundary will decrease as the size of the atomistic region is increased.

---

[2] V. V. Barkaline, personal communication.

### 3.1.1 Bond energy and electronic structure

It is known that the density of electronic states of metals and alloys $n(E)$ is connected with the bond energy $E_b$ and free energy $F_b$ [78, 379]

$$E_b = \int_0^{E_F} En(E)dE, \tag{3.1.1}$$

$$F = F_{ps} + F_{hs},$$

where $n(E)$ is the electron density of a two-component unregulated system, $E_F$ is the Fermi energy of the system. $F_{ps}$ and $F_{hs}$ are the pseudopotential and solid-spherical components of the free energy, accordingly.

Fracture energy $W_{fr}$ has an obvious estimate from above. For microscopic volumes of matter:

$$W_{fr} < E_b. \tag{3.1.2}$$

The ‘microscopicity’ is defined in each particular case depending on the type of stressed condition and structure of the material. The volume of the material is meant, sufficiently small, in order to have the dissipation processes existing in macromaterial described in microvolume by means of changing interaction potentials, and the macrostructure by changing the occupation numbers of coordination spheres. Under such presentation, only close correlations in the matter are essential for modeling mechanical processes, and the crystalline structure can be considered ‘quasi-amorphous’. By defining $E_b$ in conformity with expression (3.1.1), we can define the fracture energy by the electronic structure. Since only the closest coordination spheres are practically considered, the real non-homogeneity of the chemical or phase composition of the material can be simulated by two-component substances. The surface energy density, included in expression (4.2.1), in this approach may be interpreted as a change of energy at changing the coordination numbers in two adjacent calculated cells. The change of electronic characteristics depending on the choice of the cell meets the postulated scale noninvariance of the surface energy [44, 254]. From the viewpoint of the macrocrack propagation, it responds to the necessity to account for the effect of the loading history (i.e., in first approximation, the crack length) on the further crack propagation.

There are multiple methods to calculate the density of electronic states. A brief review of them can be found in reference [78]. It has been noted that *ab initio* the bond energy, atomic volumes and volume modules are nicely calculated. It has been shown recently that to calculate unregulated (amorphous) structures, one can broadly use the strong bond approximation [125, 353]. An advantage of such an approximation is a possibility to calculate the objects, which are within the range of intermediary dimensions of quantum systems ($10^3$–$10^5$ calculation cells). This number of cells is already enough to simulate (model) nanomaterials (for example, nanotubes and fullerenes). It means that there is a principal possibility to build a hierarchical computer model of the material in all structural levels, from the quantum up to the construction level [18].

The basics of the strong bond calculation model are connected with the provision that a complete single-electron potential in a solid $\Phi(\mathbf{r})$ can be built as superposition of atomic orbitals $\phi_\lambda(\mathbf{r}-\mathbf{n}) = \langle \mathbf{r}|\mathbf{n}, \lambda\rangle$, where $\lambda$ is the degree of degeneracy of the orbital in state $\mathbf{n}$:

$$\Phi(\mathbf{r}) = \sum_{\mathbf{n},\lambda} a_{\mathbf{n},\lambda}\phi_\lambda(\mathbf{r}-\mathbf{n}). \tag{3.1.3}$$

In this equation $a$ is the approximation parameter, which is related to weak overlapping of atomic orbitals. The Hamiltonian of the system is written as the sum of atomic potentials $V_{\mathbf{n}}^{at}$ of state $n$

$$H = T + \sum_{\mathbf{n}} V_{\mathbf{n}}^{at}. \tag{3.1.4}$$

For the $s$ states the model is characterized by matrix elements

$$\langle \mathbf{n}|H|\mathbf{n}\rangle = \varepsilon^{at} + \alpha_n; \quad \langle \mathbf{n}|H|\mathbf{m}\rangle = t_{\mathbf{nm}}, \tag{3.1.5}$$

where $\varepsilon^{at}$ is the level of atom energy, $\alpha_n$ is the field integral, $t_{\mathbf{nm}}$ is the scattering matrix. The elements of the scattering matrix characterize the probability of an electron to pass over from one atom to another. This matrix depends on distance $R = |\mathbf{m}-\mathbf{n}|$ and in general case can be taken in different forms, for example $t_{\mathbf{nm}} = t(R) \cong \exp(-qR)$ or $1/R^n$. In this approximation, for the density of electronic states of unregulated two-component alloys, an expression can be obtained [11]

$$n(E) = n^0(E) + \frac{2}{\pi}\sum_{l\le 2}(2l+1)\frac{d\delta_l}{dE}\{1+\pi \mathrm{Im}[Y-(1-Z)]\} + 32\pi \mathrm{Im} XYZ, \tag{3.1.6}$$

where

$$X = \sum_{l\le 2}(2l+1)i^{l_1-l_2-l}C_L^{L_1L_2}\left[e^{i\delta_1}\frac{d\delta_l}{dE} - t_l\right](h_l g),$$

$$Y = \sum_{l_1\le 2}(2l_1+1)i^{l_2-l_3-l_1}C_{L_1}^{L_2L_3}\sqrt{E}t_l(h_{l_1}g),$$

$$Z = \left[1-(4\pi)^2 iN\sum_{l_2\le 2}(2l_2+1)i^{l_1-l_2-l'}C_{L_2}^{L_1L'}\sqrt{E}t_{l_2}\right](h_{l_2}g),$$

$$(h_l g) = \int_{rws}^{\infty} h_l^+(\sqrt{E}R)g(r)R^2 dR.$$

In this case,

$$C_{L_1}^{LL'} = \int Y_{L_1}^*(\hat{x}_1)Y_L^*(\hat{x})Y_{L'}(\hat{x}')dQ_x$$

are Gaunt coefficients, where

$$t_l = -\sqrt{E}^{-1} \sum_l (2l+1) \exp(i\delta_l) \sin \delta_l$$

is the single-particle isoenergy $t$-matrix, $\delta_l$ is the phase shift, $h^+(r)$ is the Hankel function, and $r_{WS}$ is the Wigner–Seiz radius. On the basis of expression (3.1.6), calculations were made of the density of electronic states (Table 3.1).

We have also defined the bond energy of two-component material Al–Fe for different concentrations of Al in the material and for pure aluminum and iron. Analysis indicates that for a mix containing about 60% of aluminum, the bond energy has a sharp minimum. It means that the composition materials of this microscopic composition possess the worst physical-mechanical characteristics. From the macroscopic viewpoint, this result is probably explained by the optimum character of this percentage relation of the components for formation of intermetallic compounds in the system. This situation is observed, for example, in production of bimetallic steel–aluminum electrodes by the methods of explosive welding for aluminum electroplating baths. In technological practice, to suppress the growth of intermetallic compounds in such systems, an intermediate layer is introduced in the boundary zone, for which the bond energy between the layers is higher than in the initial two-component system.

For pure elements, the calculation results are close to the existing experimental data. Thus, for the bond energy of pure Al, experimentally the value of 18.8 eV/atom was obtained [45], which differs only by 0.5 eV from the calculated one.

Transition from a discrete (atomic) model of the material to the continual (classical mechanics of the continuous medium) demands additional studies. At what spatial scale does the transition pointwise-continual consideration essentially change the behavior of the object? McCoy and Markworth [201] have demonstrated that for iron for dimensions length $\geq$ 1 nm, the continual consideration introduces only negligible corrections.

Table 3.1
Fermi energy $E$, density of electronic states $n(E_F)$ (states/atom) and bond energy $E_b$ (Ryd/atom) in Al–Fe alloys.

| %Al | $E_F$, Ryd | $n(E_F)$ | $E_b$ |
|---|---|---|---|
| 0 | 0.614 | 25 | 2.39 |
| 10 | 0.590 | 11.95 | |
| 20 | 0.595 | 11.0 | |
| 30 | 0.605 | 5.95 | |
| 40 | 0.615 | 5.785 | 2.63 |
| 50 | 0.655 | 5.7 | 2.42 |
| 60 | 0.695 | 5.55 | 0.821 |
| 70 | 0.755 | 5.45 | 1.63 |
| 80 | 0.805 | 5.25 | |
| 90 | 0.825 | 5.05 | |
| 100 | 0.689 | 2.75 | 1.41 |

### 3.1.2 Thermofluctuation fracture initiation

In the general case, accumulation of energy in the isolated microscopic volume is a probabilistic process, with a predefined probability distribution [238, 295] $P_m$, where $m$ is the cell number. For the exponential distribution, the probability of accumulation by the cell of energy sufficient for the fracture, is given by the expression:

$$P_m = C \exp\left(-\frac{E_b}{\alpha T}\right), \tag{3.1.7}$$

where $C$ is the normalization factor as usually defined from the standard requirement $\sum_m P_m = 1$; and $\alpha$ is a parameter, depending on the properties of the material and loading form.

As a result of thermal fluctuations, interatomic bonds in the local domain can be deformed and weakened. Under the thermofluctuation approach, the domain of deformed bonds is called a dilaton [388, 389]. A dilaton can be characterized as a short-living microdynamic fluctuation defect with linear dimension $\Lambda$ and life span of $\tau_d$, inside which the interatomic bonds have been stretched by the value of $\epsilon_d \gg \epsilon_*$, where $\epsilon_*$ is the average deformation of the interatomic bond at thermal equilibrium, $\Lambda$ is the phonon path range, $\tau_d = \Lambda/s$, and $s$ is the characteristic sound speed. Value $\epsilon_d$ has the physical sense of a critical deformation of interatomic bonds in a dilaton, at which the energy pumping and a local micro-break with formation of a nucleating microcrack or pore are ensured.

For the given distribution (3.1.7), the survival equation can be written down in the form:

$$\tau = \tau_0 \exp\frac{U_b - \gamma\sigma_c}{kT}, \tag{3.1.8}$$

where $k$ is the Boltzmann constant, $\gamma$ is the activation volume (a fracture cell size at the given level of consideration), $\sigma_c$ is the applied load, and the value of $\tau_0$ is of the order of the period of thermal oscillations $\sim 10^{-13}$s. Important is the fact that the applied external loading ***does not break*** the interatomic bonds, but activates the fracture processes. Hence, equation (3.1.7) enables binding together several hierarchical levels of consideration: the quantum-mechanical (value $E_b$), mesoscopic (value $\gamma$) and macroscopic ones (value $\sigma_c$). Besides, the notion of dilaton by itself involves inclusion of collective effects in the consideration of microscopic fracture, underlining the synergetic character of the process.

According to the thermofluctuational strength theory, for the fracture energy of the preset cell we have:

$$U_m = \frac{\sigma_c \tau}{1 - \exp(-2\tau/t)}, \tag{3.1.9}$$

where $\tau$ is the time of residing in the excitation state, and $t$ is the action time of the power pulse.

With energy from the volume of the material, sooner or later (the real time depends on $P_m$) a moment arrives of cell fracture. The simplest analogue of the process is a break of an elastic element upon reaching by the system of springs a critical deformation. Eventually, the fracture passes over to the next cell, and the system generates a wave, which takes away a part of the energy connected with the cell.

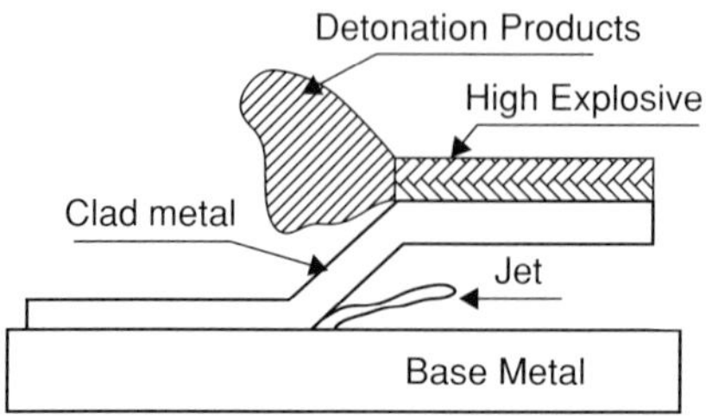


Fig. 3.2. Explosive welding process.

Because of the fact that quantum-mechanical processes and microscopic thermofluctuational processes are reversible, fracture reversibility should be observed. The processes of spontaneous healing of microscopic cracks have long been observed in experiments [23, 128, 265, 266, 271]. Like all other quantum processes, the probability of realization of the healing process is related to the size of the fracture cell.

A specific technological example realizing reversibility of interatomic bonds is explosive welding (Fig. 3.2). This is a solid state joining process. When an explosive is detonated on the surface of a metal, a high pressure pulse is generated. This pulse propels the metal at a very high rate of speed $10^2 \div 10^4$ ms$^{-1}$. If this piece of metal collides at an angle with another piece of metal, welding may occur. For welding to occur, a jetting action is required at the collision interface. This jet is the product of the surfaces of the two pieces of metals colliding. This cleans the metals to juvenile state and allows pure metallic surfaces to join under extremely high pressure. Due to energetic restrictions near the lower limit of the process the blankets of collided metals do not melt as by standard welding, they are atomically bonded. Due to this fact, any metal may be welded to any metal (e.g., copper to steel; titanium to stainless steel).

## 3.2 Influence of material defect structure on crack propagation

In a definite sense, an ideal continuum is a maximum symmetrical object, since it has an endless number of symmetry axes, shifts and other group operations. In this connection, from symmetry considerations, in the ideal material, the crack propagation can be only rectilinear. At any reduction of this infinite-dimensional symmetry group, the crack trajectory should acquire a more complicated character, and this is observed in the case of real materials, which have fluctuations of mechanical parameters.

### *3.2.1 Crack–defect interaction in classical elasticity and plasticity theory*

The mathematical description of interaction of a crack and dislocation in the 2-D interpretation within the classical theory of anisotropic elastic media was given in 1966 by R. Armstrong [8, 350]. The description was provided by analogy with consideration of the classical physical problem of interaction of a power source with images. The solution of the problem for the crack and dislocation was based on definition of interaction of the dislocation and its image arising on the free surface of the crack. This force depends on the distance from the crack surface as $1/r$. Additionally to the forces of the image, there appear the forces of elastic interaction of the dislocation and the crack, proportional to $1/\sqrt{r}$, and the dislocation–dislocation interaction, proportional to the relative distance between the dislocations as $1/(r - r')$. An additional and essential term is made up by the image interaction forces at dislocation–dislocation interaction [349, 373]. This term is proportional to factor $\sqrt{(r/r')}$ and arises because of mapping of the dislocation on singularity near the crack tip. The existence of interaction of images results in screening of dislocation fields by the crack. Introduction of these fields was caused by the physical requirement of appearance of the plastic zone near the crack tip as the sum of plastic deformations of individual dislocations. An elegant mathematical justification of screening was given in the well-known work by Bilby, Cottrell and Swinden [26] (BCS theory).

The principal difficulty in using both the BCS theory and the classical theory of cracks is the presence of singular solutions in the dislocation core and crack tip, which does not allow one to use the elasticity theory.

The mathematically justified presence of singularity near the crack tip causes an obvious protest from the physical viewpoint—in reality, endless stresses cannot exist in the body. This leads to reformulation of the basic concepts of the crack propagation theory [193, 242, 261, 262, 267] with the aim to obtain physically better justified results. These attempts, as a rule, are not satisfactory enough, since individual special cases are considered. Generalization of results and methods is a rather non-trivial problem.

Besides, the experiments have demonstrated that despite the influence of the defect structure on the crack, in many cases dislocations are absent in the vicinity of the propagating crack [68, 159, 160, 268]. It is possible in the case when near the crack tip a barrier appears for generation of dislocations, while the dislocation, already present in the material, has the free surface of the crack as a drain (the fracture acts as the attraction area for mobile dislocations). The appearance of this zone is impossible within the BCS theory. A number of possible approaches exists to explain this effect [22, 68].

This contradiction calls for the development of new approaches to studying the crack–defect interaction, based, for example, on the Lagrangian formalism of the continual theory of defects [76]. The geometric theory of the crack–defect interaction also holds much promise [226]. The reason is that from the physical viewpoint, the variation problem of crack propagation (determination of the optimal crack-propagation trajectory) can be considered as the formation and disappearance of virtual free surfaces in the bulk of the material. The energy of these processes is determined by the metric properties of the continuum in the region where the virtual surfaces are formed. Since the energy, like other invariant factors, depends on the defect structure, the metric properties are functions of the defect structure and the geometric parameters are functions of the energy related to the defect structure [306].

### *3.2.2 Defect fields and intrinsic metric of continua*

Depending on the adopted viewpoint, there exist dual interpretations of the evolution of the internal structure of the deformed body. From the geometrical viewpoint, expression

(2.4.23) sets the relative curvature tensor, since it has terms of the type $\partial\xi^l/\partial x^h$, where $\xi^l = \dot{x}^l$. For particular applications, the presence of derived coordinates in the expression for the curvature tensor leads to the dependence of the density of disclinations. Thus, we obtain that the evolution of the intrinsic metric is causing the appearance of *geometrically necessary dislocations and disclinations*.

On the other hand, it is taken that physical *carriers* of deformation are dislocations and disclinations. In this case, the evolution of the defect structure should cause a change of the internal properties of the material. This bond is usually linked by the functional dependence of the type:

$$\alpha^K_{ML} = F^{eK}_k \partial_{[M} F^{eK}_{L]}, \tag{3.2.1}$$

where $\alpha^K_{ML}$ is the tensor of dislocation density, $F^{eK}_k$ is the elastic gradient, and $[\cdot]$, as usually, defines anti-symmetrization operation.

**Remark 3.2.1.** Expression (3.2.1) is connected with the basic contradiction of the physical plasticity theory, when two ***physical independent values***—the gradient of elastic deformation and the dislocation density—are linked, without sufficient grounds, by a unique functional dependence.

Since the field of defects influences the metric properties of the continuum, the crack should *feel* this change. In the mechanics of the deformed solid body, the properties of the continuum are related with the metric tensor $g_{ik}$. With account of definitions (2.1.14) and (2.3.3), the tensor of small deformations can be related with the metric tensor as

$$\varepsilon_{ij} = \frac{1}{2}\left(g_{ij} - \delta_{ij}\right), \quad i, j, = 1, 2, 3, \tag{3.2.2}$$

where $\delta_{ij}$ is the Kronecker tensor.

In most of the cases, when considering geometrical deformation models (for example, the cycle of works [118, 244, 245]), the metric tensor of the medium is considered to be preset. The defect structure is considered only as a 1-dimensional (dislocation) one [172]. At the same time, it is well known that a complete description should also include as minimum 0-dimensional ones (point defects and vacancies). Consideration of 0-dimensional defects is required for explaining the deformation and fracture processes at non-zero temperatures [172].

A solution of the problem as postulated in Remark 3.2.1 may be setting of the geometrical structure of the medium (tensors of curvature, torsion and segmental curvature) irrespective of setting the metric. It is possible according to Remark 2.4.1. **These geometrical properties are induced by the defects present in the material.** Distribution of disclinations is set by the curvature tensor, and of point defects by the tensor of segmental curvature.

With account of definition of the rotation vector (2.3.5), similar to dislocations, disclinations are set by the vector called the Frank vector

$$\Theta_i = \epsilon_{ijk}\Omega^{jk}, \tag{3.2.3}$$

where the Cartan curvature is:

$$\Omega^{jk} = \oint_C dx^{\mu} \partial_{\mu} \omega^{ij}, \tag{3.2.4}$$

and integration is made by a random closed contour, embracing the disclination line. The length of the Frank vector is equal to the complete rotation angle of the field of dedicated director bypassing around the disclination.

The compatibility condition (2.3.6) can be written down by means of the components of the rotation vector in the following form:

$$\partial_{\mu} \omega_{\nu}^{ij} - \partial_{\nu} \omega_{\mu}^{ij} = 0.$$

This condition is equivalent to the condition of absence of disclinations. With account of expression (2.4.5), we can interpret the curvature tensor as the surface density of the Frank vector [143]

$$\Omega^{ij} = \iint dx^{\mu} dx^{\nu} \wedge R_{\mu\nu}^{ij}. \tag{3.2.5}$$

Taking into account tensor of disclination density $\beta_{ML}$ for example it is supposed that:

$$\beta_{ML} \sim \tilde{K}_{jhk}^{i},$$

where $\tilde{K}_{jhk}^{i}$ is the curvature tensor (2.4.23).

Since we have free[3] tensor fields of torsion and curvature, we have a possibility to bind various types of defects [109, 251, 342] with various curvature tensors. For example, wedge dislocations can be associated with the Riemannian curvature [71] as:

$$R_{1212} = \sqrt{I_2}\beta, \tag{3.2.6}$$

where $I_2 = 1/2\left(\mathrm{t}r^2\mathbf{G} - \mathrm{t}r\mathbf{G}^2\right)$, $\mathbf{G}$ is the metric tensor of deformed configuration (possibly, identical to the deformation tensor, p. 65), and $\beta$ is the density of wedge dislocations.

### 3.2.3 *Crack trajectory and characteristics of fracture space*

Description of the behavior of a fissured material and crack evolution is a permanent problem of the fracture theory. Therefore, the number of proposed and used empirical fracture criteria is permanently growing. Along with that, these criteria, as a rule, do not get sufficient theoretical justification. Apparently, such justification can be made only on the basis of microscopic consideration with the use of the machinery and methods of micromechanics. Among the fracture criteria, the presentation about crack propagation along geodesic energy emissions is rather well known [236]. In fact, it makes the basis of the variation principle of the crack theory [284].

Accounting for all essential fracture factors in the equation of crack trajectory is a rather complicated problem. For the media which possess the internal structure, the

[3] Free—in the sense that they are not connected with a particular type of defect.

processes that take place near the crack tip have a complex character and cannot be described as a direct sum of the operators of plastic and elastic deformation [306]. Since the account of defect structure requires introduction of all the three non-zero Cartan curvature tensors [109, 118], the space associated with the fracture should have a more general character than the Euclidean and Riemannian spaces [294, 317]. Besides, on the basis of the physical sense of the task, the geometrical properties depend also on the motion direction in this space [223]. Additionally, as is well known from the experience, the fracture process depends not only on the defining characteristics, but also on their speeds [282].

In principle, two ways of building the generalized space are possible [118, 294, 303]. One of the ways is connected with setting of the metric tensor (or metric function) in the manifold; the second with introduction into the space of preset connectivity coefficients [316]. The connectivity coefficients, in the general case, must not have any affine character.

It appears from the general elaborations that when considering fracture processes the second way (introduction of connectivity coefficients) is preferable, since in this case the defect structure becomes the factor which is initially defining the geometrical structure of the space. In this case, the space metric in the initial state becomes the calculated, not *a priori* adopted parameter. Naturally, this situation seems more satisfactory. Deviation of the metric induced by the connectivity coefficients from the metric, generated by the metric function (the Euclidean one, in the simplest case), is the deviation of the metric of the material [118].

*Metric in defect medium*

Since the defect structure is defined locally, the Finsler space is decomposed into two three-dimensional Euclidean spaces

$$\mathcal{B}_A = B \times M \subset \mathbb{E}^3 \times \mathbb{E}^3,$$

where $B$ is the object (body) considered, and $M$ is the microstructure in the reference configuration; the multiplication sign $\times$ denotes the Cartesian product of the spaces.

This representation corresponds to the well-known description of bodies with microstructures by means of the Cosserat media (position and director vectors). This decomposition is attributed to the fact that the action of the defect field occurs in the tangent space, whereas the action of the macrostructure occurs in the Euclidean space related kinematically to the initial non-deformed structure.

In this case, the metric function $F(x, \dot{x})$ written as $F(x, y)$ ($x = (x_i)$ forms the macroscopic coordinates of the point and $\dot{x} = (y^i)$ forms the microscopic coordinates). The metric tensor (2.4.14) has the form

$$g^{ij}(x, y) = \frac{\partial^2 H^2(x, y)}{\partial y_i \partial y_j}. \tag{3.2.7}$$

As in spaces with the Berwald–Moore metric, the metric tensor depends on $x$ through $y$ only. Moreover, it can be assumed that the geometrical characteristics of these spaces in the neighborhood of the tangency point are small.

Now, we consider crack propagation in the medium and build the metric tensor of the medium with microstructure. It is well known that the plastic deformation is realized by the evolution and dynamics of defects (for example, dislocation movement) and takes into account the relation between the crack propagation and the evolution of the defect structure and ignores other processes of energy dissipation. In the simplest case, it is sufficient to consider the development of defects in the region of plastic deformation caused by the crack growth. It is known that, as the crack propagates, the plastic deformation occurs in the vicinity of the crack tip and in the region of plastic-shear localization (slip bands, etc.).

In fact for the two-dimensional case, the region of deformations initiated by the crack can be regarded as superposition of a circular region of radius $R$, the center of which lies at the crack tip and rectilinear segments of slip bands (see Fig. 3.3). It is worth noting that the exact shape of the plastic-deformation zone is of no importance for the algorithm proposed. Since the space is decomposed into vertical and horizontal subspaces, the total energy concentrated in the dislocation field is a function of microscopic coordinates and is located in the tangent space. This means that, at this point of the continuum, the crack 'feels' only a certain part of the defect field of the material, which plays an important role in energy variation (precisely this variation is responsible for the crack-growth direction). Physically, this 'feeling' is the realization of the thermofluctuation mechanism of the crack origin. In the material volume formed, the field of virtual fracture area—the area where the quantum bonds disappear—recommences. Then, the physically realized surface on which the extremum of the given functional is satisfied.

We assume that $E \sim \exp(-\mu\sqrt{(}y_1^2 + y_2^2))$. At the same time, the energy should be a function of macroscopic coordinates. The reason is that the distance between the point of the basic and tangent spaces is a function of their radius vectors; the determination of this distance is a separate problem. Obviously, it is possible on the basis of the remark by G. Capriz [46, 47], that for a weak non-local body of the $n$-th order we can use the Colemann–Noll theorem of remotability [64] for the tangent space. In this case, the relative 'remoteness' $\hat{f}$ of the two values is the functional in the Banach space with

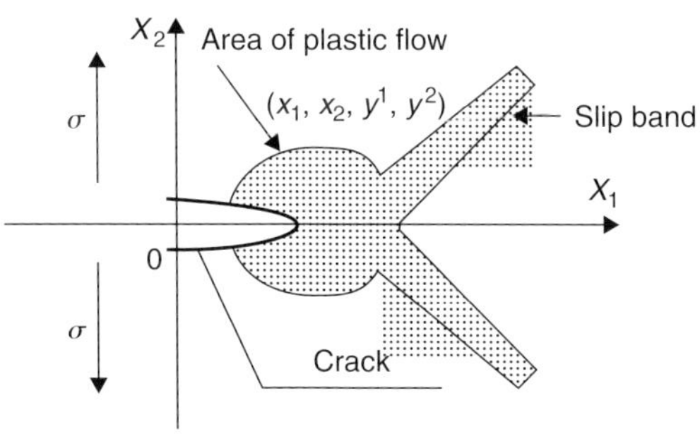


Fig. 3.3. Schematic of the plastic deformation area in the vicinity of the crack tip. The point condition in the plastic zone is described by coordinates $(x_1, x_2, y^1, y^2)$.

the norm, which is weakened by the distance growth. It is equal to the existence of an asymptotically tending to zero ***influence function***.

**Circular zone of deformation.** Apart from the influence function, which in fact is the continuum characteristics, the energy depends on the dislocation distribution function in the volume. We assume that, for a circular region, the dislocation distribution function depends on macroscopic parameters (coordinates $x^1$, $x^2$) and is uniform inside the region of radius $F$; outside this region, the influence of the defect field decreases exponentially. We introduce the notation $K = R^2 - \left((x^1)^2 + (x^2)^2\right)$. The macroscopic dependence can be written as:

$$n_1(x, y) = n_{01}\left[\boldsymbol{\theta}(K) + \boldsymbol{\theta}(-K)\exp\left[-\lambda_1\left((x^1)^2 + (x^2)^2\right)\right]\right], \tag{3.2.8}$$

where the characteristic of the medium $\lambda_1 > 0$ is the coefficient that takes into account the defect–continuum interaction (for example, deceleration forces and processes of generation and annihilation of defects) and $\boldsymbol{\theta}$ is the Heaviside function. With allowance for (3.2.8), the total energy concentrated in the defects of the circular region with the macroscopic coordinates $x^1$ and $x^2$ can be written as an integral over the region:

$$E_1 = E_0 \int_S n_1(x, y)\exp(-\mu\sqrt{(}y_1^2 + y_2^2))dS. \tag{3.2.9}$$

Here, $E_0$ is the well known elastic energy of unit dislocation and $dS$ is the element of the tangent-space area, which is the region where the defect distribution affects the energy of defects at the given tangency point of the spaces. Since the material is heterogeneous in the general case, we consider an elliptic elementary area with semiaxes $a$ and $b$, whose orientation depends on the heterogeneity. In this case, it follows from (3.2.9) that the full energy of defect continuum is

$$\begin{aligned} E_1 &= E_0\pi ab\exp(-\mu\sqrt{(}y_1^2 + y_2^2)) \\ &\quad\times\left[\boldsymbol{\theta}(K) + \boldsymbol{\theta}(-K)\exp\left[-\lambda_1\left((x^1)^2 + (x^2)^2\right)\right]\right]. \end{aligned} \tag{3.2.10}$$

**Slip band.** Real slip bands in materials have certain widths, minimally equal to the diameter of the defect. Any wide band can be represented as a set of thin slip bands. We consider an infinitely thin slip band now. Starting from the reasoning similar to the circle area, we assume that the density of defects in the band has the form of

$$\begin{aligned} n_2(x, y) &= n_{02}\delta\left(\pm k(x_1) - B - (x_2)\right)\exp\left[-\lambda_2\left((x^1)^2 + (x^2)^2\right)\right] \\ &= n_{02}\boldsymbol{\delta}(\pm t - (x^2))\exp\left[-\lambda_2\left((x^1)^2 + (x^2)^2\right)\right], \end{aligned} \tag{3.2.11}$$

where $n_{02}$ is the initial density of the defect distribution in the slip band, $\boldsymbol{\delta}$ is the Dirac delta function, $k = \tan\alpha$ ($\alpha$ is the slope angle of the slip band), $B$ is the coordinate of the slip-band origin, and $\lambda_2 > 0$. The plus and minus signs in the argument of the delta function refer to the slip bands in the upper and lower halfplanes, respectively. We denote the radius-vector of microstates by $\rho^2 = y_1^2 + y_2^2$ and the radius-vector of macrostates by $\tilde{\rho}^2 = (x^1)^2 + (x^2)^2$. Assuming that interactions in the tangent space are the same for the circular region and the band, with allowance for (3.2.10) and (3.2.11) [226]:

$$E = E_1 + E_2 = \pi a^2 b^2 e^{-\mu\rho} \left[\theta\left(R^2 - \tilde{\rho}^2\right) + \theta\left(\tilde{\rho}^2 - R^2\right)e^{-\lambda_1\tilde{\rho}} + \delta\left((\pm kx^1 + b) - x^2\right)e^{-\lambda_2\tilde{\rho}}\right]. \tag{3.2.12}$$

Since $a = a(y_1; y_2)$ and $b = b(y_1; y_2)$, we can calculate the components of the metric tensor of the Finsler space $g_{ij}$ by using (3.2.7) and (3.2.12). Because of the awkwardness of the obtained expressions, let us write down only certain components of the tensor:

$$g^{11} = \left\{\left[\left(\frac{\partial a}{\partial y_1}\right)^2 b^2 + \left(\frac{\partial b}{\partial y_1}\right)^2 a^2\right] + ab\left[b\frac{\partial^2 a}{\partial y_1^2} + a\frac{\partial^2 b}{\partial y_1^2}\right] + 4ab\frac{\partial a}{\partial y_1}\frac{\partial b}{\partial y_1} - 3ab\mu\frac{y_1}{\rho}A - a^2b^2\mu\frac{y_2^2}{\rho^3} + a^2b^2\mu^2\frac{y_1^2}{\rho^2}\right\} \times 2e^{-2\mu\rho} \times B, \tag{3.2.13a}$$

where for brevity we denote

$$A = b\frac{\partial a}{\partial y_1} + a\frac{\partial b}{\partial y_1},$$

$$B = \pi\left[\theta\left(R^2 - \tilde{\rho}^2\right) + \theta\left(\tilde{\rho}^2 - R^2\right)e^{-\lambda_1\tilde{\rho}} + \delta\left((\pm kx^1 + b) - x^2\right)e^{-\lambda_2\tilde{\rho}}\right],$$

and taking into account that $\dfrac{\partial\rho}{\partial y_1} = \dfrac{y_1}{\rho}$, $\dfrac{\partial^2\rho}{\partial y_1^2} = \dfrac{y_2^2}{\rho^3}$,

$$g^{12} = \left\{2b^2\frac{\partial a}{\partial y_2}\frac{\partial a}{\partial y_1} + 2b^2 a\frac{\partial^2 a}{\partial y_2\partial y_1} + 4ab\frac{\partial a}{\partial y_1}\frac{\partial b}{\partial y_2} + 2a^2\frac{\partial b}{\partial y_2}\frac{\partial b}{\partial y_1} + 2a^2 b\frac{\partial^2 b}{\partial y_2\partial y_1} + 4ab\frac{\partial a}{\partial y_2}\frac{\partial b}{\partial y_1} - \frac{4ab\mu}{\rho}\left[by_2\frac{\partial a}{\partial y_1} + ay_2\frac{\partial b}{\partial y_1} + by_1\frac{\partial a}{\partial y_2} + ay_1\frac{\partial a}{\partial y_2}\right] - a^2b^2y_1y_2\frac{\mu}{\rho}\left[1 - \frac{\mu}{\rho}\right]\right\} \times 2e^{-2\mu\rho} \times B. \tag{3.2.13b}$$

Components $g^{11}$ and $g^{12}$ are linked with the influence of microscopic parameters of the crack. After long but also easy calculations we can obtain the 'macroscopic' components of the metric tensor. For example:

$$g^{34} = 2a^2b^2e^{-2\mu\rho^2}\Bigg\{\Bigg[-2\delta(K)(x^2)\,[1+C] - 2\frac{\theta(K)\lambda_1(x^2)C}{\tilde{\rho}}$$
$$- \delta(1, k(x^1)+b-(x^2))B - 2\frac{\delta(k(x^1)+b-(x^2))\lambda_2(x^2)B}{\tilde{\rho}}\Bigg]$$
$$\times\Bigg(-2\delta(K)(x^1)\,[1+C] + \delta(1, k(x^1)+b-(x^2))kB$$
$$- 2\frac{\theta(K)\lambda_1(x^1)C}{\tilde{\rho}} - 2\frac{\delta(k(x^1)+b-(x^2))\lambda_2(x^1)B}{\tilde{\rho}^2}\Bigg)$$
$$+ \left[\theta(K) + \theta(K)C + \delta(k(x^1)+b-(x^2))\,B\right]$$
$$\times\Bigg[4\,\delta(1,\ K)\,(x^2)\,(x^1)\,[1+C] + +8\frac{\delta(K)(x^1)\lambda_1(x^2)C}{\tilde{\rho}}$$
$$+ 2\frac{\theta(K)\lambda_1(x^1)C(x^2)}{\tilde{\rho}^3}[1+2\lambda_1\tilde{\rho}] - \delta(2, k(x^1)+b-(x^2))kB$$
$$- 2\frac{\delta(1, k(x^1)+b-(x^2))\lambda_2 B}{\tilde{\rho}}\left[k(x^2)-(x^1)\right]$$
$$+ 2B\frac{\delta(k(x^1)+b-(x^2))\lambda_2(x^1)(x^2)}{\tilde{\rho}^3}[1+2\lambda_2\tilde{\rho}]\Bigg]\Bigg\}, \tag{3.2.13c}$$

where $B = e^{-2\lambda_2\rho}$, $C = e^{-2\lambda_1\tilde{\rho}}$.

### *3.2.4 Crack trajectory in heterogeneous medium with defects in the general case*

According to the ideology of hierarchical systems, essential regularities of processes at the lower level should find their mapping at the upper level. Since at the atomic level (§1.3) the behavior law (the bond break process or wave function collapse) is represented by a random function, the mapping of these random processes by the operator of hierarchical growth $\rho^+$ at the mesoscopic level should also bring to the behavior law of the random character. In this case, the most physically justified description of crack trajectory is represented by the probability mechanism of distribution of those or other states. It brings us to the concept of crack trajectory ensemble and path integrals [62, 104].

In conformity with the microscopic-probability hierarchical description, the macroscopic object—the forming crack—is 'sorting out' possible trajectories $a, b, \ldots,$ Fig. 3.4, stochastically forming new interfaces with the aim to minimize the total energy spent to fracture the sample as a whole. Possible new trajectories are formed by the elements of the lower hierarchical level. The principle of minimum action (for example, in the Gaussian form) is the law of hierarchical growth. From the physical viewpoint, self-healing of cracks [265], as already noted, is a realization of inversibility of quantum processes. From the viewpoint of hierarchical systems, healing is the result of control by the coordinator.

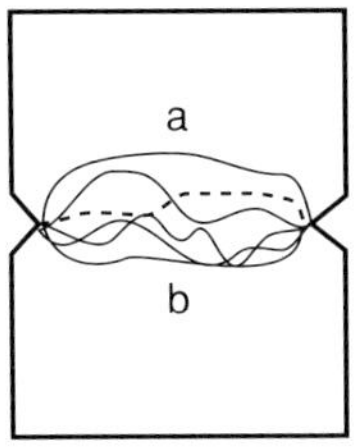


Fig. 3.4. Virtual field of trajectories. Real fracture is highlighted by the broken line.

It is known that the mathematical expression of the minimum action principle is the condition of extremum of certain functional. This condition is written down as the Lagrange–Euler equation

$$\frac{d}{dt}\frac{\partial \mathfrak{L}}{\partial \dot{q}} - \frac{\partial \mathfrak{L}}{\partial x} = Q, \tag{3.2.14}$$

where $\mathfrak{L}$ is the density of Lagrangian and connected with the energy in the given space with generalized coordinates $q, \dot{q}$. The solution of equation (3.2.14) is the line of minimum action called a geodesic. In a uniform plane in the Euclidean space the geodesic is a straight line (= crack trajectory in the ideal material in the uniform field); in more complex cases geodesics are not straight lines.

While assuming that a crack is propagating along the geodesic of energy release, with account of particular dependence of the metric tensor (3.2.7) from the structure according to (3.2.13), for the general equation of the crack trajectory (2.2.3), we have:

$$\frac{dy_i}{ds} - \gamma_{ihk}(x, x')x'^h x'^k = 0, \tag{3.2.15}$$

where $s$ is the parameter of Finsler arc length, $x'^i = \dot{x}^i(dt/ds)$ connected with $y_i$ according to expression (2.4.15), and $\gamma_{ihk}(x, x')$ are Christoffel symbols of the first kind (2.4.19).

Thus, our problem of definition of the equation of crack trajectory can be formulated as a problem of building of a geodesic for the fracture energy. In generalized space, the equation of geodesic can be written down as a function of the parameter of the curve length, actual coordinate and connectivity coefficients, accordingly $(s, x^\mu, \Gamma^\mu_{i\lambda})$:

$$\frac{d^2x^\mu}{ds^2} + \Gamma^\mu_{\sigma\lambda}\frac{dx^\mu}{ds}\frac{dx^\lambda}{ds} = 0, \tag{3.2.16}$$

where $\Gamma^\mu_{\sigma\lambda}$ are connectivity coefficients of the generalized space.

We remind the reader that the Finsler connectivity $\Gamma^\mu_{\sigma\lambda}$ is the Finsler metric connectivity only in the case where condition (2.4.11) is fulfilled.

Thus, if we build connectivity coefficients $\Gamma^\mu_{\sigma\lambda}$ according to (2.4.18) with account of (2.4.21) and metric of the defect medium, defined according to (3.2.13a), we can build the geodesic equation for the defect medium. One should remember here that the basis of the states, for which the geodesic is built, is the extended basis, for example, of the type (1.2.5).

## 3.3 Driving force acting on crack

The issue of interaction of cracks and defects of various nature has been known for a long time and rather nicely studied within the classic elasticity theory. The idea of these calculations is connected with the fact that a defect is distorting the elastic field in proximity of the crack peak. This distortion can be estimated and taken as a definite measure of the crack–defect interaction. In another presentation, in the presence of defects in the medium, they act as energy concentrators and, consequently, in the ray approximation can be viewed as additional centers of attraction or repulsion, which affect the crack with a certain force. This force can be accounted, within the classic crack theory, as superposition of the effects of individual defects [193, 242]. However, such superposition does not account for the collective effects caused by the presence of the field of defects. Collective effects cannot be obtained by a simple summing up, since, according to the MHST presentation, they have qualitative distinctions.

Let us consider a possibility to define the interaction of the crack and the defective structure from geometrical positions. This approach includes collective interactions, since the parameter defining any geometry—the metric tensor—is defined by the whole continuum, and not only by local properties of the matter in the given point. In other words, the metric tensor is the result of the combined ‘long-range action’ of the fields. In principle, the metric tensor of the medium with defects was built in §3.2.3. Now, we need to define the forces which act from the part of the defective structure on the micro- (or macro-) crack.

We shall base ourselves on the definition of structural group $G$ (3.5.14) and use the Lagrangian formalism of the continual defect theory. For the first time, this formalism was systematically developed as applied to the defect theory by Aida Kadic in her doctorate thesis [142].

*Lagrangian of fields system*

Introduction to the theory of free classic fields can be found in any textbook on the field theory, for example [32, 33]. Hereafter we shall consider the local Lagrangian, i.e., the Lagrangian which depends only on local potentials and the final number of their first derivatives

$$\mathfrak{L} \sim \int dy F\left(u(x), u(y), \frac{\partial u(x)}{\partial x}\frac{\partial u(y)}{\partial y}\right). \tag{3.3.1}$$

When considering the processes connected with interaction of different fields, experts proceed from the density of Lagrange function:

$$\mathfrak{L} = \mathfrak{L}_0 + \mathfrak{L}_{int},$$

where $\mathfrak{L}_0$ is the density of Lagrange function of free fields and $\mathfrak{L}_{int}$ is the density of Lagrange function of interaction. The Lagrangian is obtained from the density of Lagrange function by means of integrating by the four-dimensional volume. Hereafter, we shall sequentially use the term Lagrangian, since it does not result in any sort of untidiness.

The interaction Lagrangian $\mathcal{L}_{int}$ should meet a basic physical requirement—to possess the feature of relativist invariance. It can therefore represent any *invariant* algebraic or integral combination of functions of interacting fields. Integral Lagrangians lead to non-local theories and will not be considered hereafter. Thus, $\mathcal{L}_{int}$ can be designed from field functions by means of contraction of products of two values of the same tensor dimensionality; for example: two scalars, two pseudo-scalars, or two vectors.

The symmetry of the ideal continuum by defects (group $G$ action), it is necessary to take steps to restore the symmetry. According to the standard procedure for a medium without disclinations, it is achieved by prolongation of the derivative [142] (by building a covariant derivative similar to the procedure described in §2.2.1). The nonlinear connectivity $N$ is not a random value. Since $\mathfrak{X}(\mathcal{M})$ is Lie's algebra, the connectivity coefficients should be expressed through the algebra basis $\mathfrak{X}(\mathcal{M})$. In this case:

$$N_a = W_a^\alpha(\mathbf{r})\gamma_\alpha, \tag{3.3.2}$$

where $W_a^\alpha$ are gauge fields of the type of Yang–Mills fields, related to the action of the $G$ structural group. In the general case, Lagrangians of gauge fields can be built in conformity with the variation procedure [156]. In this case, gauge fields need not be the Yang–Mills type only [66, 174]. Matrices $\gamma_\alpha$ comply with the commutation rules

$$[\gamma_\alpha, \gamma_\beta] = \gamma_\alpha\gamma_\beta - \gamma_\beta\gamma_\alpha = C_{\alpha\beta}^\varepsilon \gamma_\varepsilon,$$

where $C_{\alpha\beta}^\varepsilon$ are structural constants of Lie's algebra $\mathfrak{X}(\mathcal{M})$. Structural constants comply with relation:

$$C_{\alpha\beta}^\varepsilon = -C_{\beta\alpha}^\varepsilon$$

and Jacobi identities (2.1.52). Structural constants define the Cartan–Killing metrics: [85]

$$C_{\alpha\beta} = C_{\alpha\gamma}^\varepsilon C_{\beta\varepsilon}^\gamma. \tag{3.3.3}$$

An additional term appears in the material Lagrangian, linked with potentials of gauge fields:

$$\mathcal{L} = \mathcal{L}_0 + S_2\mathcal{L}_2, \tag{3.3.4}$$

where $\mathcal{L}$ is the material Lagrangian and $\mathcal{L}_2$ is the Lagrangian of gauge fields:

$$\mathcal{L}_2 = \frac{1}{2} C_{\alpha\beta} F_{ab}^\alpha g^{ac} g^{bd} F_{cd}^\beta, \tag{3.3.5}$$

in which $g^{AB} = -\delta^{ab}$, $g^{44} = \frac{1}{\zeta}$, $g^{ab} = 0$ for $a \neq b$, $S_1$ is the bond constant, $\zeta$ is the transfer speed of the field affection, and $C_{\alpha\beta}$ are the components of the Cartan–Killing metric of subgroup $SO(3)$.

For the medium in which the Frank vector exists (it means the presence of disclinations), the connectivity depends on the additional gauge field $\Phi^\alpha$. This field is connected with breaking of the translational invariance. This type of field should be implemented, for example, for carbon nanotubes because of the presence of chirality. Consequently, (3.3.2) acquires the form of [76, 109, 142]:

$$N_a = W_a^\alpha(\mathbf{r})\gamma_\alpha + \Phi_a^\alpha. \tag{3.3.6}$$

Then, the total Lagrangian of the continuum can be written down in the form of Lagrangians of the fields connected with defects, and material Lagrangian in the form of [76, 109, 223]:

$$\mathcal{L} = \mathcal{L}_0 + S_1\mathcal{L}_1 + S_2\mathcal{L}_2, \tag{3.3.7}$$

where $\mathcal{L}_1$ is the Lagrangian of the gauge fields, connected with disclinations.

The variation procedure for potential $\mathcal{L}_1$ gives:

$$\mathcal{L}_1 = \frac{1}{2}\delta_{\alpha\beta} D^{\alpha}_{ab} g^{ac} g^{bd} D^{\beta}_{cd}. \tag{3.3.8}$$

It has been noted that values $F$, $D$ in (3.3.8) and (3.3.5) are directly connected with gauge fields. In this case the fields connected with the translational group (potentials $\phi$) are functions of the density of dislocations and can be connected with the second form of torsion [142, 180]. In the case of teleparallelism, Lagrangian $\mathcal{L}_1$ in (3.3.8) is represented as a sum of three independent torsion tensors [180]:

$$\mathcal{L}_1 = C_1^{\star}\left(-{}^{(1)}T_a + 2^{(2)}T_a + \frac{1}{2}T_a\right), \tag{3.3.9}$$

where $\star$ is Hodge dual (or Hodge star operator), which correlates the dual $p$-$n$ form [364] to the preset $n$-form.

In the case of the presence in the body of an electromagnetic field, the total Lagrangian (3.3.7) should obtain the terms connected with the electromagnetic field. The Lagrangian of the electromagnetic field can be chosen in various forms [32, 178], in the gradient-invariant example. In the Gaussian system of units, it has the form:

$$\mathcal{L}_3 = -\frac{1}{16\pi} H_{kl} H^{kl}, \tag{3.3.10}$$

where

$$H_{kl} = \frac{\partial A_l}{\partial x^k} - \frac{\partial A_k}{\partial x^l} \tag{3.3.11}$$

is the skew-symmetric tensor of electromagnetic field, and $\mathbf{A}_n, n = 0, 1, 2, 3$ is the 4-vector potential,

$$\mathbf{E} = \text{grad } A_0 - \frac{\partial \mathbf{A}_0}{\partial x^0}, \quad \mathbf{H} = \text{rot}\mathbf{A}.$$

Here $\mathbf{E}$ is the vector of electric field intensity and $\mathbf{H}$ is the vector of magnetic field intensity. The electromagnetic field brings essential corrections into the behavior of piezoelectric or magnetostriction materials and influences the crack propagation (§4.5).

*Definition of values F and D*

The procedure of building gauge fields is connected with covariant differentiation. It is known that there is a theorem in relation to the 2-form connectivity curvature [364]:

$$\Omega(X, Y) = d\omega(X, Y) + \frac{1}{2}[\omega(X), \omega(Y)], \quad X, Y \in \Omega_n \quad n \in \mathcal{M}. \tag{3.3.12}$$

With account of equation (3.3.2), expression (3.3.12) takes the form:

$$\Omega(X, Y) = F^{\alpha}\gamma_{\alpha}, \quad F^{\alpha} = \frac{1}{2}F^{\alpha}_{ab}dX^{a} \wedge dX^{b}, \tag{3.3.13}$$

where $\{F^{\alpha}_{ab}\}$ are components of the gauge field. With account of expression 2-form curvature (3.3.12), we obtain an explicit expression for the field components:

$$F^{\alpha} = dW^{\alpha} + \frac{1}{2}C^{\alpha}_{\beta\gamma}W^{\alpha} \wedge W^{\beta}. \tag{3.3.14}$$

In a similar way, for coefficients $D^{\alpha}$ we have:

$$\mathbf{D} = F^{\alpha}\gamma_{\alpha}\chi, \tag{3.3.15}$$

where $\chi$ is the state vector.

In the coordinate form, the coefficients are represented as follows:

$$D^{\alpha}_{ab} = d\Phi^{\alpha} + \gamma^{\alpha}_{\beta\gamma}\left[W^{\beta}_{a}\Phi^{\alpha}_{b} - W^{\beta}_{b}\Psi^{\beta}_{a} + F^{\sigma}_{ab}x^{\beta}\right]. \tag{3.3.16}$$

### *3.3.1 Gauge crack theory*

Crack propagation can be viewed as infinitely small translations of a certain form (object) in the direction of its propagation. For example, the crack tip may be taken as such a form. Since we study the preservation of symmetry properties, it is natural to assume Noether's theorem is fulfilled. It is known that for translations Noether's theorem leads to the law of preservation of the four-dimensional vector of the field energy-momentum [33]. By executing the standard procedure of varying the field equations and assuming that the electric field does not depend on the defect field, from equation (3.2.14) we have:

$$\begin{aligned} &\left(\mathfrak{L}\delta_{ak} - \frac{\partial\mathfrak{L}}{\partial x^{\alpha}_{,a}}x^{\alpha}_{,k} - \frac{\partial\mathfrak{L}}{\partial W^{\alpha}_{b,a}}W^{\alpha}_{b,k} - \frac{\partial\mathfrak{L}}{\partial\phi^{\alpha}_{b,a}\phi^{\alpha}_{b,k}}\right)_{,a} \\ &= \frac{\partial}{\partial t}\left(\frac{\partial\mathfrak{L}}{\partial x^{\alpha}}x^{\alpha}_{,k} + \frac{\partial\mathfrak{L}}{\partial W^{\alpha}_{b}}W^{\alpha}_{b,k} + \frac{\partial\mathfrak{L}}{\partial\phi^{\alpha}_{b}\phi^{\alpha}_{b,k}}\right). \end{aligned} \tag{3.3.17}$$

By integrating expression (3.3.17), we obtain:

$$\begin{aligned} &\int_{\Sigma}\left(\mathfrak{L}\delta_{ak} - \frac{\partial\mathfrak{L}}{\partial x^{\alpha}_{,a}}x^{\alpha}_{,k} - \frac{\partial\mathfrak{L}}{\partial W^{\alpha}_{b,a}}W^{\alpha}_{b,k} - \frac{\partial\mathfrak{L}}{\partial\phi^{\alpha}_{b,a}\phi^{\alpha}_{b,k}}\right)n_{a}ds \\ &= \frac{d}{dt}\int_{V}\left(\frac{\partial\mathfrak{L}}{\partial x^{\alpha}}x^{\alpha}_{,k} + \frac{\partial\mathfrak{L}}{\partial W^{\alpha}_{b}}W^{\alpha}_{b,k} + \frac{\partial\mathfrak{L}}{\partial\phi^{\alpha}_{b}\phi^{\alpha}_{b,k}}\right)dV. \end{aligned} \tag{3.3.18}$$

The integration surface in the left-hand part of (3.3.18) is a certain closed surface around the crack tip (Fig. 3.5). This closed surface can be represented as the sum of the free surface of crack $\pi_{a}$ and the surface of the plastic zone in proximity of tip $\pi$. Strictly speaking, we should not consider the zone of plastic deformation only, but also

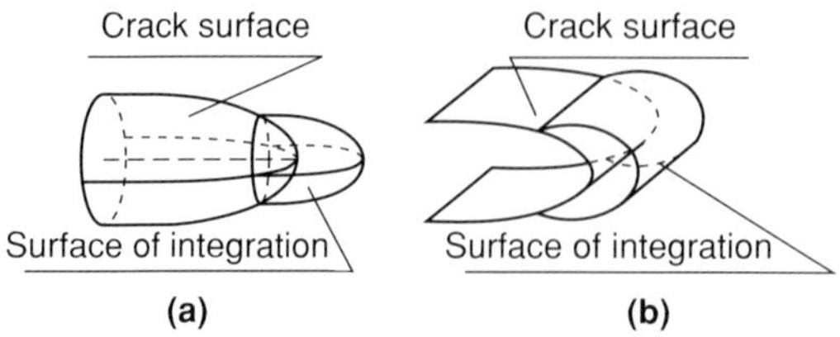


Fig. 3.5. Representation of the 'influence surface' (a) for the initial microcrack and (b) for the real crack.

take into account the long-range forces generated by defects. The precise form of this 'influence surface' should be defined additionally. Generally speaking, the influence surface or zone is connected with thermofluctuation character of atomic fracture and existence of the minimum fracture cell. An additional argument in favor of existence of the minimal scale of fracture lies in established contradictions arising at a thorough energy analysis of the continual approach to crack propagation [167, 168]. Similar results were subsequently obtained in analysis of crack propagation in the elastic–plastic medium by the method of finite elements [146, 147]. In case of additional effect of the electromagnetic field in the Lagrangian (3.3.18), additional terms of electromagnetic nature appear.

The existence of the minimum scale is also connected with the microscopic fracture reversibility fracture. It is known that quantum-mechanical processes are reversible. Therefore, the fracture processes–restoration of atomic bonds acquires the asymmetrical character only as the result of imposition of external conditions. The invariance in relation to the time sign appears in the macroscopic level as a consequence of irreversibility of statistical mechanics and thermodynamics [155]. However, because of hierarchical presentations about interaction of different levels, even at the level of microscopic cracks we should observe multiple healing. This fact is known experimentally, but it is believed that it is not enough assessed theoretically [128, 265, 266].

As follows from (3.3.18), the general force acting on the crack tip from the defective structure and external fields is

$$F_k = \int\limits_{\Sigma} \left( \mathfrak{L}\delta_{ak} - \frac{\partial \mathfrak{L}}{\partial x^{\alpha}_{,a}} x^{\alpha}_{,k} - \frac{\partial \mathfrak{L}}{\partial W^{\alpha}_{b,a}} W^{\alpha}_{b,k} - \frac{\partial \mathfrak{L}}{\partial \phi^{\alpha}_{b,a} \phi^{\alpha}_{b,k}} \right) n_a ds.$$

In respect that the crack surface is a free surface, we have an opportunity to integrate only by surface $\pi$. Then:

$$F_k = \int\limits_{\pi} \left( \mathfrak{L}\delta_{ak} - \frac{\partial \mathfrak{L}}{\partial x^{\alpha}_{,a}} x^{\alpha}_{,k} - \frac{\partial \mathfrak{L}}{\partial W^{\alpha}_{b,a}} W^{\alpha}_{b,k} - \frac{\partial \mathfrak{L}}{\partial \phi^{\alpha}_{b,a} \phi^{\alpha}_{b,k}} \right) n_a ds. \quad (3.3.19)$$

With account of expressions (3.3.8), (3.3.5), (3.3.14) and coefficients (3.3.15), (3.3.16), total force $\Phi_k$ can be elementarily transformed to the type:

$$\Phi_k = \int_{\pi} \left[ \left( \mathfrak{L}\delta_{ak} - \frac{\partial \mathfrak{L}_0}{\partial B^{\alpha}_{a}} x^{\alpha}_{,k} \right) - \frac{S_1}{2} \delta_{\alpha\beta} K^{ac} K^{bd} D^{\beta}_{cd} \Phi^{\alpha}_{b,k} \right. \\ \left. + x^{\beta} \gamma^{\alpha}_{\sigma\beta} W^{\alpha}_{b,k} - \left( \frac{S_2}{2} C_{\alpha\beta} g^{ac} d^{bd} F^{\beta}_{cd} W^{\alpha}_{b,k} \right) \right] n_a ds, \tag{3.3.20}$$

where $\Phi_k = F_k + F^{el}_k$ and

$$F^{el}_k = \int_{\pi} T^{ik} n_i ds, \tag{3.3.21}$$

is the ponderomotive force, in which

$$T^{ik} = \frac{1}{4\pi} \left( -F^{il} F^{k}_{l} + \frac{1}{4} g^{ik} F_{lm} F^{lm} \right) \tag{3.3.22}$$

is the energy-momentum tensor. Space components of (3.3.22) form the 3-D Maxwell stress tensor [178]

$$\sigma_{\alpha\beta} = \frac{1}{4\pi} \left\{ -E_{\alpha} E_{\beta} - H_{\alpha} H_{\beta} + \frac{1}{2} \delta_{\alpha\beta} (E^2 + H^2) \right\}.$$

In the case of absence of defects and electromagnetic field in the material, expression (3.3.20) gives:

$$F^{id}_k = \int_{\pi} \left( \mathfrak{L}\delta_{ak} - \frac{\partial \mathfrak{L}_0}{\partial x^{\alpha}_{,a}} x^{\alpha}_{,k} \right) n_a ds. \tag{3.3.23}$$

By its structure, expression (3.3.23) is identical to the energy-momentum tensor [1, 90]. It is rather obvious that since (3.3.23) was obtained from the Noether theorem, $F_k$ is the preserved value [76]. By simple transformations we can show that in this case expression (3.3.23) coincides with the $J$-integral of Cherepanov–Rice [52, 53, 297]. In the general case, the force acting on the crack in the defective material can be presented as superposition of forces in the ideal material (Barenblatt–Dugdale model) and corrections connected with the defective structure. Then $\Phi_k = F^{id}_k + F^{def}_k$, and for corrections we have:

$$F^{def}_k = \int_{\pi} \left[ \frac{S_1}{2} \delta_{\alpha\beta} K^{ac} K^{bd} D^{\beta}_{cd} \Phi^{\alpha}_{b,k} + x^{\beta} \gamma^{\alpha}_{\sigma\beta} W^{\alpha}_{b,k} \right. \\ \left. - \left( \frac{S_2}{2} C_{\alpha\beta} g^{ac} d^{bd} F^{\beta}_{cd} W^{\alpha}_{b,k} \right) \right] n_a ds. \tag{3.3.24}$$

These corrections define the internal force tensor of defective material, which allows one in principle to build canonically conjugate values—the deformation and stress tensors. The force deformation tensor defines the deviation of the internal metric from the metric of external space.

## 3.4 Connection of defective material structure with crack surface shape

In the previous discussion we have obtained an expression for the forces arising between the defective structure and the existing crack. It is of both theoretical and practical interest to consider the shape of the developing fracture surface and crack front line.

### *3.4.1 Equation of crack front as function of metric and field of defects*

Let us consider the positively defined Finsler space $X_n$. In this $n$-dimension space, we can set in the standard manner a hypersurface with dimensionality $m = n - 1$. In this hypersurface we can set both tensors in individual points $P$ of the field and the field tensor [294]. Let us call this hypersurface the *discontinuity surface* for variable $u^i$ of value $Z$ in the case if in this point the derivative has singularity:

$$Z_{,u^i}(P) = \frac{\partial Z(P)}{\partial u^i} = \text{SING}.$$

The set of variables $u^i$ is the extended set $\left(x^i, \dot{x}^i, t\right)$. In this case, the form

$$D\left(Z_{,u^i}(P)\right) = \lim_{u^i \to 0}\left(Z_{,u^i}(P) - Z_{,u^i}(P + \Delta P)\right) \neq \text{SING}. \tag{3.4.1}$$

is called discontinuity of the quantity $Z$ in point $P$.

Analogously, the procedure of renormalization of singular quantities is conventionally introduced into the field theory. In this case, the difference between the values of the quantity on both sides of the discontinuity surface is defined by the expression [212, 348]:

$$D\left(Z_{,u^{(2m+1)}}(P)\right) = \lim_{\Delta t \to 0}\left(Z(t) - Z(t + \Delta t)\right) = [Z].$$

**Compatibility conditions.** It is known that in the general case the compatibility conditions express an additional connection between the displacement vector and deformation tensor [316]. Since it is possible to consider the initial state, in which it is impossible to introduce the transition into the real state, the compatibility conditions in the classic sense (of the type of the Saint-Venan compatibility conditions) will not exist. These conditions can be implemented only in a limited case of the Euclidean space.

However, with account of the properties of the Finsler space and conditions of each particular task, we can also obtain additional restrictions, which have the sense of compatibility equations.

Let us consider vector field $X^i(t)$ defined along curve $C$ and assume that it describes a discontinuity surface. To consider changes in the field along the curve, the principal fact connected with the geometry of the manifold should be taken into account. To find the difference between tensor and vector fields at different points of the manifold, we must take into account that vectors $X^i$ and $X^i + dX^i$ belong to different tangent spaces $T_n(P)$ and $T_n(Q)$, respectively. Therefore, two factors must be taken into account: (a) the change $dX^i = \left(dX^i/dt\right)dt$ of the vector $X_i(t)$, which depends only on the field $X^i$ and is independent of the space metric; and (b) the difference between the metrics of tangent spaces $T_n(P)$ and $T_n(Q)$. As applied to the actual crack motion, the factor (a) will be responsible for changes in the moving crack parameters, and the factor (b) will be responsible for changes in the parameters of the medium through which the crack moves. The crack parameters associated with the field $X^I$ are, for example, fractal dimensions of the cracking surface and the crack velocity in the material. The parameter of the medium can change, for example, due to the presence of defects in the material and changes of the mechanical constants of the continuum. Therefore, the tensor differential of the vector field is represented by the sum of two terms determined by factors (a) and (b). As a result, with allowance for the transformation properties of derivatives and the necessity of obtaining a tensor quantity, for the derivatives of the field with respect to the parameter (time) (the total derivative) we have [303]

$$D(X^i) = \frac{\delta X^i}{\delta t} = \frac{dX^i}{dt} + P^{.i.}_{h.j}(x,\dot{x})\, X^h \dot{x}^j, \tag{3.4.2}$$

where

$$P^{.i.}_{k.j}(x,\dot{x}) = \gamma^{.i.}_{k.j}(x,\dot{x}) - C^{.i.}_{k.l}(x,\dot{x})\,\gamma^{.l.}_{p.j}(x,\dot{x})\,\dot{x}^p,$$

in which $\gamma_{hjk}(x,\dot{x})$, $\gamma^{.i.}_{k.j}(x,\dot{x})$ are the Christoffel symbols of the second kind and first kind defined conventionally. The quantity $C_{ijk}$ is the characteristic tensor of the Finsler geometry (the Cartan torsion (2.4.21)) caused by the dependence of the metric on the derivatives of coordinates with respect to the parameter. This tensor is

$$C^{.i.}_{k.j} = C_{ljk} g^{li}. \tag{3.4.3}$$

As follows from the physical sense of terms included in (3.4.2), the term $dX^i/dt = V^i$, where $V^i$ is the rate of change of the vector $X^i$. Since the vector $X^i$ describes the discontinuity surface, $V^I$ represents its propagation velocity. The velocity can be represented [306] as:

$$V^i = v_t g^{ij}(x,\dot{x}) e_j, \tag{3.4.4}$$

where $v_t(x,\dot{x},t) \equiv \partial X(x,\dot{x},t)/\partial t$, $e_j$ is the unit vector in the direction of discontinuity propagation, and narrowing of the tangent space of vectors $X^i(t)$ such that mapping $R \mapsto C(\wp_f)$ exists is also presented, where $\wp_f$ is the Finsler body. The Finsler body is a paired manifold and space of measurements plus a diffeomorphic mapping of the manifold onto the Euclidean space: $\chi : t \mapsto \chi_t(x,\dot{x}) = X(x,\dot{x},t)$.

With allowance of the representation (3.4.4), equation (3.4.2) takes the form:

$$\frac{\delta X^i}{\delta t} = v_t(x,\dot{x},t)\, g^{ij} e_j + P^{.i.}_{h.j}(x.\dot{x})\, X^h \dot{x}^j. \tag{3.4.5}$$

Then the first-order discontinuity operation given by (3.4.1) for field derivatives acquires the form:

$$D(P) = \lim_{\delta t \to 0} \left( \frac{\delta X^{(1)i}}{\delta t} - \frac{\delta X^{(2)i}}{\delta t} \right) \tag{3.4.6}$$

$$= \left[ v_t (x, \dot{x}, t) \, g^{ij} e_j + P_{h.j}^{.i.} (x.\dot{x}) \, X^h \dot{x}^j \right]^{(1-2)} . \tag{3.4.7}$$

The upper index in brackets means the difference of values on the two sides of the discontinuity surface. This expression considers the variations of the field $X^i$, of the discontinuity velocity $V^i$ and of the metric accompanying the discontinuity motion. In this case the expression (3.4.6) defines the Lie derivative along the hypersurface of the discontinuity field $X^i$ [303, 315]:

$$D(P) = \underset{L}{D}(P),$$

and in coordination in form of variation derivative of action integral [82, 307]. Expression (3.4.6) is the general kinematic condition of compatibility imposed on discontinuities.

*Generalized ruptures and crack motion*

Mathematically, the crack can be represented (§1.2.1) as a media rupture. Let us investigate in more detail the application of the discontinuity theory to the crack propagation.

Displacements, density of the material, particle velocities, and pressures (internal stresses) on both sides of the crack surface undergo discontinuous changes when the crack moves. Let us consider, for example, conditions of particle displacements (when the vector $X^i$ represents the displacement field).

According to local symmetry principle, a *local* discontinuity (microcrack) arises under the action of normal forces initiating it, $e_i \equiv \eta_i$. In inhomogeneous media the direction of the orthogonal vector is determined by the indicatrix. It is also known that the Lie derivative of a tensor forms a tensor of the same rank [315]. Then with allowance for orthogonality condition (2.4.34) and definition (3.4.6), we have equation of the crack front in a medium

$$g_{ij}\left(x^k, \xi^k\right) \xi^i \underset{L}{D}\left(X^j(P)\right)$$

$$= g_{ij}\left(x^k, \xi^k\right) \xi^i \left[ v_t(x, \xi, t) \, g^{jl} e_l + P_{k.r}^{\cdot j}(x, \xi) \, X^k \xi^r \right]^{(1-2)}$$

$$= \left[ g_{ij}\left(x^k, \xi^k\right) \xi^i v_t(x, \xi, t) \, g^{jl} e_l + g_{ij}\left(x^k, \xi^k\right) \xi^i P_{k.r}^{\cdot j}(x, \xi) \, X^k \xi^r \right]^{(1-2)}$$

$$= \left[ v_t(x, \xi, t) \, g_{ij}\left(x^k, \xi^k\right) \xi^i g^{jl} e_l + g_{ij}\left(x^k, \xi^k\right) \xi^i P_{k.r}^{\cdot j}(x, \xi) \, X^k \xi^r \right]^{(1-2)} = 0. \tag{3.4.8}$$

Condition (3.4.8) is satisfied when:

$$\text{(a)} \begin{cases} \left[v_t(x,\xi,t)g_{ij}(x^k,\xi^k)\xi^i g^{jl}e_l\right]^{(1-2)} = 0, \\ \left[g_{ij}\left(x^k,\xi^k\right)\xi^i P^{\cdot j}_{k.r}\left(x,\xi\right)X^k\xi^r\right]^{(1-2)} = 0. \end{cases}$$

$$\text{(b)}\ v_t\left(x,\xi,t\right)g_{ij}\left(x^k,\xi^k\right)\xi^i g^{jl}e_l = -g_{ij}\left(x^k,\xi^k\right)\xi^i P^{\cdot j}_{k.r}\left(x,\xi\right)X^k\xi^r.$$

$$\text{(c)}\ \left[v_t\left(x,\xi,t\right)g_{ij}\left(x^k,\xi^k\right)\xi^i g^{jl}e_l + g_{ij}\left(x^k,\xi^k\right)\xi^i P^{\cdot j}_{k.r}\left(x,\xi\right)X^k\xi^r\right]^{(1)}$$

$$= \left[v_t\left(x,\xi,t\right)g_{ij}\left(x^k,\xi^k\right)\xi^i g^{jl}e_l + g_{ij}\left(x^k,\xi^i\right)\xi^i P^{\cdot j}_{k.r}(x,\xi)X^k\xi^r\right]^{(2)}.$$

Condition (a) describes the independent changes of the field and metric; condition (b) characterizes correlated changes of the field and metric, when the change of the metric compensates for the field change as the crack moves; and condition (c) describes the crack motion in the Euclidean space when $g^{ij} = g^{ij}(x)$.

### *3.4.2 Break propagation in a medium*

Non-Riemannian approximation of media allows us to apply theory of breaks to the problem of crack propagation in media taking into account the real properties of defects field [212, 224, 228]. Inasmuch as the Riemannian space is the tangent space to the Riemannian space the results may be local identical.

We consider that the fracture processes are determined by the applied external stress rather than by the structural boundaries in the material. The fracture is considered as motion of the discontinuity of the characteristics of the material. A physical realization of the discontinuity is the crack front. After the discontinuity (front) formation, it evolves into the crack surface. Thus, by the crack growth is meant the dynamic process of discontinuity formation and evolution. The internal defective structure of the material may influence the fracture process via the distortion of the front shape and surface of the propagating crack. Such distortions complicate the surface geometry, thereby leading to the appearance of nonlinear and fractal properties [15]. The evolution of the fracture surface is a subject of special research. The classical solution of the problem of the elasticity theory for the crack near its tip demonstrated that the stress field obtained has a singularity. Moreover, stress fields near the crack tip are bounded under actual conditions, and defect fields near the distributed crack tip are absent [274], for example, because of defect sinks on free crack edges (the so-called dislocation-free zone arises) [159, 194, 268]. Thus, the stress fields are renormalized, and the defect structure is the physical reason for the occurrence of gauge fields.

We take the functional dependence between the stress and strain tensors in the form[4]

$$\sigma_{im} = \lambda_1 \varepsilon_{kk}\delta_{im} + 2\mu_1\varepsilon_{im}, \tag{3.4.9}$$

where $\lambda$, $\mu$ are elastic constants.

The equations of motion of the medium in Lagrangian coordinates have the form [130]:

$$\rho(\mathbf{w} - \mathbf{Q}) = \nabla_i \mathbf{S}^i = \frac{\partial \mathbf{S}^i}{\partial x^i} + \mathbf{S}^n \gamma^i_{ni}, \tag{3.4.10}$$

where $\rho$ is the density of the medium, $\mathbf{Q}$ is the mass force, $\gamma^i_{ni}$ is the Christoffel symbol of the first kind (2.4.19), $w_j = \ddot{u}_j$ and $\mathbf{S}^j$ denotes the $j$th component of the vector representation of the stress tensor $S^{ij}$, $\mathbf{S}^i = S^{ij}\mathbf{e}_j$. In the given section, we distinguish between the stress tensor $S^{ij}$ and the instantaneous true stress tensor $\sigma_{ij}$ [130]. For an arbitrary metric space

$$\sigma_{\beta k} = g_{ik} g_{\alpha\beta} S^{mn} A^{\alpha}_{m} A^{i}_{n},$$

where $g_{ik}$ is the metric tensor of the space and $A^{\alpha}_{m}$ is the matrix of transition from the coordinate system $x$, $y$ to the coordinate system $x'$, $y'$.

For simplicity, we assume that the mass forces are absent in the bulk of the material[5]. Thus, since the Stokes theorem is fulfilled for any space and does not depend on the metric, in the Finsler space we obtain the equation of motion

$$\rho(\mathbf{w}) = \nabla_j \mathbf{S}^i = \frac{\partial \mathbf{S}^i}{\partial x^j} + P^{*i}_{h\ j}\mathbf{S}^n, \tag{3.4.11}$$

where $P^{*i}_{h\ j}$ are determined according to (2.4.17), and the quantity $P^{i}_{k\ j}(x, \dot{x})$ characterizes the parallel translation of vectors (the connectivity of the space, (2.4.18)). According to results of §2.4.5, orthogonality condition, which determines the direction of the crack propagation, is the condition of the touch of the hypersurface and indicatrix (2.4.34):

$$g_{ij}(x^k, \xi^k)\xi^i S^k = 0. \tag{3.4.12}$$

Taking into account (3.4.12) we can find local direction of the crack propagation.

Stress tensor versus metric takes the form:

$$\sigma_{ij} = \lambda \nabla_k + \mu(\nabla_i u_j + \nabla_j u_i), \tag{3.4.13}$$

where $\nabla_i = g_{ik}\nabla^k$. Since the deviation from the Euclidean metric is associated with the properties of the continuum and the free crack surface is in air, the rare surface of the

---

[4] Strictly speaking, in the general case the functional dependence of stress and strain tensors for micropolar body is given by a set of 45 relations [270]. For visibility of the analytical results we must use more or less reasoned simplification.

[5] In reality the defective structure of the body is producing the mass forces, but we assume these forces are included in the generation metric of the continuum.

discontinuity (the crack) can be considered located in the conventional Euclidean space. Then equation of motion (3.4.11) in terms of discontinuities [289] assumes the form

$$\left[\mathbf{S}^{i}_{;j}\right] + \left[P^{*i}_{h\ j}\mathbf{S}^{n}\right] = \rho\left[\mathbf{w}\right]. \tag{3.4.14}$$

Since we examine the deformation of a solid, the particles of the medium before the crack front will move continuously. Due to the fact that the discontinuity function is induced by the vector field $\dot{u} = v \in C^2$, this field generates an infinitesimal transformation of the type:

$$\overline{x}^{i} = x^{i} + v^{i}(x)d\tau, \tag{3.4.15}$$

and discontinuity function is equal to taking Lie's derivative [303, 289]. We note that for stationary propagation of the crack, (3.4.15) represents the equation of motion of an elementary volume of the medium. Then the equation of crack surface motion (3.4.14) assumes the form:

$$\underset{L}{D}\,\mathbf{S}^{i}_{;j} + \underset{L}{D}(P^{*i}_{h\ j}\mathbf{S}^{n}) = \rho\,\underset{L}{D}\,\mathbf{w}. \tag{3.4.16}$$

Since the convolution of geometric objects of different natures is under the derivation sign, according to the rules of operation with the Lie derivative, in calculations we can consider the convolution of the Lie derivatives.

## 3.5 Macroscopic group properties of deformation process and gauge fields introduction procedure

The mechanism of group theory broadly used in physics of the twentieth century for a long time found no real application in solid mechanics. However, after the works of L. V. Ovsyannikov [272, 273] the methods of group theory have found their constructive application in solid mechanics [5]. In the last decade, the necessity of group analysis has become almost obvious. This is related to the fact that, according to F. Klein, the geometry of any space can be built on the basis of the symmetry group of this space. The basic definitions and theorems of the group theory are presented in Appendix C. Let us consider a possibility to build an abstract group related with a deformed solid body.

*Theory of hierarchies and group of deformation operators*

Since from the viewpoint of the MHST, the structure of solid mechanics is a *multi-echelon hierarchy*, the group structure of solid mechanics is a plurality of the group structure of echelon elements. In most cases, when applying the group analysis to the mechanics of solids, one considers group properties of various *equations* of solid mechanics [5]. This allows one to make important conclusions in relation to a possible structure of equation solutions, and to make certain invariant solutions. This consideration level corresponds to studying the structure of the *education layer*. Along with that, there is a plurality of solutions, connected with the most possible structure of the continuum, in which the mechanical processes are realized, that is, the structure of the *selection layer*. This direction, founded by Walter Noll in 1958, was subsequently called by Clifford Truesdell the *rational mechanics*. It is usual to consider that all the types of purely mechanical continuum possible within the suppositions made are known [5].

Consideration of the group DSM structure in the *self-organization layer* was practically never made. This study corresponds to consideration of symmetry properties of solid mechanical processes (dynamic structures) and their links with the properties of the medium.

*Initial Galilean symmetry*

For a classic ideal continuum, the spatial-temporal symmetry leads to the choice of the Galilei group as a symmetry group [200]. The Galilei group is a ten-parameter group of linear uniform transformations of coordinates of the Euclidean 3-D space and 1-D time, which leave invariant the space distance and the interval and direction of time:

$$\mathbf{r}' = \mathbf{r} + \mathbf{V}t, \quad t' = t. \tag{3.5.1}$$

The invariance in relation to the Galilei group leads to the Euclidean nature of the classic ideal continuum. The Galilei group includes rotations in relation to random axes, Galilei transformations (boosts), mappings in relation to random plane, and all possible products of these transformations.

### *3.5.1 Kinematics*[6]

In approximation of the continuum, its properties are described by continuously differentiable functions of spatial coordinates and time. It is supposed that any body $\mathcal{B}$ can be viewed as a certain manifold, having the power of the continuum $\mathfrak{R}$:

$$\mathcal{B} = \{X_{(\alpha)}\}, \quad \alpha \in I, \quad I \subset \mathfrak{R} \tag{3.5.2}$$

where $I$ is a certain continual manifold of indices. The elements are endowed macroscopic properties of this body and are called particles of the medium or material points.

The physical properties of this particle are defined by macroscopic characteristics of the respective physical infinitely small volume of real medium, comprising, therefore, a macroscopic number of atomic—molecular units of the matter. Its dimensions should be essentially smaller than the characteristic dimensions of the fields under study. The macroscopic parameters of a material point are introduced by means of averaging of a respective microscopic characteristic by a physically infinite small volume and time interval, which is much smaller than the characteristic time of the motion under study. In doing so, we make a large-grain roughening of the special-temporal description of the macroscopic body, enabling us to exclude small-scale fluctuations of its characteristics

[6] This subsection was written with the participation of V. V. Barkaline.

from consideration, and establish the spatial and temporal scales, which define the minimum distance and minimum time interval, for which the macroscopic values still have sense. Particular values of these scales depend on the nature of the body and its condition, and are defined in the kinetic description level. It is possible that the roughening itself is made under different scales, which are independent. For example, for *physically independent* fields of stresses and deformations, independent averaging procedures should be introduced [106].

The infinite divisibility of power manifolds of the continuum introduces continuous distributions of the additive matter characteristics into consideration: mass, energy, charge, electromagnetic spin, etc. From the mathematical viewpoint, division (3.5.2) defines the determinacy of the respective additive measures on the manifold $\mathcal{B}$, which represents the macroscopic body.

The physical properties of the continual model define the mathematical structure of the geometry, connected with the medium. I. Kunin asserts [173] that the postulate of existence of internal geometry of the medium makes probably the most general equivalent of the notion of elasticity. Respectively, it is expedient to define the non-elastic phenomena (plasticity and creeping) as changes of the internal geometry. Along with that, one usually neglects the fact that the elastic deformation by itself causes a deviation of the internal geometry from the Euclidean nature, which in principle may be defined by experimental methods [143, 154]. The defective continuum is defined similar to (3.5.2) as:

$$\mathcal{B} = \{Y_{(\beta)}\}, \quad \beta \in I', \quad I' \subset \mathfrak{R} \tag{3.5.3}$$

where $Y_{(\beta)}$ are its material points or simple uniform body. To obtain a description of a defective structure, separation (3.5.3) should be correlated with separation (3.5.2) of the ideal continuum. It means that we must mentally bring each element of separation to the natural condition according to Kondo. This transformation for the defect body is not all times continuous transformation [48, 316]. Since there are no grounds to assume that $I = I'$, the number of elements in the defective state should not inevitably coincide with the number of elements in the natural state. In this case, for every element of separation, the motion equation of the elements of the ideal continuum is valid, and the elements are invariant in relation to the Galilei group.

It means that if we apply the Galilei transformation (3.5.1) to solution of these equations, we obtain the solution thereof again. Hereafter, out of the obtained motions of these ideal elements, we need to compile the motion of the defective body, invariant in relation to the Galilei group. One of the possible variants of group representation of ideal continuum microscopic separation can be found in [48]. According to G. Maughin [200], it means invariance of this motion in relation to superposition of the motion of an absolutely solid body. Thus, a group of motions of the defective body is a subgroup of a direct product of Galilei groups of each natural element, to isomorphic Galilei group[7]:

$$G = \prod G_{(\alpha)}.$$

[7] V.V. Barkaline, personal communication.

However, the group of motions is not a gauge group. The gauge group is linked with the choice of model of the medium (in fact, by the links that exist in the medium, i.e., the law of correlation of separations (3.5.2) and (3.5.3)). From the viewpoint of mathematics, possible motions of the points are forming a layer, while the plurality of motions of all the points of the body forms a fibering, in which the Galilei group is taken as a structural group. The correlation law is defined by correlations between the elements of the layers, i.e., by the connectivity in this fibering. In fact, we need to select this connectivity, in order to have the body with defects that move in conformity with the preset group of motions [174].

The choice of this subgroup can be interpreted as localization of the global Galilei group in the terms of gauge symmetry. In this case, one usually uses a certain expansion of the group, for example increase of its dimensionality. Thus, as was shown in [190], to describe the final deformation of ideally plastic media, the defining equations can be written down in the form of the Lie-type system:

$$\dot{X} = AX, \quad A \in SO(5, 1) \tag{3.5.4}$$

where $X$ is defined in the Minkowski space-time $\mathbb{M}^{5+1}$ with an indefinite signature metric (5,1). In this case, the Ilyushin stress space

$$\left\{3/2s^{11}, \sqrt{3}/2s^{11} + \sqrt{3}s^{22}, \sqrt{3}s^{23}, \sqrt{3}s^{13}, \sqrt{3}s^{1}\right\},$$

where $s$ is the deviator of stresses with the Euclidean norm $\|s\| := \sqrt{s \cdot s}$, is only a sufficiently restricted special case of space $X$.

### 3.5.2 Dynamics

As a rule, when studying plastic deformation, one considers the dynamics of certain invariant structures (for example, yield surfaces [132]), where the topology of these structures does not change. The topology conservation is connected with disregard of generation–annihilation defects of various kinds (ideal plasticity). In the case of an ideal plastic medium, the yield surface is generally fixed. In this connection, the deformation process is connected with some route, which is trying to reach the plasticity surface. A characteristic of this route may be, for example, the Odquist parameter, proportional to the arc length of the trajectory of plastic deformation.

This sort of criterion can be formulated for all the existing plasticity theories. It follows from the principle of macrodeterminism that the topological invariance of the routes is correct both in the space of deformations and in the space of stresses. However, the *frontal* comparison of two different deformation ways—one straight and another zigzag-shaped one, but not surpassing the borders of flux tube—results in a well-known contradiction, when the difference of closer ways produces the final value, different from zero. In this connection, we can approach a bit differently the mathematic structure of plasticity theories. It is known that the notion of the group is closely connected with motion of invariant objects of this or that sort. Therefore, if we are interested in the symmetry group of motions of a solid body, we need to establish the presence of invariant objects, connected with the solid body and consider their possible motions.

It follows from the general reasoning that such objects should be certain minimal elements, which do not change in the process of deformation, and only move inside the body. In the case of an absolutely solid body, the body itself is such an element, and possible motions of the solid body are defined by the Galilei group in conformity with the law (3.5.1).

In case we choose out of all the motions of the solid body only the motions connected with plastic deformation, we need to find a group for invariants of the plasticity theory. In the principal aspect, the invariants of the motions of the body should be defined in the atomic–molecular level by the quantum-mechanical way, since the real elementary objects, which preserve their properties at deformation, are the atoms, which obey the laws of quantum mechanics.

The Hamiltonian of this motion has the invariance group, containing subgroups of permutations of identical particles of the system. This subgroup is common for the quantum and classic Hamiltonians of the given system. In the classic mechanics, where the principle of identity of similar particles is not introduced, instead of a group they apply its trivial presentation. (In this case, all the elements of the group are mapped on the unit element of presentation.) It means that in the classic theory no dynamic operator of particle permutation appears. Therefore, in principle, every particle can be awarded with a ‘name’, which does not change during further evolution. As such particle name we can use a set of its initial coordinates and impulses, set at a certain time moment $t_0$, which is the same for all the particles of the system. Since different trajectories in the phase space of the system do not intersect, the point names represent dynamic invariants of the evolution of the system, and any of their functions is also a dynamic invariant.

In the case of a deformed body with microstructure, the minimal elements from the mathematic viewpoint are the sets of coordinates characterizing individual points or structures of the body (*names of the points*). An invariant, in this case, is any function of the ‘names’. A violation of invariance means an appearance of any new object (for example, a crack, a defect, or a fracture surface). If no fracture is taking place even at the microlevel (there is no generation of microcracks, dislocations, etc.), then in the process of plastic deformation the number of names does not change, but at the same time, the mutual location of sets changes in a random manner [174].

### *3.5.3 Group structure of deformation curve*

Since our task is to describe a plastically deformed material, it is necessary to bind the internal parameters of the deformed body with experimentally defined characteristics. Since in practice we work with experimental characteristics, from the mathematics viewpoint we have a reverse task—by a preset property to restore the internal structure. It has been noted above that the full group of permutations, connected with the body, is unknown, but the well-known experimental stress–deformation curve is its integral representation. Let us demonstrate that the curve (diagram) itself possesses a group structure.

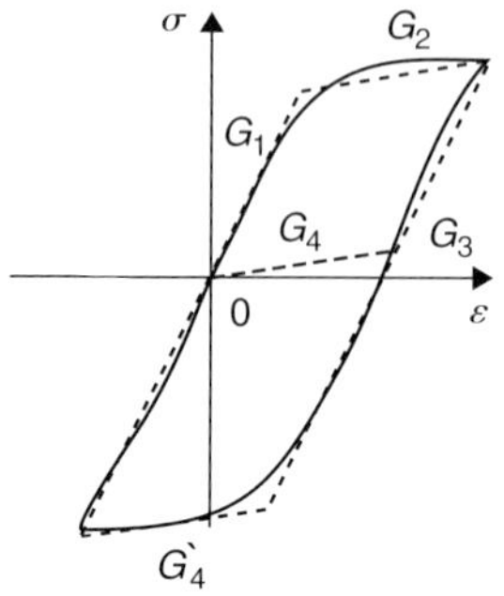


Fig. 3.6. Diagram of material behavior at loading: solid line, real; dotted line, ideal.

It is well known that the real deformation processes of solids are irreversible. This is connected with multiple factors—defect level of the structure, internal friction, energy dissipation, and others. Nevertheless, the classic $\sigma$–$\varepsilon$ diagram (Fig. 3.6) with known reservations is broadly used in material science, since it is assumed that the majority of materials operate within the zone of linear elasticity. The whole mechanism of the classic mechanics of deformed solid body is also built within these assumptions. At the same time, modern materials, working in critical loading areas, demonstrate a rather sharp deviation from the classic behavior (see Fig. 3.6) and require, for their description, a more adequate consideration of deformation processes.

*The $\sigma$–$\varepsilon$ diagram*

Let us consider the macroscopic deformation process. For description of deformation processes, as the elements of the point symmetry group we can introduce the operators which affect the physical-mechanical characteristics of the material and describing the motion in the state space within the stress–deformation diagram (Fig. 3.6). From the physical sense of the sections of the classic stress–deformation diagram for ideally plastic bodies, we can take as the elements of the complete group, describing the deformed state of the continuum, operators $G_1$—elastic deformation; $G_2$—plastic deformation; $G_3$—elastic relaxation; $G_4$—plastic relaxation; $O$—initial state (state of dead loading). In the simplest case of point mapping, the operators describe the transition by the angular points of the diagram—from the initial non-deformed state into the state of maximum elastic deformation, from the elastic deformed state into the plastically deformed state. If we assume that the idealized deformation and relaxation processes take place without the fracture of the material, then, on the basis of physical sense of the operators, the multiplication table for the point mapping will have the form:

$$\begin{array}{cccccc} & G_1 & G_2 & G_3 & G_4 & G_5 = 0 \\ G_1 & G_2 & G_2 & 0 & G_1 & G_1 \\ G_2 & G_2 & G_2 & G_4 & 0 & G_2 \\ G_3 & G_1 & G_4 & G_4 & 0 & G_3 \\ G_4 & G_2 & G_1 & 0 & G_3 & G_2. \end{array} \tag{3.5.5}$$

In conformity with Appendix C, let us check the fulfilment of the required conditions for the elements of the group. Fulfilment of condition 1 follows out of the definition of the multiplication table. The existence of the unit element follows from the definition of the elements of the manifold $\Im$. By means of direct substitution, we may ensure the correctness of condition 3:

$$G_1^{-1} = G_3, \tag{3.5.6a}$$

$$G_2^{-1} = G_2, \tag{3.5.6b}$$

$$G_3^{-1} = G_1, \tag{3.5.6c}$$

$$G_4^{-1} = G_2. \tag{3.5.6d}$$

Fulfilment of conditions (3.5.6a) and (3.5.6b) follows from the physical sense of the operators of the multiplication table. In cases (3.5.6c), (3.5.6d), we adopt the necessity of existence of the left-hand reverse element. However, for the operators, set by the multiplication law (3.5.5), the condition of transitivity is not observed. This is explained by the violation of symmetry, arising in the deformation process. The violation of symmetry explains the necessity of definition of the standard procedure of elongation of derivatives and introduction of gauge fields [115, 116].

*Gauge transformations*

Violation of the transitivity condition for the multiplication table (3.5.5) allows correcting the structure of the operator of plastic deformation [210]. Thus, by consequently applying the multiplication table for combination $G_1G_2G_3$, we have:

$$G_1(G_2G_3) = G_1, \tag{3.5.7}$$

$$(G_1G_2)G_3 = G_4. \tag{3.5.8}$$

From equations (3.5.7) and (3.5.8), taking $G_2 = G_2' + G_2''$ and assuming the action of the operator to be linear, we have:

$$(G_1(G_2' + G_2))G_3 = (G_1G_2')G_3 + (G_1G_2'')G_3, \tag{3.5.9}$$

$$G_1((G_2' + G_2)G_3) = G_1(G_2'G_3) + G_1(G_2''G_3). \tag{3.5.10}$$

By uniting the right-hand parts of (3.5.9) and (3.5.10), we can obtain the conditions imposed on the structure of operator $G_2$. Let us note here that in fact we considered a point group connected with the action of deformation operators. For a correct transition to continuous groups, we need to define in a stricter way the deformation operators and conditions of their action. However, it is clear that expressions (3.5.9) and (3.5.10) are also the defining conditions for the procedure of introduction of gauge fields [115] of the deformation process.

In the case of satisfaction to equations (3.5.9) and (3.5.10), manifold $\Im$ forms the fifth-order group.

*Group parameterization*

Let us consider a possibility of parameterization of group $\Im$. Usually, for plastic deformation with account of defective structure, the internal energy of the material $s$ in the stationary case is written down as:

$$s = s(S, \varepsilon_{\alpha\beta}(S^{\delta}_{\alpha\beta\gamma}, R^{\varepsilon}_{\alpha\beta}, K_{\beta\gamma}), S^{\delta}_{\alpha\beta\gamma}, R^{\varepsilon}_{\alpha\beta}, K_{\beta\gamma}),$$

where $S$ is the entropy and $\varepsilon_{\alpha\beta}$ is the deformation tensor, characterizing the structure of the material, where the Greek indices are numbering the components of the tensors. In this connection, in the case of plastic deformation, the deformation tensor is represented as a sum of three components:

$$\varepsilon_{\alpha\beta} = {}^{el}\varepsilon_{\alpha\beta} + {}^{elpl}\varepsilon_{\alpha\beta} + {}^{pl}\varepsilon_{\alpha\beta},$$

where upper indices are not tensor indices, but characteristics of the states. Thus, ${}^{el}\varepsilon_{\alpha\beta}$ is the tensor of elastic deformation, ${}^{pl}\varepsilon_{\alpha\beta}$ of plastic deformation of non-defective material, and ${}^{elpl}\varepsilon_{\alpha\beta}$ is the tensor of joint elastoplastic deformation, connected with the defects of the material. $S^{\delta}_{\alpha\beta\gamma}$, $R^{\varepsilon}_{\alpha\beta}$, $K_{\beta\gamma}$ are, accordingly, the tensors of curvature, torsion and segment curvature.

Actually, ${}^{elpl}\varepsilon_{\alpha\beta}$ defines a deviation of the deformation curve of the real material from the ideal one. If we consider isothermal deformation of the material, the entropy has only configuration character and is also the distribution function of defects, that is, $S^{\delta}_{\alpha\beta\gamma}$, $R^{\varepsilon}_{\alpha\beta}$, $K_{\beta\gamma}$. It is known that in the Lagrange formalism the elastic deformation (that is, ${}^{el}\varepsilon_{\alpha\beta}$) correlates to the affine connectivity [6, 130], while term ${}^{pl}\varepsilon_{\alpha\beta}$ may correspond to torsion tensor $R^{\varepsilon}_{\alpha\beta}$, which is identical to the tensor of dislocation density $\alpha^{\varepsilon}_{\alpha\beta} = R^{\varepsilon}_{\alpha\beta}$ [6]. Direct link with the defects field can be chosen according to Fig. 3.6. In this connection, the torsion tensor responds to the change speeds of the defining variables [223]. Tensor $K_{\beta\gamma}$ corresponds to linear defects in the material, and the presence of all three tensors of Cartan curvature generates the affine-metric space, more general than the Riemannian one [109]. The presence of the speeds of the defining variables in the transformation law causes the change of the structure of the transformation law. In this connection, the continuous mapping:

$$X^{*} = GX + A, \tag{3.5.11}$$

where $G$ is the complete group of translations and rotations of the solid body, $G = SO(3) \triangleright T(3)$, transits into the $t$ time-dependent mapping:

$$X* = \Im\,(x, y, t)\, X + A\,(x, y)\,. \tag{3.5.12}$$

The dependence of the mapping (3.5.12) on time is connected with introduction of speeds into the mapping and corresponds to the description of the loading history (that is, corresponds to the fact that the tensor of dislocations density $\alpha^{\varepsilon}_{\alpha\beta}$ is the tensor, depending on the history of loading). In mapping (3.5.12), $x = x^{i}$ is the set of defining variables (coordinates), $y$ is the tangent vector in point $x = x^{i}$:

$$\mathbf{y} = y^{i}\left(\frac{\partial}{\partial x^{i}}\right)_{x}. \tag{3.5.13}$$

In such a case, mapping (3.5.12) defines the Finsler space [306], and $\Im$ is the group symmetry of this space. For this group three Cartan curvature tensors will be the defining variables, that is, $\Im$ is the three-parametric final group. Continuity $\Im$ will depend on the continuity of the set of defining variables, which is connected with the deformation character of the body. In this connection, the compatibility tensor in the Finsler representation disappears, and the complete deformation decomposes into the horizontal one (index $h$) and the vertical one (index $v$) [306] in conformity with decomposition of the deformation gradient (2.5.8).

In this case, deformation gradient $F$ represents a fibered manifold of dimension $2n$, and the transformation of the fibering, which translates the layers into itself, has the form:

$$\exp(F) = \exp({}_vF_i^j)\partial_j \otimes D\, l^i, \tag{3.5.14}$$

and structural group $G$ can be represented in the form of (3.5.14) [206, 364]. The introduction the structural group allows transiting to new coordinates $G_{\alpha\beta}$, where $G_{\alpha\beta}$ are elements of the matrices making up the group. These matrices can form, for example, a complete set of deformation operators [214]. These operators act as the right-hand shift operators (set of fundamental fields):

$$D_a^{(R)} = D_{a\alpha\beta} G_{\alpha\beta} \frac{\partial}{\partial G_{\alpha\beta}}, \tag{3.5.15}$$

where $D_{a\alpha\beta}$ are group generators.

Decomposition into the vertical and horizontal components cannot be considered as decomposition into a simple sum of plastic and elastic deformations because of a complex internal structure of the components. We note here that the division of variables in the transformation law into two sets—set $x = (x^i)$, defining the properties of the space, and set $y$, defining the properties of the moving object (field of speeds) (tangent space to the manifold)—is ideologically close to description of the motions of a test particle in the gravitation field.

Fibering of the Finsler space allows introducing, hereafter and in the correct manner, gauge fields and the respective Lie group [206, 364]. The gauge invariance is introduced by other ways into the fracture theory, since it is necessary to explain the incompatibility effects at plastic deformation of materials, when the total deformation cannot be described as a direct sum of plastic and elastic deformations [115, 116, 306, 376].

*Group parameterization and damage criterion*

It follows from the experiments that in the process of real deformation, even in the case of small elastic deformations, a plastic component is present in the process. It means that the composition of representations $G_1 \circ G_2$ is continuous. Field of elastic energy $\mathcal{E}$ is connected with the elastic deformation, while the plastic deformation is physically realized as the process of dynamic origination–elimination of defects. Consequently, the parameters of complete group $\Im$ are the fields of torsion $T$, $C$, fields of curvature $R$, $S$ (p. 56), connected with the field of defects in the usual way, and the elastic energy:

$$\mathcal{E} = \mathcal{E}(T, C, R, S), \quad \Im = \Im(T, C, R, S, \mathcal{E}(T, C, R, S)). \tag{3.5.16}$$

The connectivity and other metric and topological properties of the internal space of the deformed solid body with microstructure can be found from solutions of the equations of the Cartan structure [161]. If we take the assumption, usual in the microscopic fracture theory, that the plastic deformation is a function of the tensor of dislocation density, then the infinitesimal group generators can be obtained as the function of the invariants of the deformation energy [67]. In this case, the torsion fields are not included in the number of group parameters, the obtained theory is applicable to isotropic material, and a complete group is reduced down to the Poincare group [67]. For reducing the notation, let us introduce certain designations for the group parameters $a^n_{(i)}$, $i = 1, 2$, $n = 1, \ldots, N$, where $i$ numbers the starting and final state of the mapping, and $n$ numbers the group parameters. Then, the volume of the group is the absolute value of determinant:

$$\left| \frac{\partial a^n_{(1)}}{\partial a^n_{(2)}} \right| = V_{\Im}. \tag{3.5.17}$$

It follows from expression (3.5.17) that the current damage of the material is:

$$\mu(t) = \frac{V_{\Im}(t)}{V^{\lim}_{\Im}(t)}, \tag{3.5.18}$$

where $V^{\lim}_{\Im}$ is the group volume corresponding to destruction of the material. This volume can be defined experimentally by the values of the respective defect fields at fracture. Respectively, the representation volume of group $\Im$ has the form:

$$\mu^S(t) = \frac{S_{\Im}(t)}{S^{\lim}_{\Im}(t)}, \tag{3.5.19}$$

where $S$ is the area limited by representation operators in the space state $\sigma - \epsilon$. In conformity with the selected representation of mappings and the structure of group $\Im$, both the tensor and scalar criteria of damage follow from group criterion (3.5.18).

## 3.6 Four-dimensional formalism and conservation laws

Earlier, we have considered a quasi-stationary crack. In this context, the quasi-stationary nature should be understood as invariance in relation to the time of the processes, connected with the growth of the crack. In the general case, invariant processes are broadly studied in relativistic physics. It is known that the laws of conservation of physical values follow from the properties of symmetry of space-time or the presence of internal symmetries. Thus, the time homogeneity is related to the possibility of shifting the time count start $t \Rightarrow t + \Delta t$, which leads to energy conservation. The space homogeneity follows from the possibility to transfer the start of coordinates $x \Rightarrow x + \Delta x$, which entails the momentum conservation. The space isotropy (absence of dedicated directions) is connected with a possibility to realize the transformation of rotation $\phi \Rightarrow \phi + \Delta\phi$ and leads to the angular momentum conservation law. On the contrary, the conservation law of the electric charge is caused by the invariance of the motion equations in relation to the internal (local gauge) symmetry. It is less well known that the law of mass conservation, as the final result, is a consequence of invariance in relation to the Galilei group [290].

The spatial-temporal symmetry for a classic ideal continuum results in selection of the Galilei group symmetry as the group symmetry of the ideal medium.

A study of invariant values should be based on the formalism of the special relativity theory [43, 178]. The special relativity theory is formulated for the 4-D Minkowski space. It means that now we can have the time coordinate treated in the same way as the spatial coordinate, and the state of the system is described by a 4-vector, which includes spatial and temporal components $\mathbf{x} = \mathbf{x}x^0 = -ict, +x^1, +x^2, +x^3$. The choice of signs in front of spatial and temporal components is defined by the metric signature. Since the temporal dependencies are included into the state vector, it automatically means a description of dynamic systems (for example, a dynamic crack [101]).

A fundamental role is played by the Lorentz transformation from one system of spatial-temporal coordinates $x^\alpha$ to another system $x^\beta$

$$x^\beta = \Lambda_\alpha^\beta x^\alpha + a^\beta,$$

where $a^\beta$ and $\Lambda_\alpha^\beta$ are the constants, restricted by the condition of the metric tensor invariance

$$\Lambda_\alpha^\beta \Lambda_\gamma^\delta g_{\beta\delta} = g_{\alpha\gamma}. \tag{3.6.1}$$

The role of the distance in the Minkowski space is played by the *interval*

$$c^2\tau^2 \equiv (dx^1)^2 + (dx^2)^2 + (dx^3)^2 - c^2dt^2$$

where $\tau$ is the 'proper time', which measures the distance of the neighboring points on the world line.

Equation (3.6.1) is the equation of the light cone, which is the boundary of a possible evolution of the system, i.e., the evolution of any dynamic system, a crack included, cannot leave the boundaries of this cone.

Further, one can introduce the four-velocity $u^\mu = dx^\mu/d\tau$, analogously to the classical velocity definition and velocity of light as:

$$u^0 = \frac{c}{\sqrt{1 - \frac{v^2}{c^2}}}, \qquad u^k = \frac{v^k}{\sqrt{1 - \frac{v^2}{c^2}}}.$$

The four-momentum $p^\mu$ is defined by extension of the classical notion of mass times velocity:

$$p^\mu = mu^\mu,$$

where $m$ is the rest mass of the particle (body).

Similar to the classical mechanics, the relativity theory introduces the action integral $A$. For a local Lagrangian (3.3.1), the action integral taken over the spatial-temporal domain

$$A = \int \mathfrak{L} d^4\mathbf{x}$$

with the requirement $\delta A = 0$ will result in the Euler–Lagrange equations of the associated variational problem (3.2.14).

In an arbitrary Finsler space-time, however, the set $\{(x, y)\} \in \overset{\circ}{T}_z \,|L(x, y) < 0$ may have many arbitrarily connected components; correspondingly, there may be arbitrarily many 'light cones' [287]. Finsler space-times with two or more future light cones at each point are probably not of physical interest. Note that a plastic deformation in different ways (it is equal to a birefringent medium) is not described by one Finsler structure with two future light cones, but rather by two Finsler structures with one future light cone for each. However, there is no mathematical reason to exclude them. For the formulation of the Fermat principle for the ray path (that is, crack trajectory), we need to just select only one light cone from the additional mechanical or physical consideration.

The equation of the energy conservation law (conservation of energy-momentum tensors) is the requirement that a certain vector should be divergence-free in a particular 'volume' of space-time, that is, it represents a mechanical conservation law [152, 177].

In the general case, the studies of the conservation and balance laws are based on the Noether theorem [255]. In the fracture mechanics, the study of preserving values has caused an increased interest in connection with broad application of invariant $J$-integrals.

Let us consider transformation of a random value $f(x, y)$ under the effect of the translational group (details of the group theory can be found in Appendix C):

$$f^*(x, y) = f(x, y) + a. \tag{3.6.2}$$

The function $f$ is said to be (globally) invariant under the group operation in $f^* = f$ for all the values of parameter $a$. In the neighborhood of the identity transformation, $\tau_0$, which is sometimes denoted $I$, the function $f^*$ may be written, to first order in $a$ as

$$f^*(x, y) = (1 + a\mathbf{v}) f(x, y),$$

where $\mathbf{v} = \partial/\partial x$ is called the infinitesimal generator of the group. Taking into account the independence of variables of the Finsler space and equation (3.5.13), the first prolongation [151, 269] of $y$ is equal to:

$$pr^{(1)}\mathbf{v} = \mathbf{v}, \quad \mathbf{v} = {}^{\{1\}}\chi_i \frac{\partial}{\partial x_i} + {}^{\{2\}}\chi_i \frac{\partial}{\partial y_i}, \tag{3.6.3}$$

where

$$\mathbf{x}^* = \mathbf{x} + {}^{\{1\}}\xi\,(\mathbf{x}, \mathbf{y}), \quad \mathbf{y}^* = \mathbf{y} + {}^{\{2\}}\chi\,(\mathbf{x}, \mathbf{y}),$$

is the one-parameter infinitesimal transformation of the independent and dependent variables and Lie's infinitesimal invariance condition on the action integral can be stated succinctly as [151]

$$\left(pr^{(1)}\mathbf{v} + {}^{\{K\}}D_i\, {}^{\{K\}}\chi_i\right)\mathfrak{L} = 0, \tag{3.6.4}$$

where $D$ is the covariant derivative (2.4.22). When the condition (3.6.4) is fulfilled, then the Noether theorem asserts the existence of $p$ functions $P_i$ on the jet bundle space such that

$$D_i P_i = \operatorname{Div} \mathbf{P} = 0,$$

for every solution of the Euler–Lagrange equations.

It means that quantity $P_i$ satisfies the conservation conditions. By solving jointly (3.6.4), (3.6.3) in relation to $P_i$, we can obtain a complete set of preserved values, specific for the given problem [151]. In this connection, non-traditional conservation laws appear. Thus, when building a covariant four-dimensional description of elastodynamics [152, 153], and also in applying differential operators div, curl, there arise additional connections and relations of mutuality between the physical and mathematical moments, the law of energy and mass conservation.

At the same time, the affect of differential operators grad, div, curl on the energy-momentum tensor of a defective non-homogeneous deformed body $\mathcal{B}$, which theoretically enables us to obtain specific preservation laws, is currently unknown and requires additional studies.

## Certain outcomes

- The introduction of right-hand shift operators by formula (3.5.15) allows building the gauge fields of the processes of plastic deformation. These fields are built as coefficients of a set of fields commutating with fundamental fields.
- The gauge fields arising at plastic deformation are the function of the defective structure of the medium, since dislocations define the deformation gradient according to expression (3.2.1), and disclinations define the curvature tensor according to expression (2.4.23).
- The metric of the medium, generated by defects, allows building a correct description of deformation without attracting any additional postulates. In this connection, the equation of crack trajectory and the shape of the fracture surface can be calculated with the use of the break theory. Unlike the restrictions that were known earlier, conditions (a)–(c) according to the results of §3.4.1 define a direct dependence of the direction of the crack propagation and the front shape of the defective structure of the medium. Besides, since the parameters in the conditions (3.4.8) depend on the respective speeds, the energy dissipation is also affecting the shape of the surface and crack front.
- Singularities of the fields of stresses and deformation in the classic solution of the problem of interaction of the defect and the crack do not appear when using the gauge description. In this connection, the parameters of the gauge potential are the curvature tensors and torsions of the space, defined through the metric.
- The introduction of the group of the space symmetry, connected with the macroscopic deformation, allows correcting the geometrical properties of the space process, associated with fracture. The conditions, from the requirement of the group law of the transformation law of linear deformation operators, allow constructing gauge potentials without attracting any additional postulates.
- The knowledge of gauge potentials makes it possible to obtain the Lagrangian of the deformed solid body. The precise knowledge of the Lagrangian allows defining the thermal-mechanical properties of the material [6], the peculiarities of the defect dynamics [109], forecasting the trajectory of crack propagation [223]. The knowledge of the defect dynamics is especially important for forecasting the behavior of materials in critical conditions (for example, at increased temperature). A possibility to fix the defects by affecting the field variables in conformity with separation of the general Lagrangian

into the field and massive parts, $\mathfrak{L}_t + \mathfrak{L}_m$, will enable regulating the mechanical properties of the material. This can be especially important when designing materials with preset fracture zones in operation [211]. To achieve this, we need to create the defect distribution, at which the metric of the Finsler space, functionally connected with $\Gamma^{*\beta}_{\delta\alpha}$, has the structure under which the crack propagation is restricted within the preset zones.

- An additional study is required into the problem of transition from the point symmetry group of the deformed body to the Lie group generated by the geometry of the Finsler space.

In Edelen and Lagoudas, *Gauge Theory and Defects in Solids* (North-Holland, Amsterdam 1988) it is mentioned that in most cases connection, torsion and curvature quantities are associated with the complete differential system generated by the additional state variables required in order to describe materials with defects. In this case the connection, torsion and curvature forms are simply systems of exterior forms defined on the space of reference histories $\mathbb{R}^4$ and $\mathbb{R}^4$ does not take part in the dynamics of state variables. From our opinion we can operate in real material space $\mathbb{M}_3 \times \mathbb{R}$ and associate connection, torsion and curvature with real physical properties of intrinsic space of deformed and fissured body.

# 4 Application of General Formalism in Macroscopic Fracture

## Contents



It follows from the general consideration of hierarchy systems that the control level and the objects located therein should not possess the whole plenitude of information about the structure and behavior of its slaving level. Therefore, a transition from micro- to macroscopic consideration is a rather non-trivial problem and cannot be reduced to a simple summing up of microscopic interactions. In an ideal variant, we could suppose that in some future, by using super-large and super-rapid computers, it will be possible, for example, by using the molecular dynamics methods, to perform a 'determinate' (non-probabilistic) calculation of a real steel cube. However, the number of states that may be realized in such a cube is expressed by A. N. Kolmogorov's *superhuman* numbers, which, naturally, prevents one from treating such suppositions seriously. Thus in any case certain averaging models of the media will be under consideration. Additionally by physically satisfactory 'determinate' calculation we may take into account the probabilistic character of the wave function which leads to the macroscopic indeterminacy.

The basic difficulty of transition to upper hierarchy levels of the fracture process lies in evaluation of the effects of microscopic variables on the macroscopic state of the system. In this case, the hierarchy growth operator $\rho^+$ is an operator of continual averaging, while the complication of the internal structure of the hierarchy layer (action of the operator $\rho^*$) corresponds to complication of real media models. It is well known from continual theory that the problem of averaging strictly depends on media model and solution problem. Thus, the earlier considered microscopic methods should be accordingly modified to describe the plastic deformation and fracture in particular macroscopic problems. Let us consider the problem of a mainline crack propagation in a non-uniform body in the random strain–stress deformation condition.

## 4.1 Macroscopic variational approach to fracture

It is known that the variational approach is successfully applied for consideration of the macroscopic fracture process [235, 284]. A variational problem for a mainline crack can be formulated as follows. Let us consider an open connected domain $\Omega$ of the Euclidean space $\mathbb{R}^3$, the border of which is $\partial\Omega$. Aggregate $\Omega^- = \Omega \cup \partial\Omega$, belonging to $\Omega$, represents the domain occupied by the body before the moment of deformation. A derivative of any function $\psi(\cdot), \xi \in \mathbb{R}^3 \to \psi(\xi \in \mathbb{R}^3)$ by any variable $\xi_i$ will be $\partial_{\xi_i}\psi$. A derivative by material coordinates $x_i$ for simplicity is written down as $\partial_i$.

Let part $\Gamma_c$ of border $\partial\Omega$ represent a section or a notch. Domain $\Gamma_c$ is briefly called a crack, and body $B$ is a fractured body. Then a *fracture* represents a geometrical change of $\Gamma_c$ [39]. This change can be understood as an action of a group of translations on the initial geometrical configuration. This geometrical change can be represented as a single-parameter family of continuous and reversible mappings:

$$\psi(.;\eta) : x \in \Omega^- \to x' \in \Omega_\eta^- \subset \mathbb{R}^3,$$

where $\eta \in \mathbb{R}$ and $\psi(x;0) = x$. Domain $\Omega_0^- = \Omega_{\eta=0}^- = \Omega^-$ represents the influence of the crack. Strictly speaking, for real processes, we cannot consider $\eta = 0$, the crack in its motion changes the properties of the material and elastic fields in the propagation direction. But in ideal plasticity our statement of problems is broadly appropriate. Dynamic effects will be taken into account hereafter.

Similar to § 3.3, we assume that for an idealized 2-D case $\Omega_0 = \mathbb{R}^2$, the process of crack propagation represents a translation of a geometrical object along its own direction. New coordinates are given by a single-valued reversible transformation of coordinates:

$$x'_j = \psi(x;\eta) = x_j + \eta\alpha_j \quad \text{for all} \quad x \in \Omega_0^-, \quad j = 1, 2, \tag{4.1.1}$$

where parameter $\eta \in \mathbb{R}$ defines the number of unit translations (it corresponds to the number of steps $i$ of propagation) and $\boldsymbol{\alpha}$ is a unit vector in the direction of crack growth.

In the case where certain forces or preset displacements are applied to body $B$, the body occupies the *deformed domain* $\boldsymbol{\varphi}(\boldsymbol{\Omega}_\eta)$. This domain is characterized by mapping $\boldsymbol{\varphi}(\cdot) : \boldsymbol{\Omega}_\eta^- \to \mathbb{R}$, such that it preserves orientations in $\boldsymbol{\Omega}_\eta^-$ and is mapped on $\boldsymbol{\Omega}_\eta$ in a physically defined manner. Mapping $\boldsymbol{\varphi}(\cdot)$ is called deformation, set $\boldsymbol{\Omega}_\eta$ the *initial (reference) deformed domain*. Mapping $\mathbf{u}(\cdot) : \boldsymbol{\Omega}_\eta^- \to \mathbb{R}^3$, defined by ratio $\mathbf{u}(x) = \boldsymbol{\varphi}(x) - x$, is called the *field of displacements* corresponding to deformation $\boldsymbol{\varphi}$.

*Defining equations*

If we assume that body $B$ is made from a non-uniform material, the density of the fracture energy $w(\mathbf{x})$ for $\mathbf{x} \in \Omega_0$ can be represented as a sum of elastic ${}^e w$ and dissipative parts ${}^d w$:

$$w(\mathbf{x}) = {}^e w + {}^d w. \tag{4.1.2}$$

Strictly speaking, the dissipative part of the fracture energy does not coincide with the fracture energy connected with plastic deformation. Note that well-known representation of energy decomposition

$$w(\mathbf{x}) = {}^{el}w + {}^{pl}w$$

follows from the traditional polar decomposition, representation (4.1.2) arises from fibering structure of internal geometry of a real body.

In a similar way, displacement field $\mathbf{u}(\cdot)$ is split into the elastic ${}^{e}\mathbf{u}$ and dissipative parts ${}^{d}\mathbf{u}$ [184]:

$$\mathbf{u}(\cdot) = {}^{e}\mathbf{u}(\cdot) + {}^{d}\mathbf{u}(\cdot), \tag{4.1.3}$$

and the following relation is correct:

$$w(\mathbf{x}) = {}^{e}w[\mathbf{x};\, \partial_x {}^{e}\mathbf{u}(\mathbf{x})] + {}^{d}w[\mathbf{x};\, \partial_x \mathbf{u}(\mathbf{x})]. \tag{4.1.4}$$

On the basis of assumptions made, we can build the defining equation of the problem.

**Elastic behavior of material.** The full elastic energy of the fracture, stored in the volume of the material, is the function of the field of elastic shears ${}^{e}\mathbf{u}$ and can be calculated as:

$${}^{e}W[{}^{e}\mathbf{u}(\cdot)] = \int_{\Omega^-} {}^{e}w[\mathbf{x};\, \partial_x {}^{e}\mathbf{u}(\mathbf{x})]d\Omega. \tag{4.1.5}$$

In this connection:

$${}^{e}W[{}^{e}\mathbf{u}(\cdot) + \delta^{e}\mathbf{u}(\cdot)] = {}^{e}W[{}^{e}\mathbf{u}(\cdot)] + \delta_u {}^{e}W + \delta_u^2 {}^{e}W + o(\|\delta^{e}\mathbf{u}\|^3), \tag{4.1.6}$$

where $o(\|\delta^{e}\mathbf{u}\|^3)$ is an infinitely small (i.s.) value of the same order as $\|\delta^{e}\mathbf{u}\|^3$, and $\|\cdot\|$ is the norm in the Euclidean sense. By unwinding expression (4.1.6) with account of (4.1.5), we have:

$$\begin{aligned} {}^{e}W[{}^{e}\mathbf{u}(\cdot) + \delta^{e}\mathbf{u}(\cdot)] &= {}^{e}W[{}^{e}\mathbf{u}(\cdot)] + \int_{\Omega^-} \partial_{\partial_i u_j} {}^{e}w \partial_i \delta^{e} u_j d\Omega \\ &\quad + \frac{1}{2}\int_{\Omega^-} \partial_{\partial_i u_j} \partial_{\partial_k u_l} {}^{e}w \partial_i \delta^{e} u_j \partial_k \delta^{e} u_l d\Omega + o(\|\delta^{e}\mathbf{u}\|^3). \end{aligned} \tag{4.1.7}$$

In respect that $\sigma_{ij} = \partial_{\partial_i u_j} {}^{e}w$, and denoting the tensor of rigidity coefficients $E_{ijkl} = \partial_{\partial_i u_j} \partial_{\partial_k u_l} {}^{e}w$, equation (4.1.7) takes the form:

$$\begin{aligned} {}^{e}W[{}^{e}\mathbf{u}(\cdot) + \delta^{e}\mathbf{u}(\cdot)] &= {}^{e}W[{}^{e}\mathbf{u}(\cdot)] + \int_{\Omega^-} \sigma_{ij} \partial_i \delta^{e} u_j d\Omega \\ &\quad + \frac{1}{2}\int_{\Omega^-} E_{ijkl} \partial_i \delta^{e} u_j \partial_k \delta^{e} u_l d\Omega + o(\|\delta^{e}\mathbf{u}\|^3). \end{aligned} \tag{4.1.8}$$

**Non-elastic behavior of material.** In the general case, the dissipative processes in the proximity of the crack tip include a number of effects. These are the plastic deformation of the material, and the effects connected with a break of interatomic bonds. The effects connected with breakage of interatom bonds are usually accounted for by introduction of the surface crack energy (parameter $\gamma$ is the density of surface energy). The plastic deformation is accounted for in different ways. For example, we can assume that the behavior of the material in the plasticity zone responds to the linearized stress–deformation law:

$$\delta_u \sigma_{ij} = E_{ijkl} \partial_k \delta^e u_l = E_{ijkl} (\partial_k \delta u_l - \partial_k \delta^d u_l). \tag{4.1.9}$$

On the other hand, we may consider that the non-elasticity effects are connected with the corrections introduced into the characteristics of the material. This allows rewriting expression (4.1.9) in the form:

$$\delta_u \sigma_{ij} = (E_{ijkl} + \Delta E_{ijkl}) \partial_l \delta u_k, \tag{4.1.10}$$

where value $\Delta E_{ijkl}$ is a non-elastic correction to the tensor of rigidity moduli. This way is thought to be more promising, since it allows one in principle to account in structure $\Delta E_{ijkl}$ not only the effects of plasticity, but also the corrections connected with the processes of other origin.

Thus, since integration when defining the work of fracture takes place by a closed loop in the area of the crack tip, an important factor can be the change of strength of the material as a result of dynamic propagation. This can be accounted for by introducing the notion of the weakened bond zone (or cohesive zone) in the proximity of the tip. The length of this zone can be defined [101, 240, 298] as a function of the speed of crack propagation $v$:

$$R = R_{v=0}/f(v), \tag{4.1.11}$$

where $R_{v=0}$ is the length of the influence zone [123, 192] at a quasi-static crack growth, $f(v) = (1 - v^2/c^2)^{-1/2}$ is the function of the speed effect for a scalar wave equation, and $c$ is the speed of sound in the material (unique for the scalar equation). The relative length of this cohesive zone $\lambda = R_{v=0}/(2\varepsilon) \sim 10^{-4}$–$10^{-1}$.

The influence function $f(v)$, written down in this form, reflects the fact that for a crack, similar to many other dynamic problems, a quasi-particle can be introduced, related with the motion of the end of the crack. This quasi-particle is called a *crackon*. The notion of crackon was introduced into the fracture theory for the first time in [237]. The results, connected with the crackon, are considered in the finalizing work [235] (see the respective references therein). Structurally similar expressions can be written for the vector wave equation. An additional contribution into the energy dissipation process is made by 'engagement' of dissipative configuration forces, connected with the flow of energy [1]. These forces come into play upon achievement by the crack of the critical propagation speed and depend on $K_I$, $K_{II}$.

On the basis of expression (4.1.9), with the precision down to third-order infinitely small values, we can write down similar to (4.1.7) for non-elastic deformations:

$$^dW[^d\mathbf{u}(\cdot)] = \int_{\Omega^-} {}^dw[\mathbf{x}; \partial_x {}^d\mathbf{u}(\mathbf{x})] d\Omega. \tag{4.1.12}$$

We obtain:

$$\begin{aligned}
{}^dW[\mathbf{u}(\cdot)+\delta\mathbf{u}(\cdot)] &= {}^dW[\mathbf{u}(\cdot)] + \delta_u{}^dW + \delta_u^{2d}W + o(\|\delta^e\mathbf{u}\|^3) \\
&= {}^dW[\mathbf{u}(\cdot)] + \int_{\Omega^-} \partial_{\partial_k u_l}{}^e w[\mathbf{x}, \partial_x{}^e\mathbf{u}]\partial_k\delta^d u_l d\Omega \\
&\quad + \frac{1}{2}\int_{\Omega^-} E_{ijkl}\partial_i\delta^e u_j\partial_k\delta^d u_l d\Omega + o(\|\delta^e\mathbf{u}\|^3) \\
&= {}^dW[\mathbf{u}(\cdot)] + \int_{\Omega^-} \sigma_{kl}\partial_k\delta^d u_l d\Omega \\
&\quad + \frac{1}{2}\int_{\Omega^-} E_{ijkl}\partial_i\delta^e u_j\partial_k\delta^d u_l d\Omega + o(\|\delta^e\mathbf{u}\|^3).
\end{aligned} \tag{4.1.13}$$

By uniting the equations for the elastic and dissipative parts, we obtain an expression for the full fracture energy, accumulated and dissipated in body $B$:

$$\begin{aligned}
W[\mathbf{u}(\cdot)+\delta\mathbf{u}(\cdot)] &= W[\mathbf{u}(\cdot)] + \int_{\Omega^-} \sigma_{kl}\partial_k\delta u_l d\Omega \\
&\quad + \frac{1}{2}\int_{\Omega^-} E_{ijkl}\partial_i\delta^e u_j\partial_k\delta u_l d\Omega + o(\|\delta^e\mathbf{u}\|^3) \\
&= W[\mathbf{u}(\cdot)] + \int_{\Omega^-} \sigma_{kl}\partial_k\delta u_l d\Omega \\
&\quad + \frac{1}{2}\int_{\Omega^-} (E_{ijkl} + \Delta E_{ijkl})\partial_i\delta u_j\partial_k\delta u_l d\Omega + o(\|\delta^e\mathbf{u}\|^3).
\end{aligned} \tag{4.1.14}$$

In equation (4.1.14), obvious transformations have been made:

$$\begin{aligned}
E_{ijkl}\partial_i\delta^e u_j &= E_{ijkl}(\partial_i\delta u_j - \partial_i\delta^d u_j) = E_{ijkl}\partial_i\delta u_j - E_{ijkl}\partial_i\delta^d u_j \\
&= E_{ijkl}\partial_i\delta u_j - \Delta E_{ijkl}\partial_i\delta u_j = (E_{ijkl} - \Delta E_{ijkl})\partial_i\delta u_j,
\end{aligned}$$

where transformation $\Delta E_{ijkl}\partial_i\delta u_j = -E_{ijkl}\partial_i\delta^d u_j$ has been obtained from comparison of expressions (4.1.9) and (4.1.10).

**Variational problem.** With the use of the above formalism, we can formulate a variational problem of crack growth (VPCG). Let the following conditions be observed:

- There exists a fixed displacement $\mathbf{u}^0(\cdot) : \mathbf{\Gamma}_u \to \mathbb{R}^3$ on a non-empty submanifold $\mathbf{\Gamma}_u$ of border $\partial\mathbf{\Omega}$ with $\mathbf{\Gamma}_u \cap \mathbf{\Gamma}_c = \varnothing$ and $\mathbf{\Gamma}_u \cup \mathbf{\Gamma}_c = \partial\mathbf{\Omega}$. This is equivalent to splitting the complete displacement in conformity with the principle of superposition of the linear elasticity theory on the sum of displacements of the body without cracks, which is under loading, and the body without loading with a crack, on the surface of which the load is acting [284, 311].
- There are no volume forces in volume $\mathbf{\Omega}$ or surface forces on $\mathbf{\Gamma}_c$.
- The body is in equilibrium, that is, the stress equilibrium condition is observed on $\mathbf{\Omega}^-$.

Also, there exists a solution of the variational problem of the crack growth. The solution of the problem is understood in the sense of existence of the displacement field $\mathbf{u}(\cdot)$. Let in these conditions value $d\mathbf{u}_0$ increase by $\mathbf{\Gamma}_c$. Our task is to define the growth of displacement $d\mathbf{u}$ of domain $\mathbf{\Omega}$, caused by change $d\mathbf{u}_0$, at the assumption that the growth is infinitesimally small and all the control equations are linearized (infinitesimal VPCG).

Like before, we take $\delta\mathbf{u} = \delta^e\mathbf{u} + \delta^d\mathbf{u}$; the displacement variation complies with condition $\delta\mathbf{u} = d\mathbf{u} = +d\mathbf{u}$ on $\mathbf{\Gamma}_c$. Since $\mathbf{u}(\cdot)$ is the solution of VPCG, the first variation $\delta_u W[\mathbf{u}(\cdot)]$ is equal to zero according to the principle of virtual work. In this case, according to expression (4.1.14) the change of total fracture work, induced by displacement variation $\delta\mathbf{u}(\cdot)$ with precision down to terms $o(\|\delta^e\mathbf{u}\|^3)$, is [39, 284]

$$W[\mathbf{u}(\cdot) + \delta\mathbf{u}(\cdot)] - W[\mathbf{u}(\cdot)] \approx \delta_u^2 W = +\frac{1}{2}\int\limits_{\Omega^-} \delta\sigma_{ij}\partial_i\delta u_j d\Omega. \tag{4.1.15}$$

**Remark 4.1.1.** The absence of volume forces in $\Omega$ corresponds to representation of the material as ideal. When considering real materials (for example, with dislocations), the variational problem should be respectively changed.

**General conditions of crack growth.** On the basis of the considered general provisions, it is possible to set up the conditions of crack growth.

Let us consider a case of an endless plate with a semi-endless crack, located along the negative semi-axes $OX$, and let us consider the VPCG. We assume that the field of elastic displacements ${}^e\mathbf{u}(\cdot)$ as a result of the fracture process $\boldsymbol{\phi}$ is transformed into a new elastic field ${}^e\mathbf{u}'(\cdot)$ according to the law:

$${}^e u_k'(\mathbf{x}') = {}^e u_k'[\phi^{-1}(\mathbf{x}'; \eta)], \quad \text{for} \quad \forall\, \mathbf{x}' \in \Omega'. \tag{4.1.16}$$

In this expression $\eta$ is the parameter characterizing the onset of fracture. Condition (4.1.16) is the condition of self-consistency. According to assumptions of the previous discussion, the actual increase of displacement $d\mathbf{u}$ minimizes value $\delta_u^2 W$. Since we are considering a thermodynamically isolated system, at the start of the fracture (at $\eta = 0$) this minimum should be preserved. Thus:

$$\frac{\partial}{\partial\eta}\delta_u^2 W[\mathbf{u}'(\cdot); \eta] \,|_{\eta=0} = 0. \tag{4.1.17}$$

It has been shown in [39] that condition (4.1.17) can be transformed into the condition for work of elastic and dissipative forces:

$$\frac{\partial}{\partial\eta}\delta_u^2 W[\mathbf{u}'(\cdot); \eta] \,|_{\eta=0} = J_{j'}\alpha_{j'} - R_{j'}\alpha_{j'} = J_\alpha - R_\alpha = 0, \tag{4.1.18}$$

where $R_{j'} = R_{j'}^i + R_\alpha^d$ is the sum of non-uniform $R_{j'}^i$ and dissipative $R_{j'}^d$ resistance to fracture:

$$R_{j'}^i = -\int\limits_{\Omega} \partial_{j'}\left[\frac{1}{2}E_{ijkl}\partial_i d^e u_j \partial_k d^e u_l\right] |_{\partial_{x'}u'} \, d\Omega,$$

$$R^d_{j'} = -\int\limits_{\Omega} \partial_{j'} \left[ \frac{1}{2} E_{ijkl} \partial_i d^e u_j \partial_k d^d u_l \right] d\Omega.$$

It is seen from the expressions that the dissipative resistance appears at a permanent deformation. Value $R_{j'}$ is the full resistance to fracture.[1] Vector $\mathbf{R}$ is the vector of energy dissipation. In (4.1.18) value $J_{j'}$ is $j'$ component of vector of the Cherepanov–Rice $J$-integral [52, 53]:

$$\mathbf{J} = \int\limits_{\Omega} -\partial_{l'} \left( E_{ijk'l'} \partial_i d^e u_j \,|_{x'}\, \partial^e_{j'} u_{k'} + \left[ \frac{1}{2} E_{ijkl} \partial_i d^e u_l \right] |_{\partial_{x'} u'}\, \delta_{l'j'} \right) d\Omega.$$

### *4.1.1 Thermodynamics of crack growth and influence of weakened bond zone on crack equation*

In the preceding paragraphs, we have considered an idealized crack propagation supposing a lack of energy exchange with the external medium. In reality, this factor can cause an essential, if not the defining influence on the crack propagation [168]. Besides, an important factor affecting the crack growth is the weakened bond zone that exists in front of the crack. The origin of this zone is clear: unlike the ideal crack model of Barenblatt–Dugdale, the fracture does not occur instantly. The atomic bonds are deformed from the equilibrium condition until the break occurs. Strictly speaking, when considering the crack propagation, it is necessary to introduce two characteristic speeds into the theory: the speed of atomic bond break and the speed of crack growth [343]. Since part of the energy stored by the system at deformation is spent on deformation and breaking of atomic bonds, this leads to changing elastic constants, energy dissipation and other effects. A crack model with such a zone was introduced into consideration by Leonov and Panasyuk.

Undoubtedly, a precise account of the effects connected with the existence of the weakened bond zone is possible only in the successive quantum fracture theory, which, unfortunately, is still far from complete [54, 55, 219]. However, the effect of weakened bonds can be partially accounted for by the methods of the classic theory of deformed solid body [101, 123, 192].

Let us investigate the problem of the cohesive zone length (Fig. 4.1). The weakened bond zone idea provides a simple yet useful device for examining crack tip phenomena at a level of observation for which deviations from linearity in material response are included. On the basis of energy concepts, stress intensity factor in a mode I deformation field can be related to the constitutive characteristics of the cohesive zone. However, it was not possible to determine the size of the cohesive zone in terms of the remote loading and the constitutive properties of the zone by means of the global energy arguments. Instead, the determination of the size $R$ must follow from a study of the stress distribution. The main ideas of cohesive zone models are simpler to convey for mode I deformation

[1] Similar to the values introduced in electricity: resistance is on what the electromotive force (EMF) makes its work.

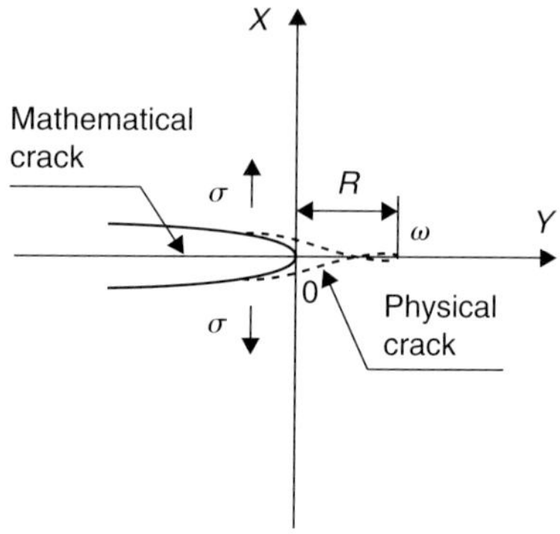


Fig. 4.1. Model of crack without cohesive zone.

than for mode II deformation [101]. We believe that our system is under the action of concentrated force. The stress distribution has a form

$$\sigma_{yy}(\varepsilon, 0) = \frac{p^* \sqrt{l}}{\pi \sqrt{\varepsilon(\varepsilon + l)}} \tag{4.1.19}$$

for normal stress on the crack line $\varepsilon > 0$, where $p^*$ is the magnitude of each concentrated force acting at $\varepsilon = -l$ to open the crack. The corresponding stress intensity factor is

$$K_I = p^* \sqrt{\frac{2}{\pi l}}. \tag{4.1.20}$$

The stress intensity factor (SIF) by the complex loading due to the cohesive stress is determined by superposition of partial SIF. The concentrated force (4.1.19) is first replaced by traction $-\sigma(l)$ distributed over the infinitesimal interval $-l < \varepsilon < (-l + dl)$, and the result is then integrated over the range $0 < l < R$ to yield

$$K_{Icoh} = -\sqrt{\frac{2}{\pi}} \int_0^R \frac{\sigma(l)}{\sqrt{l}}\, dl. \tag{4.1.21}$$

The total stress is non-singular or $K_I = 0$ if $R$ is chosen to satisfy

$$K_{Iappl} = \sqrt{\frac{2}{\pi}} \int_0^R \frac{\sigma(l)}{\sqrt{l}}\, dl \tag{4.1.22}$$

for given $\sigma(l)$. To make further progress with the cohesive zone idea, it is necessary to be more specific about the nature of the cohesive stress $\sigma(l)$. The simplest case of a uniform cohesive stress was introduced by Dugdale to represent a plastic zone near a crack tip in a thin sheet of an elastic–plastic non-hardening material under tension. He assumed the cohesive stress to be the tensile flow stress $\sigma_0$ of an ideally plastic material. Then, from (4.1.21), the traction on the plane $y = 0$ will be less than or equal to $\sigma_0$ everywhere demands that

$$R = \frac{\pi}{8}\left(\frac{K_{Iappl}}{\sigma_0}\right)^2. \tag{4.1.23}$$

From a physical point of view the cohesive stress $\sigma(l)$ originates from long-distance collective interactions of two metallic surfaces. After nuclear forces break over divergence of crack edges this interaction is caused by Van der Waals forces [72]. Thus the cohesive stress can be calculated *ab initio* by the quantum-mechanical methods.

With $R$ determined, the crack line loading is completely specified. It should be additionally noted that the physical crack tip is at $\varepsilon = R$ and not at $\varepsilon = 0$. The fact that the crack tip is initially taken to be at $\varepsilon = 0$ is simply a mathematical artifice of the approach. This is usually by the construction of different models, when the idealized case is generalized. For example, with the aim of abstraction from the problem of the shape of the crack surface, we can first formally consider a zero-thickness crack, and then consider corrections to the theory, arising from the existence of the crack thickness.

*Energy balance*

Study of the energy balance can be made on the basis of thermodynamic considerations [181]. We think that the external medium performs the work $W_{ext}$ of separating the body. Then, on the basis of the first and second laws of thermodynamics, we can write down:

$$dW_{ext} = dW + \mathbf{R}d\mathbf{u}. \tag{4.1.24}$$

From the formal viewpoint, expression (4.1.24) coincides with the Griffith criterion—the required condition of crack growth:

$$dW_{ext} = dW + 2\gamma dl, \tag{4.1.25}$$

where $\gamma$ is the surface energy of the formed crack. However, the physical interpretation of expressions (4.1.24) and (4.1.25) is essentially different. Term $2\gamma$ is a change of internal energy, while $\mathbf{R}d\mathbf{u}$ is the increase of the entropy of the body. The non-equivalence of the Griffith criterion and the thermodynamic consideration was analyzed in detail by B. V. Kostrov and L. V. Nikitin [168]. They considered the energy balance (Fig. 4.2) in a small area $V$ around the tip of the propagating crack and analyzed the equation of the power spent on the fracture, in the form:

$$\dot{W} = \int_{\partial\Omega_g} (q_g - g_i\dot{u}_i)\,dS - \lim_{\varepsilon\to\infty}\int_{S_\varepsilon}\left\{\sigma_{ij}\dot{u}_i - q_j + \left[\int^t \left(\sigma_{ik}\dot{\varepsilon}_{ik} - q_{i,j}\right)dt + \frac{1}{2}\rho\dot{u}_i\dot{u}_i\right] v_j\right\} n_j dS, \tag{4.1.26}$$

where $\varepsilon$ is a small domain in the proximity of the crack tip, $q_i$ is the vector of heat flow in the area of the crack motion, $q_p$ is the density of the heat flow, $\partial\Omega_p$ is the crack surface in the area of volume $V$, which is outside the weakened bond zone, $v_j$ is the vector of

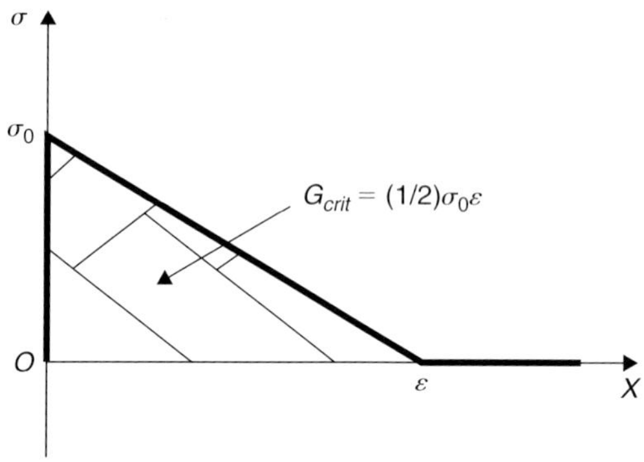


Fig. 4.2. Correlation of the crack opening and applied stress $\sigma$. Fracture energy $G_{crit}$ is proportional to the area under the curve. The weakened bond zone is schematically shown with a thick line along the direction of crack propagation.

the surface motion speed $S_\varepsilon$, $\partial\Omega_g$ is the surface inside volume $V$, on which the cohesive forces $g_i$ are acting, $q_g$ is the density of the heat flow through the surface $\partial\Omega_g$, and $n_j$ is the external normal to the surface embracing $V$.

Analysis of expression (4.1.26) allows one to write down the following for the energy balance under definite restrictions on the process characteristics:

$$\dot{w} = 2\gamma v \int_{-d}^{0} g_i \langle \frac{\partial u_i}{\partial t} \|_{\xi_j=\text{const}} \rangle d\xi, \tag{4.1.27}$$

where $u_i$ is the crack opening, $\xi$ are new local coordinates, selected in such a way that $\xi = -R$ is the end of the weakened bond zone, where the crack opening $u$ reaches its limit value, and $\langle\rangle$ means a jump in the crack line.

We note here that expression (4.1.15) was obtained in supposition of varying only the displacement (length of the crack $a$). In the general case, it is possible also to vary the loading $\mathfrak{L} \equiv \{u, T\}$, where $\mathbf{T}$ is the random power parameter. In this case, equation (4.1.24) can be written down as:

$$\delta W_{ext} - \delta W = \left(\delta W_{ext} - \delta W\right)_{\mathfrak{L}=C_{ext}} + \left(\delta W_{ext} - \delta W\right)_{a=C_{ext}}. \tag{4.1.28}$$

The procedure of varying in the equation is performed in extremal $C_{ext}$. From well-known suppositions [181], we assume:

$$\delta(W_{ext})_{\mathfrak{L}=C_{ext}} = d\Phi \equiv \int_{\partial\Omega} \mathbf{T} \cdot \mathbf{u} ds. \tag{4.1.29}$$

In this case, we can write down:

$$\delta W_{ext} - \delta W = \left(\delta W_{ext} - \delta W\right)_{\mathfrak{L}=C_{ext}} = G d\mathbf{u}, \tag{4.1.30}$$

$$G \equiv -\left(\frac{\partial\,(W - \Phi)}{\partial a}\right)_{\mathfrak{L}}.$$

In this expression, $G$ is the energy release rate. With account of the fact that in the extremal, the first variation of fracture energy is equal to zero (in conformity with § 4.1), equation (4.1.30) can be rewritten as follows:

$$(\delta W_{ext} - \delta W)_{\mathfrak{L}=C_{ext}}$$

$$= \int_{\partial\Omega} \mathbf{T}\left(\frac{\partial \mathbf{u}}{\partial \mathfrak{L}}\right)_{a=C_{ext}} d\mathfrak{L}ds - \frac{\partial}{\partial \mathfrak{L}}\left(\int_{\Omega} \frac{1}{2}\sigma\varepsilon dS\right)_{\mathfrak{L}=C_{ext}} d\mathfrak{L}$$

$$= \left[\int_{\partial\Omega} \mathbf{T}\left(\frac{\partial \mathbf{u}}{\partial \mathfrak{L}}\right)_{a=C_{ext}} ds - \int_{\Omega} \sigma\left(\frac{\partial \varepsilon}{\partial \mathfrak{L}}\right)_{a=C_{ext}} dS\right] d\mathfrak{L} = 0. \tag{4.1.31}$$

By comparing equations (4.1.24) and (4.1.29) with equation (4.1.30) and taking into account condition (4.1.30), we have condition of equilibrium:

$$Gda = Rda. \tag{4.1.32}$$

*Expression for fracture energy*

Condition (4.1.30) allows obtaining a corrected equation for the fracture energy [181]:

$$G = \frac{1}{2}\int_{\partial\Omega} \left\{\mathbf{T}\left(\frac{\partial \mathbf{u}}{\partial a}\right) - \left(\frac{\partial \mathbf{T}}{\partial a} \cdot \mathbf{u}\right)\right\} ds. \tag{4.1.33}$$

We underline here that the integration is made on the crack length. When calculating particular states, we have to set an obvious type of power value **T**, the geometry of loading and crack.

Let us consider, for example, a flat stressed state, for which $\sigma_{22} \neq 0$, $\sigma_{ij} = 0$ (see Fig. 4.3). In this case the elementary work of external forces is $\delta W_{ext} \equiv \sigma_{22}^{\infty} L_1 d(\Delta L_2)$,

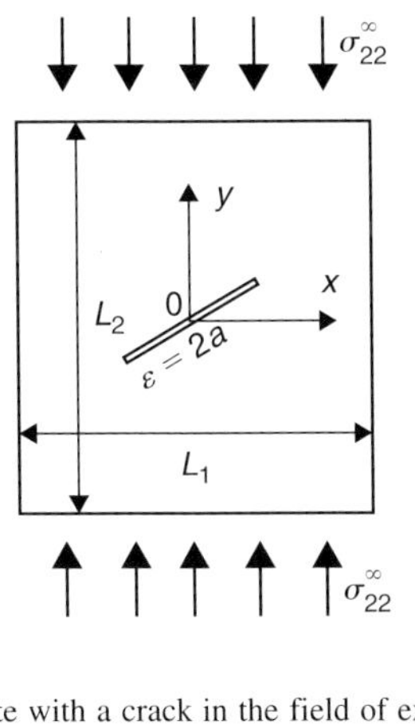


Fig. 4.3. Plate with a crack in the field of external loads.

where $\Delta L_2$ is the change of the plate height from the action of the applied stretching load $\sigma_{22}^{\infty} > 0$. Then, with account of the known expressions for SIF, we have:

$$G = \frac{1-\nu^2}{E}\left(K_I^2 + K_{II}^2\right) = \frac{1-\nu^2}{E}(\sigma_{22}^{\infty})^2\frac{\pi a}{2}\cos^2\phi.$$

This allows finding the relative elongation of the plate as a result of the crack growth:

$$\frac{\Delta L_2}{L_2} = \frac{1-\nu^2}{E}\left(1+\frac{\pi}{2}\frac{a^2}{L_1 L_2}\cos^2\phi\right)(\sigma_{22}^{\infty})^2. \quad (4.1.34)$$

It was already mentioned that in the case of dynamic fracture, corrections appear in equation (4.1.33), connected with the action of configuration forces [1, 101]. In this case:

$$G(t) = \frac{1}{2\mu}\left[A_I(v)K_I^2 + A_{II}(v)K_{II}^2\right],$$

$$A_I(v) = \frac{a(1-b^2)}{D(v)}; \quad A_{II}(v) = \frac{b(1-b^2)}{D(v)};$$

$$D(v) = 4ab - (1+b^2)^2; \quad a(v) = \sqrt{1-\frac{v^2}{c_d^2}}; \quad b(v) = \sqrt{1-\frac{v^2}{c_s^2}};$$

where $c_d, c_S$ are the speeds of compression and shear waves, accordingly.

*Curvilinear crack propagation*

Characteristic objects, illustrating the necessity of a synergetic approach to description of the phenomenon as a whole, are the crack and fracture process of the material at propagation of a mainline crack.

The difficulties connected with forecasting crack propagation and the dependence of the trajectory from multiple random parameters have caused a presentation about the propagation trajectory as about a stochastic process [336]. In this connection, the propagation process is represented as random and discrete in time. For a strict consideration of a stochastic crack propagation, we need to write down for the trajectory one of the known equations of stochastic processes (for example, the Fokker–Planck–Kolmogorov (FPK) equation) and to solve it. However, this strict approach is rather complicated. To study the basic regularities, we adopt the Markov character of the process and consider step-by-step crack propagation. Thus, the characteristics of the applied stress are included ('incorporated') into the random counter of the propagation process $N(t)$. The process counter is connected with discretization of the crack propagation process and characterizes the distribution of probabilities of angle $\Theta$, the deviation angle of the direction of crack development in step $i+1$ against the direction in the $i$-th step, and $X_i$, the increase of the crack length in the $i$-th step. Nevertheless, this forced discretization is more physically justified than the continual approximation, since in this case, we indirectly account for the already mentioned concept about the existence of the minimal fracture cell [146, 147, 261, 262]. From the physical viewpoint, the step-by-step consideration allows meeting the condition of self-consistency of the process of crack propagation [211, 227], while the initial angle of crack growth (propagation) is defined by the Sih criterion [330, 331] where the function of energy density should have its minimum. This criterion coincides with the VPCG (see p. 118). The function of energy density has the form:

$$S = \frac{\sigma^2 A}{16\mu}(3 - 4\nu - \cos\Theta)(q + \cos\Theta), \tag{4.1.35}$$

where $A$ is the crack length, $\sigma$ is the applied stress (having determinative character), $\mu$ is the shear module, and $\nu$ is the Poisson coefficient.

The Sih criterion requires:

$$\frac{\partial S}{\partial \Theta} = 0, \quad \frac{\partial^2 S}{\partial \Theta^2} > 0. \tag{4.1.36}$$

However, introduction of only parameters $N(t)$ does not allow one to correctly describe the stochastic crack propagation. Moreover, conclusions are sometimes made on the impossibility to present a crack trajectory as a random process [327]. Therefore, the authors of [336] were forced to additionally introduce an assumption on the random character of the argument of values $\Theta$, $N$, $X_i$ in such a way that $\Theta = \Theta(\gamma_i)$, $N = N(\gamma_i)$, $X_i = X_i(\gamma_i)$. Unfortunately, the obtained stochastic characteristics of crack propagation are related with empirical parameters and have insufficient theoretical justification. In principle, the physical reason of change of direction of crack propagation is not justified evidently, although it is obvious that in the proposed approach it can be connected only with a sharp change of mechanical properties of the material in structural borders. The change of properties can be accounted for by means of the Dundurs parameters [79, 80]. He offered two non-dimensional parameters for characterization of the refraction angle, when the crack meets the border of non-homogeneity:

$$\alpha = \frac{\Gamma(\kappa_2 + 1) - (\kappa_1 + 1)}{\Gamma(\kappa_2 + 1) + (\kappa_1 + 1)}, \quad \beta = \frac{\Gamma(\kappa_2 - 1) - (\kappa_1 - 1)}{\Gamma(\kappa_2 + 1) + (\kappa_1 + 1)}. \tag{4.1.37}$$

Indices 1 and 2 define, accordingly, the first and second materials; $\kappa = 3 - 4\nu$ for a flat deformation, $\kappa = (3 - \nu)/(1 + \nu)$ for a flat stressed state, $\Gamma = \mu_1/\mu_2$, $\mu$ is the shear modulus. Equation (4.1.37) has a more 'transparent' structure, if we represent it in the form:

$$\alpha = (E_1' - E_2')/(E_1' + E_2'),$$

where $E' = E/(1 - \nu^2)$ is the strain modulus for the flat case. The physically permissible values $\alpha$, $\beta$ fill the parallelogram enclosed between $\alpha = \pm 1$ and $\alpha - 4\beta = \pm 1$ in plane $\alpha$, $\beta$. These parameters characterize the distinctions of elastic properties of two contacting materials—in the sense of the trend of the parameter to zero at disappearance of elastic distinctions between the materials.

Sometimes, as a criterion of the preferable direction of crack propagation (including a reflection from the border or crossing it), one chooses the ratio of the energy production rate (EPR) of the reflected energy, $\mathcal{W}_d$, to the production rate of the passed energy, $\mathcal{W}_p$, of the crack [122]. In this connection, the ratio does not depend on the crack length and the SIF:

$$\mathcal{W}_d/\mathcal{W}_p = \frac{1-\beta^2}{1-\alpha}(\|d\|^2 + \|e\|^2 + 2\mathrm{Re}(de)]), \tag{4.1.38}$$

where $d$ and $e$ are non-dimensional parameters depending on the values of $\alpha$ and $\beta$, which enter the expression for SIF [122, 382]:

$$K_1 + iK_2 = k_1 a^{1/2-\lambda}(d(\alpha,\beta)a^{i\varepsilon} + e(\alpha,\beta)a^{-i\varepsilon}). \tag{4.1.39}$$

The EPR criterion is not the only one. The criteria of maximum normal stresses, and also some others, are used as defining ones. The mathematic grounds for these criteria are considered in [382]. Drawing out of expressions for EPR for these criteria, as a rule, is based on the classic solution of Muskhelishvili for a crack stress fields. In reality, the criteria of such simple form are not enough satisfactorily to describe the preferred direction of crack development. Thus, for the criteria of maximum normal stresses, it was shown [382] that, especially for non-uniform materials, the direction of the turn of the end of the crack does not always coincide with the direction of maximum stresses, but depends on the strength of the material and orientation of the crack in relation to the material (basic axes of the elasticity tensor of the material). It has been shown that the crack propagation may not have a stable character. For an isotropic material when loading under mode I, the stability or instability depend only on the stress, acting along the direction of crack propagation; a non-isotropic one depends also on the stress and on the mechanical characteristics of the material in which the crack is propagating.

The parameters which have a direct sense of relations of the energy concentrated in the passed crack to the energy concentrated in the reflected crack were independently proposed in references [211, 222, 227]. These parameters were obtained on the basis of the optical-mechanical analogy in the fracture theory. The advantage of the proposed criteria is the physically better justified argumentation, not connected with the idealized presentation about the structure of the material, since the variational methods are used, not depending on the structure of the medium. Here, the observed value—the trajectory of crack propagation—is derived as a result of direct calculations, unlike the above-considered works, where the trajectory equation is not derived in the explicit form.

The formulated fracture criteria, based on mechanical considerations, can be redefined in the form of thermodynamic restrictions and conditions [181]. For an ideal plasticity, the thermodynamic essence of the fundamental fracture criteria—criteria of Griffith and Irwin—is connected with the fact that the elastic fracture energy (integrals of Cherepanov—Rice ) do not depend on the integration contour. It has been shown in [148] that the elastic fracture energy for a curving crack depends on the second Dundurs parameter (4.1.37) and can be written down as:

$$^{e}W = \frac{1}{2}(1-\beta^2)\left(\frac{1-\nu_1}{\mu_1} + \frac{1-\nu_2}{\mu_2}\right)K_{II}^2. \tag{4.1.40}$$

With account of equation (4.1.40), the trajectory of crack propagation can be defined for small distances (in the $i$-th step).

## 4.2 Crack trajectory equation as a variational problem

The theories that use energy as a variable, as well as other values, which are preserved in mechanics—a momentum, an angular momentum—allow obtaining the most fundamental results. This is the advantage of the methods developed by us and enabling us to build a crack trajectory in non-uniform media without detailed definition of the internal state of the material in the conditions of a random stressed-deformed state.

*Some peculiarities of crack propagation*

It is known [284] that in the ideal case, at an ordinary single-axis stretching along axis $y$ of a plate made from a uniform elastic material, the initial crack $-\varepsilon < x < \varepsilon$ will propagate along axis $x$ rectilinearly. In real materials, which contain numerous defects, macroscopic crack propagation has a number of peculiarities, the origin and the nature of which are not quite clear. Among these peculiarities are the following: a rather confidently established fractal character of the fracture process [16, 56], percolation effects [15, 338], stochastization of trajectories [220, 221]. Explanation of these effects requires a deeper elaboration of the physical grounds of the fracture process and crack propagation.

If the loading is complicated, the trajectory will depend on the history of loading [17, 232]. Moreover, despite the fact that operation of many articles is taking place at rather stable boundary conditions, the trajectory of the crack arising in these conditions can be curvilinear. Besides, the problems of internal stability of the material and sustainable crack propagation are of interest in connection with the necessity of creating composite materials with preset operational properties [340]. All these peculiarities are related to deviation of the crack as a real physical object from the model of an ideal crack, defined by the classic theory of elasticity and plasticity (models of Barenblatt–Dugdale). Therefore, when building a crack model, these peculiarities should be accordingly accounted for, or the model should assume a possibility of accounting for them.

In principle, it is possible, by using decomposition by small parameter and solving the problem by successive small recalculations, to calculate the mutual influence of internal stresses in the body and propagating crack on the trajectory of the crack. Such approach makes the basis of all computer-based methods of finite-element analysis and methods of analytical studies of a weakly curved (almost rectilinear) crack [242].

An additional difficulty stems from the fact that in many cases we have to take into account the stochastic behavior of the crack propagation process. This is necessary, since because of the complexity of fracture processes and a great number of complex parameters, involved in the process, the arising cracks not only have a curvilinear character, but their propagation is chaotic. This does not allow solving, in the deterministic formulation, the problem of forecasting a crack trajectory in an inhomogeneous or composite material. Generally the crack propagation has an evident synergetic character.

*Variational problem for a real crack*

Let us consider the VPCG for an unlimited plate in plane $XOY$. The plate comprises two rigidly bound semi-plates: $x \leq 0$ is a homogeneous isotropic medium, $x > 0$ is an

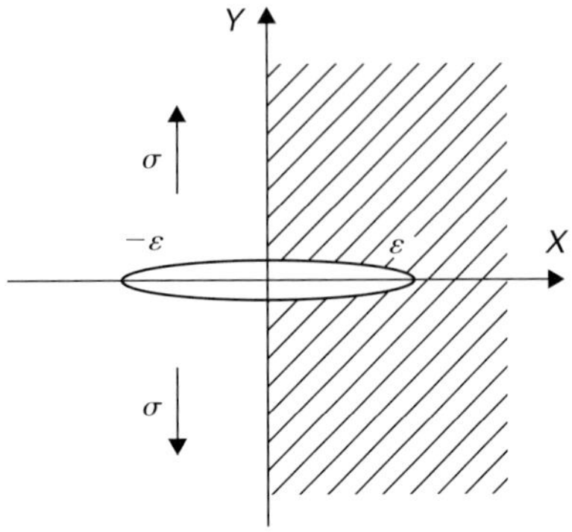


Fig. 4.4. Initial fracture in a medium.

inhomogeneous isotropic medium. The plate is stretched along axis $Y$ by the dead stress, and the initial crack is located along axis $X$, $(-\varepsilon < x < \varepsilon)$, $y = 0$ (Fig. 4.4).

Let us designate the elastic modules of the homogeneous medium as $\lambda_0, \mu_0$, of the inhomogeneous one as $\lambda(x, y)$, $\mu(x, y)$. In the considered case we have two power factors operating in the system: external applied loads and the forces of surface tension, arising at appearance of free surfaces in a solid material. Microscopically, we may additionally investigate the interatomic bond breaking and acoustical emission from crack tip but in continuum limits all these factors vanish. With account of these phenomena and because of the established variational condition (4.1.15), we shall seek the crack trajectory in the form of functional extremum for the fracture energy [284]:

$$\delta \int_{\varepsilon_A}^{\varepsilon_B} (2\gamma - P_1 u_1)\, ds = 0, \tag{4.2.1}$$

$$(2\gamma - P_1 u_1) = F\,(x, y, \dot{y}),$$

where $\gamma$ is the density of surface energy of the material, $P_i = \sigma_{ij} n_j$ are the stress tensor components on the pads, the position of which coincides with the crack surface, $n_i$ is the guiding cosine of the external normal to the crack surface, $u_i$ is the displacement of the crack edges, $\sigma_{ij}$ is the tensor of the stresses in the material, and $\varepsilon_A$, $\varepsilon_B$ are the initial and final crack lengths, accordingly. The equation of crack trajectory is written down in the form as: $y = y\,(x)$. In our VPCG, we consider a simple loading, and curvature of the trajectory in an inhomogeneous plate will depend only on distribution of inhomogeneity. It is well known that any functional extremum satisfies the Euler equation. The Euler equation for the formulated problem has the form:

$$\frac{d}{dt}\left(\frac{\partial}{\partial y'}\left[\sqrt{1 + y'^2}\,(2\gamma - P_1 u_1)\right]\right) - \frac{\partial}{\partial y}\left[\sqrt{1 + y'^2}\,(2\gamma - P_1 u_1)\right] = 0. \tag{4.2.2}$$

From formula (4.2.2), we can obtain the equation for trajectory $y(x)$ in the form as:

$$\frac{y''}{\sqrt{1+y'^2}}F - \frac{y'^2 y''}{\left(\sqrt{1+y'^2}\right)^3} + \frac{y'}{\sqrt{1+y'^2}}\left\{\frac{\partial F}{\partial x} + \frac{\partial F}{\partial y}y' + \frac{\partial F}{\partial y'}y''\right\}$$
$$+ \frac{y'y''}{\sqrt{1+y'^2}}\frac{\partial}{\partial y'}F + \left\{\sqrt{1+y'^2}\left[\frac{\partial}{\partial x}\frac{\partial}{\partial y'}F + \left(\frac{\partial^2}{\partial y'^2}F\right)y'' + \left(\frac{\partial}{\partial y'}\frac{\partial}{\partial y}F\right)y'\right]\right\}$$
$$- \sqrt{1+y'^2}\frac{\partial}{\partial y}F = 0 \tag{4.2.3}$$

**Remark 4.2.1.** As a strict consideration of the VPCG, one should take into account that for non-conservative systems the operations of integration and variation are irreplaceable. For such systems, one should request the equality to zero of the time integral from variation of the Lagrange function, as is required by the Hamilton–Ostrogradsky principle:

$$\int_{t_A}^{t_B} \delta L dt = 0, \quad \delta L = \delta \int_{x_A}^{x(t)} \Phi ds.$$

In this case, not only is the crack length varied, but also its end position depending on time. Here, we can use the variational equation of L. I. Sedov, which accounts for all possible accompanying effects. The discussion on this topic can be found in [235].

If we assume functional $F(x, y)$ and the crack trajectory to be rather smooth (without accounting for the fractal character of the crack propagation), the deviations of the crack trajectory in the following step from the direction of propagation in the previous step will be small, $y' \ll 1$. In the general case, value $\gamma$ depends on the microscopic properties of the medium. This dependence is discussed in more details in the following chapter. In our case, because of the chosen small steps of separation, in the first approximation we can take $\gamma \approx$ const. Then, equation (4.2.3) reduces [211, 227] essentially to a nonlinear equation in the form as:

$$\frac{\partial Q}{\partial y}\frac{\sqrt{1+y'^2}}{Q^2} + \frac{d}{dx}\frac{y'}{Q\sqrt{1+y'^2}} = 0; \tag{4.2.4}$$
$$y' = \frac{dy}{dx}; \quad Q = \left(\sigma_{ij}n_i u_j\right)^{-1} \equiv C(x, y),$$

where $\sigma_{ij}$ is the normal stress on the surface element of the crack surface, $n_i$ is the guiding cosine of the external normal to the crack surface, and $u_i$ is the displacement of crack edges. We note here that we could have immediately written, from the general principles [311], the variation of the fracture work as $\delta\left(\sigma_{ij}\varepsilon^{ij}\right)$. However, in practice, stresses and deformations are not taken to be independent, but are bound, for example, by means of Hooke's law. Self-consistency of equation (4.2.4), required for description of the fracture surface [282], is ensured by a step-by-step procedure of varying the fracture energy by the small vicinity near the crack tip:

$$f_1(x, y) = \frac{\partial \ln Q(x, y)}{\partial x}, \quad f_2(x, y) = \frac{\partial \ln Q(x, y)}{\partial y}.$$

With account of adopted designations for equation (4.2.4), we have:

$$y'' - y' f_1(x, y)\left(1 + y'^2\right) + f_2(x, y)\left(1 + y'^2\right)^2 = 0. \tag{4.2.5}$$

In equation (4.2.5), term $1 + y'^2$ is the length element in the Euclidean space in a 2-D case. Depending on the particular type of inhomogeneity, the equation solutions admit, in particular, the existence of stochastization [220, 221]. Let us consider the stability of solutions of equation (4.2.5).

## 4.3 Propagation stability and influence of material inhomogeneity on crack trajectory

Both theoretically and practically, crack propagation stability is an interesting issue. First, we need to define additionally what we understand as stability/instability of crack propagation. This can be a sufficient smoothness of the trajectory, consistency of the speed of crack propagation, absence of bifurcation points, and other criteria. Hereafter, as a rule, we shall study the condition of sufficient smoothness and the condition of branching (bifurcation origin).

At development of a main crack from the field of microcracks, various ways of crack evolution are possible depending on fulfilment or non-fulfilment of the conditions of the trajectory consistency. It appears that right at the ***formation of macrocracks and in the proximity of bifurcation points, the effect of microstructure is essential.*** Thus, one of the examples of macroscopic crack inconsistency is the process of self-supporting fracture. In this case, for a chain of linearly located microcracks, a situation can be realized, when the crack propagation acquires an avalanche, self-supporting character after destruction of the sole bridge between two adjacent cracks [329]. It is interesting that for explosive propagation the stress level is sufficient, which is essentially smaller than what is required at the first fracture. Realization of this sort of bifurcation transitions is dangerous in operation and requires a more detailed analysis. Another example of the importance of the microscopic structure for crack development is the fact that the presence of residual stresses can be a necessary condition for crack branching [131, 140].

### *4.3.1 Stability of crack trajectory*

Quasi-stationary and dynamic crack growth in real materials is aggravated by deviation of the material from the ideal state. This non-ideality leads to the crack's deviation from the rectilinear propagation. Deviations, in their turn, can build up in the course of time or with the travel along the trajectory—in this case the trajectory is unstable. The instability of trajectories is caused by various physical factors. This can be inhomogeneous dislocation field, inhomogeneous mechanical properties of the continuum, and also random fluctuations of applied stresses.

**Analysis of consistency by method of weak trajectory perturbations.** A study of consistency of trajectories to fluctuation of the stress field can be made within the framework of the classic elasticity theory on the basis of the methods of the perturbation theory [65, 242, 267]. It was assumed that the crack propagates in the next 'step' of its growth in such direction that the SIF

$$K_{II} = 0, \tag{4.3.1}$$

and the growth was considered of a smoothly curved crack under the effect of point forces of amplitude $A$, applied in points $x^+$, $x^-$, symmetrically in relation to the plane of crack propagation (Fig. 4.5). It was assumed that in the absence of cracks in the volume of the material, the applied stress generates the field of stresses and displacements $\boldsymbol{\sigma}^{nc}$, $\mathbf{u}^{nc}$, accordingly. When a crack appears, additional stresses and displacements appear $\boldsymbol{\sigma}$, $\mathbf{u}$, satisfying the Lame equilibrium conditions and the boundary conditions:

$$\mu\nabla^2\mathbf{u} + (\lambda + \mu)\nabla(\nabla \cdot \mathbf{u}) = 0, \quad x \in \mathbb{R}^2 \backslash C_\varepsilon(t);$$

$$\sigma_{ij}n_j + \sigma_{ij}^{(nc)}n_j + P_i = 0 \quad \text{on} \quad C_\varepsilon^\pm,$$

where $\mu$, $\lambda$ are the Lame constants, $C_\varepsilon(t) = \{\mathbf{X} : X_2 = \varepsilon\psi(X_1), -\infty < X_1 < Vt\}$ is the crack surface, $0 \leq \varepsilon \ll 1$ is a small parameter, defining the crack displacement, $\psi$ is the function, defining the shape of the crack, $n_j$ is the unit normal vector to the crack surface, and $P_i = P_i(X_1 - Vt)$ are the surface cohesive forces, a function of the crack growth speed $V$. This growth speed was taken to be small, in order to have the inertial

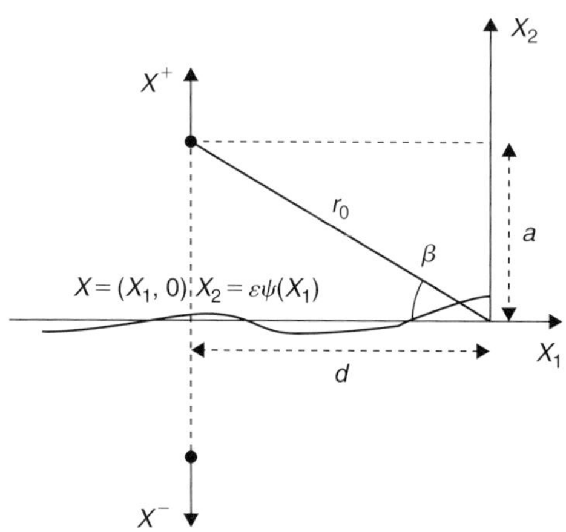


Fig. 4.5. Curvilinear crack under effect of point forces.

effects negligibly small. The surface cohesive forces can be defined by different methods, depending on the mechanisms causing cohesion. For example, the account of dislocation components on the crack surface [193] leads to a system of two bound integral equations in relation to the function of dislocation distribution by axes $X, Y$, accordingly.

Additional fields arising in the presence of the crack were searched in the form of a series by parameter $\varepsilon$:

$$\sigma \sim \sigma^{(0)} + \varepsilon\sigma^{(1)}, \tag{4.3.2}$$

where $\sigma^{(0)}$ is the stress field for an unperturbed crack and $\varepsilon\sigma^{(1)}$ is the first-order perturbation. It is obvious that physical carriers of perturbations can be, for example, dislocation, disclinations and other sources of incompatibility [193].

For an unperturbed crack in the proximity of the tip, we have an asymptotic approximation of stress field [242]:

$$\sigma_{22}^{(0)} \sim \frac{K_I^{(0)}}{\sqrt{2\pi x_1}} - \sigma_{22}^{(nc)}(0,0) - P(0) + A_I^{(0)}\sqrt{x_1}, \quad x_1 > 0,$$

where a moving system of coordinates $x_1 = X_1 - Vt$, $x_2 = X_2$ is used. For coefficient $A_I^{(0)}$, we have:

$$A_I^{(0)} = \sqrt{\frac{2}{\pi}} \int_{\infty}^{0} h(x_1)\{\sigma_{22}^{(nc)\prime}(x_1,0) + P'(x_1)\}dx_1, \tag{4.3.3}$$

where the prime defines a derivative by $x_1$ and

$$h(x_1) = \sqrt{\frac{-2}{\pi x_1}}, \quad x_1 < 0$$

is the weight function for a semi-endless crack.

In a similar way, Cottrel and Rice assume in their work [267] that perturbation $\varepsilon\psi$, induced by a small, order $\varepsilon$, load fluctuation, causes a small addition $\varepsilon k_{II}(Vt)$, when the crack tip is in position $X_1 = Vt$. The total perturbed SIF of order $\varepsilon$ can be presented as: $\varepsilon(K_{II}^{(1)}(Vt) + k_{II}(Vt)$, and the crack growth criterion (4.3.1) takes the form:

$$K_{II}^{(1)}(Vt) + k_{II}(Vt) = 0. \tag{4.3.4}$$

Value $K_{II}^{(1)}(Vt)$ is defined from the integral-differential equation, which binds together the stress tensor, the crack shape and cohesive forces.

Generally speaking, equation (4.3.4) is a linear correlation between $\psi(Vt)$ and $k_{II}(Vt)$, and its formal solution can be obtained by means of the Laplace transformation. For the solution, we introduce auxiliary functions:

$$f_k(\beta) = (2\kappa + 3)\cos\frac{\beta}{2} - \cos\frac{5\beta}{2},$$

$$f_a(\beta) = (2\kappa + 5)\cos\frac{3\beta}{2} - \cos\frac{7\beta}{2},$$

where $\kappa = 3 - 4\nu = (\lambda + 3\mu)/(\lambda + \mu)$ and $\beta$ is the angle between the negative direction of axis $x_1$ and direction towards the applied point force (we remember that the moving system of coordinates is linked with the crack tip). Then, equation (4.3.4) reduces to complex equation:

$$f_a(\beta) - i\lambda f_k(\beta) + (i - 1)M\sqrt{\lambda} + \hat{I}_D(\lambda) = 0, \tag{4.3.5}$$

where $M = 2\sqrt{2\pi}(\kappa + 1)r_0\sigma_{11}^{\infty}/A$ is a non-dimensional parameter,

$$\hat{I}_D(\lambda) = -\frac{8}{\pi}\hat{a}\int_0^{\infty}\frac{\exp(i\lambda y)}{\sqrt{y}\hat{r}^2} \times\left[\frac{2(\hat{d} - y)}{\hat{r}^2}\left(\kappa - 3 + 8\frac{\hat{a}^2}{\hat{r}^2}\right) + i\lambda\left(\kappa - 3 + 4\frac{\hat{a}^2}{\hat{r}^2}\right)\right]dy, \tag{4.3.6}$$

and introduce the non-dimensional geometrical parameters:

$$\lambda = -kr_0, \quad \hat{a} = \frac{a}{r_0} = \sin\beta, \quad \hat{d} = \frac{d}{r_0} = \cos\beta,$$

$$\hat{r} = \sqrt{\hat{a}^2 + (\hat{d} - y)^2} = \sqrt{1 - 2\hat{d}y + y^2}.$$

The standard research procedure of the consistency of solutions of equation (4.3.5) has shown that there exist critical values of losing the consistency by the trajectory. Thus, for purely imaginary roots at $\kappa = 2$ and $\beta \in [\beta_1, \beta_2]$, ($\beta_1 \approx 21°$, $\beta_2 \approx 75°$) there is at least one root of equation for any values of parameter $M$, and the crack trajectory is unstable. It means that on application of the load within the defined angle sector, the crack will always be unstable. At small angles, $\beta < \beta_1$, the crack is unstable at $M > M^{cr}$. In the case of complex roots of the equation, the study of solutions is more complicated, but here also the consistency depends on $M$, and value $M^{cr}$ can be defined.

For the case of loading the body with point forces only (distributed cohesive forces are absent), it was demonstrated that several ranges of parameter $M$ exist, at which the crack loses its stability. However, the types of inconsistencies arising in the system were not analysed.

The considered approach correlates nicely with known results, but, unfortunately, its transfer onto the system of point or continuously distributed forces looks problematic because of the necessity of correct definition of the contribution of all the forces into the highest decomposition terms of the stress and displacement fields (4.3.2).

*Wave processes and stability of ray approximation at crack propagation*

The analogy of the process of crack propagation with the process of wave propagation in a medium yields a number of fruitful results. Thus, the works [240, 241] consider a collision of propagating cracks with an isolated region of inhomogeneity of fracture energy and show that in this case a new type of elastic waves is generated, which propagate along the front of the moving crack. The authors consider propagation of the Barenblatt–Dugdale crack with a zone of weakened bonds in front of the crack tip (4.2) and analyze small perturbations of the propagation speed and crack front shape. It is adopted in this case according to [90, 167] that the energy production rate (EPR) for the moving crack $G$ is the function of the propagation speed $v$ and the EPR $G_{rest}$ for an immobile crack:

$$G = g(v)G_{rest}.$$

It is assumed that in the region with critical fracture energy $G_{crit_0}$, areas exist with fluctuation of fracture energy $\Delta G_{crit}(z, x)$, in which the crack propagation speed is fluctuating. Eventually, the profile of the crack tip $A(z, t)$ is distorted, along with fluctuations of $G_{rest}$. It is assumed that $A(z, t) = a(z, t) - v_0 t$, where the crack front lies along line $a(z, t)$ in plane $x, z$. Linear analysis of perturbations [293] is applied to the equation of perturbations, obtained by L. Freund [99, 100]. In the first approximation, the equation has the form:

$$\frac{\Delta G_{crit}(z, v_0 t)}{G_{crit_0}} = \frac{g'(v_0)}{g(v_0)} \frac{\partial A(z, t)}{\partial t} + \frac{\Delta G_{rest}[z, t; A(z', t')]}{G_{rest_0}}, \tag{4.3.7}$$

$$-\infty < z' < +\infty, \ t' \leq t.$$

In these designations, it has been stressed that $\Delta G_{rest}$ is the functional from $A(z, t)$. Computer simulation helped to show the existence in inhomogeneous media of the areas, where the crack propagation is *forbidden*. The existence of such areas in inhomogeneous materials was independently demonstrated analytically [227]. The solution of equation (4.3.7) at transition to wave numbers has the form [241]:

$$\Delta \hat{G}_{crit}(k, v_0 t) = G_{crit}\left(C_v(v_0)\frac{\partial \hat{A}(k, t)}{\partial t} + 2I(k, t)\right), \tag{4.3.8}$$

where $I(k, t)$ by the physical sense are retarding potentials, which take into account the preceding loading history, and $C_v$ is the function of the longitudinal and lateral sound speed [293]. Account of retarding potentials, that is, account of the history of loading, is necessary for the macroscopic integral principles of the crack theory, otherwise the fracture model changes essentially and can lead to erroneous conclusions [17].

The results similar to [99, 100] and also confirming the wave propagation along the front of the growing crack were obtained with computer simulation in the simplest break model of the weights on springs [103]. Superposition of two independent motions—propagation of the front and the wave along the front—in principle, can lead to development of chaotic processes in the system according to the Landau mechanism. The initial weak fluctuations of surface shape are indications of chaotization, which not always develops into a full-scale picture.

*Analysis of crack trajectory stability*

In as much as in the given problem we shall investigate a realization, but not a process (this means the stable state is under investigation), we are not going to study the time dependence. Let us denote $z = y'$ and rewrite (4.2.5) as a system of equations:

$$\begin{cases} y' = z, \\ z' = zf_1\left(1 + z^2\right) - f_2\left(1 + z^2\right)^2. \end{cases} \tag{4.3.9}$$

In the general case, the second equation of the system (4.3.9) is the equation with separable variables, but account of the complex functional dependence on coefficients $f_1, f_2$ is complicating the situation. The orbits of equation (4.3.9) in the phase plane are set by the equation:

$$\frac{dz}{dy} = \frac{zf_1\left(1+z^2\right) - f_2\left(1+z^2\right)^2}{z} = f_1\left(1+z^2\right) - f_2\frac{\left(1+z^2\right)^2}{z}. \tag{4.3.10}$$

Solution of equation (4.3.10) is essentially dependent on the type of inhomogeneity of the medium. This dependence is essential, since stressed–deformed state is the function of physical-mechanical characteristics of the medium, in which the crack propagates, $\sigma_{ij} = \sigma_{ij}\left(s_{ijkl}\right)$. If we assume that the properties of the medium are smoothly changing across the trajectory of crack propagation, then in the analysis we can neglect the terms above the second order. In this case we have:

$$\frac{dz}{dy} = f_1\left(1+z^2\right) - f_2\left(\frac{1}{z}+4+6z\right). \tag{4.3.11}$$

Direct analytical integration of equation (4.3.10) does not yield reasonable results. However, we have a possibility to regulate the type of critical points in the phase plane by means of suitable selection of inhomogeneity (parameters $f_1, f_2$). Thus, for example, a stable focus can represent contraction (coagulation) of microcracks in sole mainline crack. Unstable nodal points along the propagation trajectory will correspond to decay of the mainline crack into the field of microcracks (microdamages). This situation is interpreted as a crack trajectory stochastization [220]. However, since trajectory parameters depend on the character of inhomogeneity along both axes, from the technology viewpoint, it seems possible to regulate parameter $f_2$ (inhomogeneity along axis $Y$). It has a very strong effect on the phase trajectories of the equation. The behavior of the solutions of the second equation of the system (4.3.9) is demonstrated in Figs. 4.6–4.8. The right-hand borders of axis OX are chosen in the proximity of the point of abrupt loss of consistency by the system (values of deviation get outside the physically justified borders). With an insignificant growth of the parameter, a strong dispersion of the beams of trajectories is observed.

Value $z = 0$ is stable practically at all combinations of the parameter. A clear instability near $y = 0.5$ by initial condition $z(0) = -1$ is observed in all figures[2]. It can be explained by a possible stochastization of the trajectory behavior at the preset plurality of parameters or instability of computational methods.

Thus, when designing a composite material in such a way that

$$f_2 = \frac{z^2}{\left(1+4z+6z^2\right)}, \tag{4.3.12}$$

we have a positive attractor, the critical point is $z = 0$ provided $f_1 = 0$.

[2] Strictly we need additional analysis of stability of using computer methods, but we believe that standard packages, for example Maple, Mathematics, check the correctness.

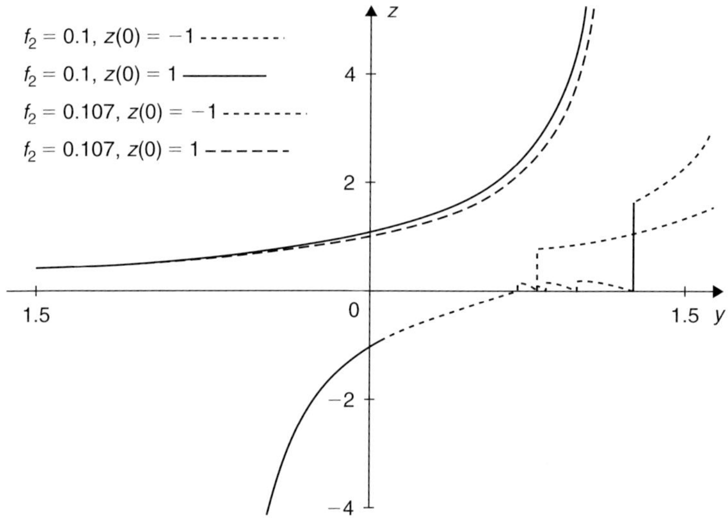


Fig. 4.6. Parameter $f_1 = 1.0$.

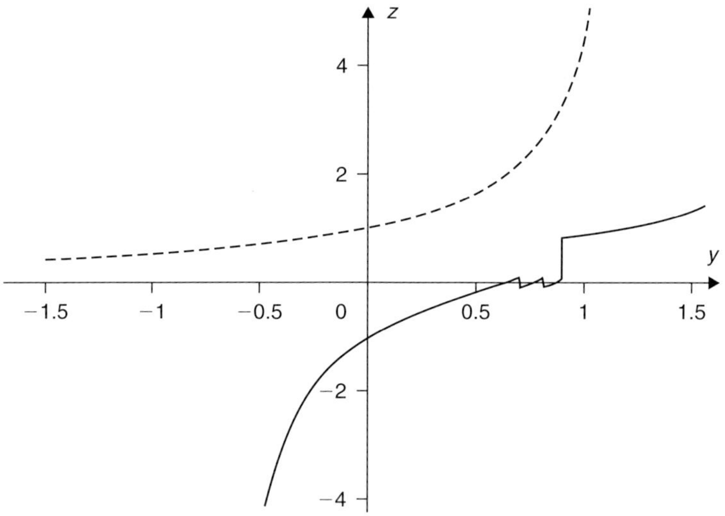


Fig. 4.7. Parameters $f_1 = 1.0$, $f_2 = 0.093$.

Strictly speaking, a conclusion about a possibility of linear analysis of critical points of nonlinear system (4.3.9) requires additional studies. Thus, for example, for fulfilment of the conditions of existence of stable and unstable manifolds, we need [359] to satisfy additional conditions:

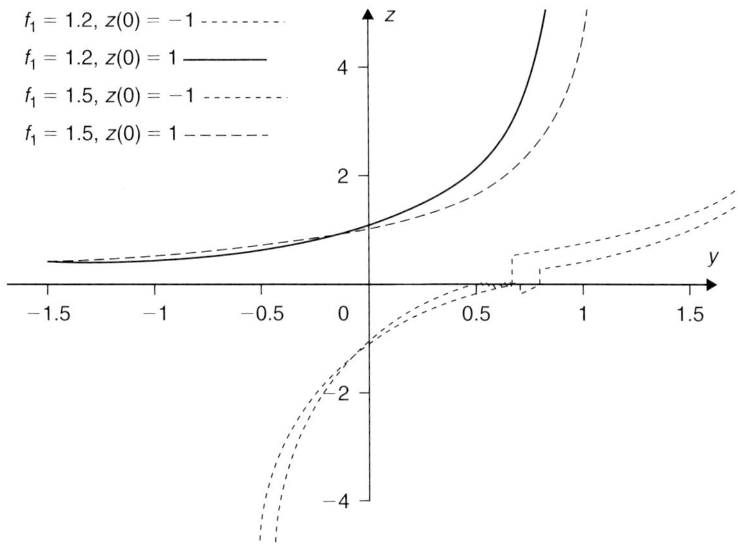


Fig. 4.8. Parameter $f_2 = 0.1$.

$$\lim_{\|z\|\to 0} \frac{\left\| f_2 \left(1 + 4z + 6z^2\right) + f_2 \right\|}{\|z\|} = 0,$$

which imposes requirement $f_2 = o\left(z^2\right)$, to which condition (4.3.12) does not satisfy. In this case, we can assume:

$$f_2 = z^{\xi} / \left(1 + 4z + 6z^2\right),$$

where $\xi = 2 - \varsigma$, $\lim_{x\to\infty} \varsigma = 0$, at additional condition $f_2 = o(z)$. By technological methods, by choosing value, we can regulate the speed of transition into the stage of fracture through propagation of the mainline crack.

### *4.3.2 Crack propagation in real media*

We have mentioned already that it is difficult to study equation (4.2.5) in the general form, therefore, let us first consider the case when the deviation of the crack in layer 2 from axis $OX$ is small, that is, $y' \ll 1$. In this approximation, we restrict ourselves only to linear terms by $y'$, and neglect the higher order terms because of their small values. Let us decompose the equation of crack trajectory $f_1$, $f_2$ into series by degrees of $y$. In this case, we restrict ourselves to decomposition of the terms not above the third order:

$$f_1(x, y) = f_1^0(x, 0) + f_{1,y}^0(x, 0)y + \frac{1}{2} f_{1,y^2}(x, 0)y^2 + \frac{1}{3!} f_{1,y^3}(x, 0)y^3 + \varepsilon_1 F_1(x, y), \tag{4.3.13}$$

$$f_2(x,y) = f_2^0(x,0) + f_{2,y}^0(x,0)y + \frac{1}{2}f_{2,y^2}(x,0)y^2 + \frac{1}{3!}f_{2,y^3}(x,0)y^3 + \varepsilon_2 F_2(x,y). \tag{4.3.14}$$

Hereafter, we take into account that, since $y = 0$ is the solution of the equation, then:

$$f_1^0(x,0) + f_2^0(x,0) = 0.$$

Besides, because of the assumption of the smallness of deviations from the rectilinear propagation in equation (4.2.5), we neglect the terms of the order above $y$. By making a standard replacement of variables $y = z\exp(\frac{1}{2}\int C(x,y)dx)$, where $C(x,y)$ is the sum of all coefficients at $y'$, we eliminate in equation (4.2.5) the terms containing $y'$. In references [42, 61], the authors use replacement $y = zU(x,y)^{\frac{-1}{2}}$, which leads to similar results. Equation (4.2.5) takes the form:

$$z'' + \frac{1}{2}f_1^1 z - \frac{1}{4}f_1^2 z + f_2^0 + \frac{1}{2!}f_{2,y^2}^0 z^2 \exp A + \frac{1}{3!}f_{2,y^3}^0 \exp^2 A = 0, \tag{4.3.15}$$

$$A = \frac{1}{2}\int f_1(x,y)dy.$$

Let us account in decomposition $f_1$, $f_2$ only zero terms, then (4.3.15) takes the form:

$$z'' + N(x,0)z + M(x,0)z^2 + L(x,0)z^3 = -\varepsilon_1 F_1(x,0) - \varepsilon_2 F_2(x,0), \tag{4.3.16}$$

$$N(x,0) = \frac{1}{2}f_1' - \frac{1}{4}f_1^2 + f_{2,y}^0, \tag{4.3.17}$$

$$M(x,0) = \frac{1}{2!}f_{2,y^2}^0 \exp\ A, \quad L(x,0) = \frac{1}{3!}f_{2,y^3}^0 \exp\ 2A.$$

Thus, assuming that in the proximity of axis $X$, inhomogeneity of the layer and stressed state can be roughly considered linearly dependent on $Y$, we have obtained an equation which complies with the crack trajectory. The supposition on the character of boundary conditions, which ensure the unchanged character of stress distribution, allows forecasting the local direction of crack propagation.

Hereafter in this discussion we shall consider the behavior of the crack in linearized formulation. The behavior will be defined by the type of coefficient $N(x,0)$; therefore, we can decompose the coefficient in more detail.

Let us decompose structure $N(x,0)$. We assume that at crack propagation, the temperature of the medium does not change (the body is at permanent temperature), and, except for a small zone in the proximity of the crack tip, the relations of the linear elasticity theory are fulfilled. Then, we have for the deformation tensor:

$$\varepsilon_{kr} = \frac{1+\nu}{E(x,y)}\sigma_{kr} - \frac{3\nu}{E(x,y)}\sigma\delta_{kr},$$

where $E(x, y)$ is the Young modulus, and $\nu$ is the Poisson coefficient. We think that conditions of loading are such that:

$$\sigma_{22} = \sigma_{22}(x, y) \neq 0, \quad \sigma_{ij} = 0, \quad i, j = 1, 2. \tag{4.3.18}$$

After rather cumbersome calculations, we obtain for coefficient $N(x, y)$:

$$N(x, y) = \frac{2\Delta\sigma_{22}}{\sigma_{22}} - \frac{\Delta E}{2E} + 2\left(\frac{\nabla\sigma_{22}}{\sigma_{22}}\right)^2 + \frac{1}{4\sigma_{22}^2}\left[\frac{2\partial\sigma_{22}}{\partial x} - \frac{\sigma_{22}}{E}\frac{\partial E}{\partial x}\right]^2, \tag{4.3.19}$$

where $\nabla = \partial/\partial x$, $\Delta = \nabla^2$.

The assumption on the character of the boundary conditions, which ensure the unchanged character of stress distribution, allows forecasting the propagation direction of a small (change of $x$ is small) crack. For calculation of a long crack, it is necessary to use the iteration procedure and recalculate the stressed state that arises at progression of the crack to a certain distance.

Our model problem (varying with fixed ends) responds to the task of definition of probable break surfaces of the construction with two known concentrators of stresses.

### *4.3.3 Trajectory in linear approximation*

Let us study the trajectory equation (4.3.15) in linear approximation. Similar to previous discussions, the linear approximation is understood as sufficient smoothness of crack behavior, enabling us to neglect the terms of the type $y'^2$. However, the properties of the medium in which the crack propagates can change in a nonlinear way. It is important in principle that these nonlinear property changes should not cause a transition to the fractal character of crack propagation (Chapter 5). The restriction, introduced earlier (4.3.18), has an essential justification from the microscopic viewpoint. According to the principle of local symmetry, a crack always propagates in such manner that the main stress is perpendicular to the front. By making a respective replacement of coordinates, we can always ensure it locally.

By neglecting all the terms of the order above the first one, and assuming the free terms to be negligibly small, we obtain the equation of the type:

$$z'' + N(x, 0)z = 0. \tag{4.3.20}$$

Depending on the properties of the medium, that is, values of coefficient $N(x)$, solutions of this equation have different forms.

*Crack trajectory in media with determined structure*

Let us consider the mechanical properties of the medium through which the crack propagates to be preset. Like earlier, we assume that at crack propagation, the stressed state does not change, $\sigma_{22}$ = const, $E = E(x)$. In this case, equation (4.3.19) will be rewritten as:

$$z'' + \left[ -\frac{1}{2E(x)} \frac{\partial^2 E(x)}{\partial x^2} + \frac{1}{4E(x)^2} \left( \frac{\partial E(x)}{\partial x} \right)^2 \right] z = 0. \qquad (4.3.21)$$

Let us consider the solution of equation (4.3.21) for media with differing mechanical characteristics.

(A) $E = E_0 \sin(kx + C)$, where C is selected from the physical conditions of non-negativeness of the elastic modulus. The medium of this type may respond, for example, to periodical distribution of dislocation density in reaction-diffusive kinetic models [2, 300]. At these assumptions, equation (4.3.21) reduces to equation of the Hill type:

$$z'' - \left( k^2 - \frac{k^2}{\tan^2(kx + C)} \right) z = 0. \qquad (4.3.22)$$

The approximated solution of equation (4.3.22) can be sought in the form of a range of exponents [210] or formal power series. Solution of equation with account of the terms up to the sixth order can be presented as:

$$\begin{aligned}
z(x) &= z(0) + \mathrm{D}(z(0))\, x + 0.25k^2 \left[2 - \tan^2(C)\right] z(0)\, x^2 \\
&+ \frac{1}{3} \frac{k^2}{\sin^3(C)} \left[ \frac{1}{2} k\, z(0) \cos(C) \sin^2(C) + k\, z(0) \cos^3(C) \right. \\
&\left. + \frac{1}{2} \mathrm{D}(z(0)) \sin^3(C) - \frac{1}{4} \sin(C)\, \mathrm{D}(z(0)) \cos^2(C) \right] x^3 \\
&- k^3 \left[ k\, z(0) \left[0.02833 \tan^2(C) + 0.114583 \tan^4(C)\right] \right. \\
&\left. - 0.0833\mathrm{D}(z(0)) \left[ \frac{1}{2} \sin(2C) - \tan^3(C) \right] \right] x^4 \\
&+ \frac{1}{5} \frac{k^4}{\sin^5(C)} \left[ \frac{1}{2} k \sin^4(C) \cos(C)\, z(0) \right. \\
&+ 0.91566 \sin^2(C)\, k\, z(0) \cos^3(C) + 0.4166 \cos^5(C)\, k\, z(0) \\
&- 0.4166 \sin^5(C)\, \mathrm{D}(z(0)) - 0.5416 \sin^3(C)\, \mathrm{D}(z(0)) \cos^2(C) \\
&\left. - 0.3654833 \cos^4(C) \sin(C)\, \mathrm{D}(z(0)) \right] x^5 + \mathrm{O}(x^6),
\end{aligned}$$

where $D(z(0))$ is the differential operator, the first derivative of the function in zero.

In more detail, the consistency zones of equations of Hill types are studied, for example, in [280].

On the other hand, a possibility is known to present the process of crack propagation as a wave process [247]. This correlates with the fact that the Fermat principle in quantum mechanics has as its analogue the principle of last action of Maupertuis [34]. Then, equation (4.3.20) can be considered as the equation analogous to the Schrödinger equation, and representation can be introduced about the coefficients of transmission $D$ and reflection $R$ for the crack. In this case, the equation can be written down as:

$$z'' + \frac{k^2}{\tan^2(kx + C)} z = k^2 z. \tag{4.3.23}$$

The coefficients of transmission $D$ and reflection $R$ express the share of the fracture energy 'passed' by the material in the direction of crack motion and 'dissipated' within the volume. According to definition [34]:

$$R = \left(\frac{k_1 - k_2}{k_1 + k_2}\right)^2, \quad D = \frac{4k_1 k_2}{(k_1 + k_2)^2},$$

where $k_1^2 = 1$, $k_2^2 = 1 - \tan^{-2}(kx + C)$. In this case, in (4.3.22) the coefficient at $z$ has the sense of the normalized energy difference before and after the potential barrier (structural border). For values $R$ and $D$, we have, accordingly:

$$R = \left(\frac{1 - \sqrt{1 - \tan^{-2}(kx + C)}}{1 + \sqrt{1 - \tan^{-2}(kx + C)}}\right)^2;$$

$$D = \frac{4\sqrt{1 - \tan^{-2}(kx + C)}}{1 + \sqrt{1 - \tan^{-2}(kx + C)}^2}.$$

The curves of the coefficients of transmission and reflection for the possible area of their existence are presented in Fig. 4.9. Breaks of the curves in the proximity of points $(0 \pm \pi n - 2C)/2k$, $n = 1 \ldots N$ are caused by singularity of the root in the expressions for the coefficients.

The obtained results allow one to suppose that 'zones' or 'regions' exist, at propagation through which the crack is not able to cross the border of this area, and either stops or reflects from the border. The role of these borders can be performed both by the borders of structural formations in the material, and by, for example, the walls of dislocations. The existence of such zones of 'enclosure' arises from the physical restriction of the positiveness of the radical expressions of the coefficients of reflection and transmission.

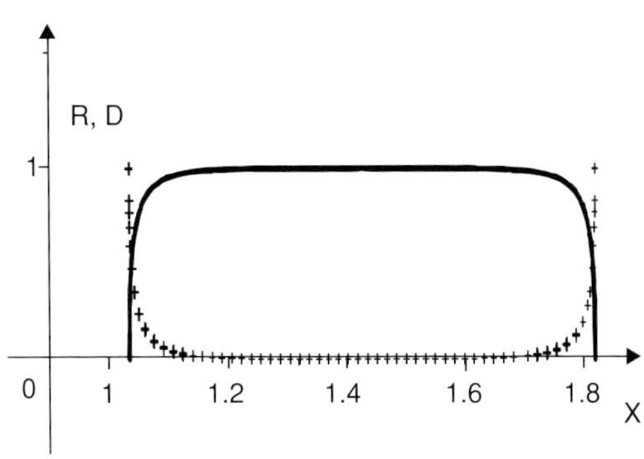


Fig. 4.9. Behavior of the coefficients of transmission (dotted line) and reflection (solid line) $C = 5$, $k = 2$.

Otherwise, the coefficients are imaginary and their physical sense corresponds to the impossibility of propagation of the mainline crack in this area. Experimentally, the crack braking on grains is rather nicely studied; however, the results obtained by us are evidence that the crack not only is not able to pass from a soft material into a solid one, but vice versa from a solid material into a soft one. In the area of imaginary coefficients, there should exist a 'sub-barrier' penetration of the crack from one area into another, similar to the quantum tunnel effect. It is probable that the physical sense of this behavior lies in disappearance of the mainline crack, its degradation down to diffusive microcracks and further organization up to the mainline crack.

(B) Let $E(x) = E_0 + kx$ (linear medium). A medium of this sort can be a metal hardened in some special way. Then, equation (4.3.20) has the form:

$$z'' - \frac{k^2}{(E_0 + kx)^2} z = 0.$$

Solution of this equation is obtained as:

$$z(x) = \sqrt{E_0 + kx}\,(C_1 + C_2 \ln(E_0 + kx))\,. \tag{4.3.24}$$

Coefficients $C_1$, $C_2$ are defined from the initial conditions. With account of the expression for $z$ we have:

$$y(x) = C_1 + C_2 \ln(E_0 + kx). \tag{4.3.25}$$

An interesting situation arises in the case of decreasing Young modulus, $k < 0$, in point $x = E_0/k$. The trajectory inclination in this point tends to infinity that corresponds to the crack propagation across the inhomogeneity (same as in the previous case, a crack non-passage zone appears). The schematic solution curve is presented in Fig. 4.10

(C) $E(x) = E_0 \exp(kx)$. A medium with this character of change in the elastic modulus can be, for example, a multilayer package with a diffuse border between the layers. Equation (4.3.20) takes the form:

$$z'' - \frac{5}{4} k^2 z = 0.$$

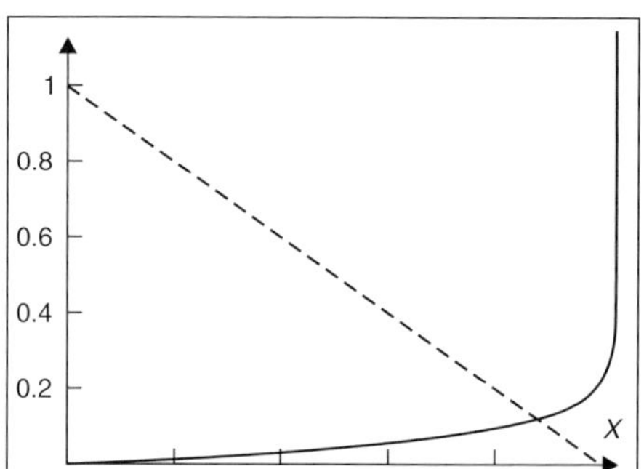


Fig. 4.10. Decrease of elastic modulus (dotted line) and deviation of the crack trajectory away from the rectilinear propagation.

Its solution is an exponent, $z = \exp(\frac{5}{2}kx)$. In this case, for $y$ we have:

$$y = \exp\left(\frac{\sqrt{5}}{2}kx + \frac{1}{2}\int C(x, y)dx\right). \tag{4.3.26}$$

According to expression (4.3.15), $C(x, y) = \frac{1}{2}\int f_1(x, y)dy$, the solution behavior depends on $f_1(x, y)$. In the case of $C(x, y) > -\sqrt{5}kx$, we have a crack propagation which is asymptotically approaching axis $OX$; in the case $C(x, y) = -\sqrt{5}kx$ we have stationary propagation $y = \text{const}$; and in the case of $C(x, y) < -\sqrt{5}kx$, the crack is infinitely deviating from axis $OX$. In this case, a variant of propagation is possible, when at initial normal propagation in relation to the layers, the crack turns around and starts to move along the border of the section.

(D) $E(x) = E_0(\exp(kx) + \exp(-kx)) = E_0 \coth kx$. The solution can be obtained rather simply in the form of a series. With precision up to the terms of the 6th order, we have:

$$z(x) = z(0) + D(z)(0)x + \frac{1}{4}z(0)x^2 + \frac{1}{12}D(z)(0)x^3$$
$$+ \frac{1}{32}z(0)x^4 + \frac{7}{480}D(z)(0)x^5 + O(x^6).$$

For the coefficients of transmission and reflection, we have:

$$R = \left(\frac{1 - \frac{1}{2}\sqrt{4 - th^2(kx)}}{1 + \frac{1}{2}\sqrt{4 - th^2(kx)}}\right)^2,$$

$$D = \frac{2\sqrt{4 - th^2(kx)}}{\left(1 + \frac{1}{2}\sqrt{4 - th^2(kx)}\right)^2}.$$

Behavior of the coefficients of transmission and reflection is presented in Fig. 4.12.

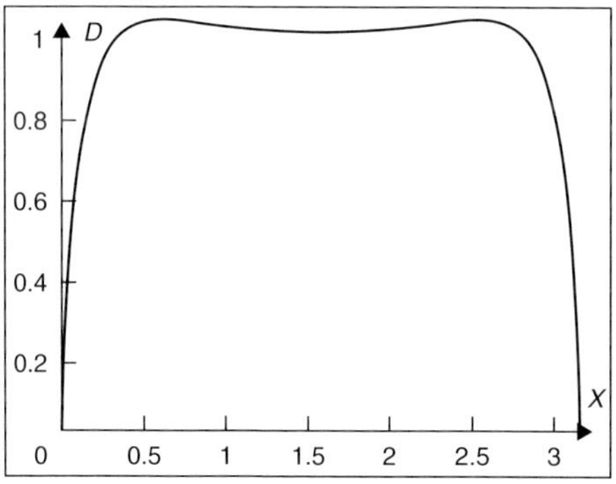


Fig. 4.11. Change of transmission coefficient along the crack trajectory for a linear medium.

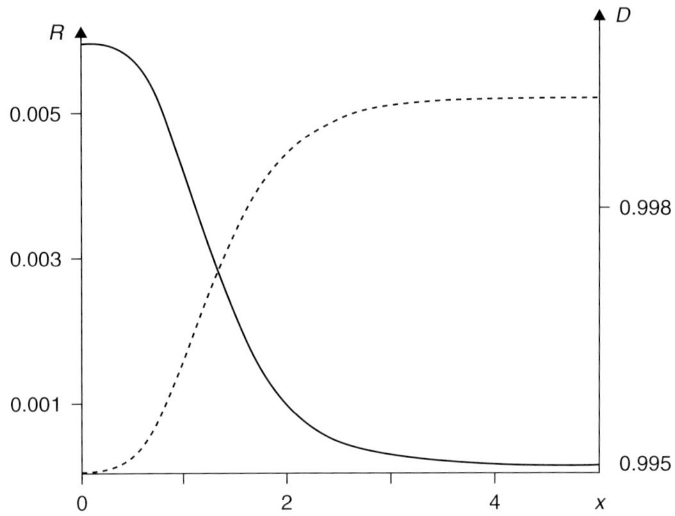


Fig. 4.12. Coefficients of transmission (solid line) and reflection (dotted line) for a hyperbolic medium.

It follows from the analysis of the transmission coefficient that the passage of the crack in the initial direction is impossible, and the coefficient is asymptotically tending to zero. This is what is expected, since a direct construction of the crack trajectory curve indicates an exponential deviation away from the direction of initial propagation. Analysis of the curves shows that $R + D = 1$, which follows from the physical sense. In general, it follows from analysis of cases (C) and (D) that a medium with the exponential dependence of properties is more favorable for development of a mainline crack, since such combinations of parameters are possible, at which the transmission coefficient never turns into zero in any point of its existence.

### *4.3.4 Influence of weakened bonds zone on crack trajectory*

The results of the preceding paragraphs were obtained in the assumption of an ideal fracture process. At the same time, as mentioned in §4.1.1, the presence of such zone brings rather serious corrections into the fracture process. Let us consider the influence of the zone of weakened bonds on the equation of the crack trajectory. It is known that the fracture energy required creating a crack of length $l$ in this case has the form of (4.1.33).

In equation (4.1.33), $\varphi$ is the angle between the current direction of crack propagation and axis $X$. Strictly speaking, value $\varphi = \varphi(x, y)$; however, in the first approximation, we can assume that $\varphi = \text{const}$. This is the condition of self-consistence and corresponds to the supposition about the crack propagation at the following step of calculations in the direction of the previous step. Along with that, in the physically justified model of crack propagation, a possibility should be present to change the direction of its development.

This is achieved by making the change $y' = \text{tg}(\varphi)$ to be a part of the trajectory equation. Then, we have:

$$f_1(x, y) = A\frac{\partial \ln\left(E^{-1}(x, y)\right)}{\partial x}; \qquad f_2(x, y) = A\frac{\partial \ln\left(E(x, y)^{-1}\right)}{\partial y}. \tag{4.3.27}$$

With account of the fact that:

$$A = \left(\left(1 - \nu^2\right)\frac{\pi \cos^2\varphi}{2}\left(\sigma_{22}^{\infty}\right)^2\right)^{-1}$$
$$= \left(\left(1 - \nu^2\right)\frac{\pi}{2}\left(\sigma_{22}^{\infty}\right)^2\right)^{-1}\frac{1}{\cos^2\varphi} = B\left(1 + y'^2\right),$$

we can rewrite expression (4.3.27) as:

$$f_{\{1,2\}}(x, y) = B\left(1 + y'^2\right)\frac{\partial \ln\left(E(x, y)^{-1}\right)}{\partial\{x, y\}}, \tag{4.3.28}$$

where indices in braces indicate the choice of the respective value. By substituting coefficients (4.3.28) into the equation of crack trajectory (4.2.5), we have:

$$y'' - B\frac{\partial\left(E(x, y)\right)}{\partial x}y'\left(1 + y'^2\right) + B\frac{\partial\left(E(x, y)\right)}{\partial y}\left(1 + y'^2\right)^2 = 0.$$

As before, by restricting the non-linearity with quadratic terms by the derivative, we have:

$$y'' - B\frac{\partial\left(E(x, y)\right)}{\partial x}y'\left(1 + y'^2\right) + B\frac{\partial\left(E(x, y)\right)}{\partial y}\left(1 + y'^2\right)^2 = 0. \tag{4.3.29}$$

We note here that in equation (4.3.29), unlike the previous cases, we did not make transformation coordinates in order to eliminate the terms containing $y'$. This is connected with the fact that consideration of the trajectory consistency and stochastization modes requires one to study the full structure of the equation. In this case, the terms at the first derivative make an essential contribution to behavior of the system.

Let us consider crack propagation in a medium with a definite value of the elasticity modulus $E(x, y)$.

Earlier [211, 222], we have considered the crack propagation in the medium with $E(x) = E_0 + kx$. In that case, solution (4.3.24) was obtained. In our case, from expression (4.3.29) we have:

$$y(x) = C_1 + C_2 \exp(kBx). \tag{4.3.30}$$

If we consider a crack in a medium with a decreasing elasticity modulus, $E(x) = E_0 - kx$, we have:

$$y(x) = C_1 + C_2 \exp(-kBx). \tag{4.3.31}$$

Comparison of expressions (4.3.30) allows asserting that the account of the weakened bonds zone causes a sharper deviation of the crack trajectory away from the rectilinear propagation. Unlike the previous case, in expression (4.3.31) $\lim_{x\to\infty} y(x) = C_1$, therefore, we can speak about stabilization of the crack trajectory in the medium with a linear decrease of the elasticity modulus. This result matches in principle the result of the general study of stability of crack trajectory [229].

### 4.3.5 *Crack propagation across singular border [216, 217]*

Crack propagation across structural inhomogeneities represents a significant practical interest, for example, for composites and structured materials. Moreover, all real materials have internal structural borders (for example, grain structure, walls of dislocations, etc.). In the elastic formulation, an idealized problem of interaction of the crack with an inhomogeneity (a border or a dislocation) can be solved, for example, by the method of complex potentials of Muskhelishvili [345]. In the variational statement, the problems of accounting for borders are connected with discontinuity of function $E(x)$ in borders 0, $L$, etc. This causes a necessity to consider, on varying the fracture energy, piecewise smooth functionals, and equation (4.2.1) in the case of crack passage across $i$ borders should be rewritten in the form of:

$$\begin{aligned}\delta U &= \delta U|_A^{0-\varepsilon} + \delta U|_{0-\varepsilon}^{0+\varepsilon} + \delta U|_{0+\varepsilon}^{L-\varepsilon} + \delta U|_{L-\varepsilon}^{L+\varepsilon} + \ldots + \delta U|_{L+\varepsilon}^{B} \\ &= \delta \int_A^{0-\varepsilon} \sigma_{ij}(x,y)\, n_i u_j ds + \delta \int_{0-\varepsilon}^{0+\varepsilon} \sigma_{ij}(x,y)\, n_i u_j ds \\ &\quad + \delta \int_{0+\varepsilon}^{L-\varepsilon} \sigma_{ij}(x,y)\, n_i u_j ds + \ldots + \delta \int_{L_i+\varepsilon}^{B} \sigma_{ij}(x,y)\, n_i u_j ds = 0,\end{aligned} \tag{4.3.32}$$

where $\lim \varepsilon = 0$. Term $\delta U_{0-\epsilon}^{0+\epsilon}$ is the additional fracture energy, connected with a sharp border. Naturally, it does not coincide with the energy of the main material, since in real samples the properties of the transition layer differ from the properties of the main material. Besides, in the border zone, the stressed-deformed state is locally observed; this state is different from the state of the main material. To simplify the calculations, let us consider a crack which crosses one border. Condition (4.3.32) can be written down as:

$$\delta U = \delta \int_A^{0-\varepsilon} dU_1 + \delta \int_{0-\varepsilon}^{0+\varepsilon} dU|_1 + \delta \int_{0+\varepsilon}^{B} dU_2 = 0, \tag{4.3.33}$$

where $dU_i$ is the fracture energy at the crack's passage of the $i$-th inhomogeneity, $dU|_i$ is the fracture energy, connected with the existence of the $i$-th border. This energy is not the energy of layer separation, since it is not connected with the crack propagation along the borders of the section. According to our problem statement, the crack approaches the border almost normally, and the fracture energy, connected with the existence of the border, is the energy required for separation of the material along the layer. In the statement of the problem of ideal fracture:

$$dU|_1 = \int_{-\infty}^{+\infty} dU\,(1)\,\mathbf{D}(x-1)dx,$$

where $\mathbf{D}(x)$ is the delta-function and $dU(1)$ is the bond energy of the border number. Since in equation (4.3.32), the coefficient in square brackets is the function of fracture energy, then with account of (4.3.32) coefficient $N$ acquires the form of:

$$\begin{aligned} N\,(x,\,y\,(x)) = &-\frac{1}{2}\frac{E\,(x,\,y(x))}{\sigma^2\mathbf{D}\,(x)}\left[2\frac{\sigma^2\mathbf{D}\,(x)\,P\,(x)}{E\,(x,\,y(x))^3} - 2\frac{\sigma^2\mathbf{D}\,(1,\,x)\,P\,(x)}{E\,(x,\,y(x))^2}\right.\\ &-\frac{\sigma^2\mathbf{D}(x)}{E\,(x,\,y(x))^2}\left(D_{[1,1]}E\,(x,\,y(x)) + D_{[1,2]}E\,(x,\,y(x))\left(\frac{d}{dx}y\,(x)\right)\right.\\ &+\left(D_{[1,2]}E\,(x,\,y(x)) + D_{[2,2]}E\,(x,\,y(x))\left(\frac{d}{dx}y(x)\right)\right)\frac{d}{dx}y(x)\\ &\left.\left.+\,D_{[2]}E\,(x,\,y(x))\left(\frac{d^2}{dx^2}y(x)\right)\right) + \frac{\sigma^2\mathbf{D}(2,\,x)}{E\,(x,\,y(x))}\right]\\ &+\frac{1}{4}\frac{\left(-\sigma^2\mathbf{D}(x)P(x) + E\,(x,\,y(x))\,\sigma^2\mathbf{D}\,(1,\,x)\right)}{\sigma^4\mathbf{D}^2(x)},\,vspace * 4pt \end{aligned} \tag{4.3.34}$$

where $\mathbf{D}(1, x)$ is the first derivative of delta-function and $D_{[1,2]}$ is the differential operator, defined by the expression:

$$\begin{aligned} P\,(x) &= D_{[1]}E\,(x,\,y\,(x)) + D_{[2]}E\,(x,\,y\,(x))\,\frac{d}{dx}y\,(x)\\ &= \frac{\partial E\,(x,\,y\,(x))}{\partial x} + \frac{\partial E\,(x,\,y\,(x))}{\partial y}\frac{d}{dx}y\,(x)\,. \end{aligned}$$

In expression (4.3.34), like before, it has been adopted that $\sigma = \sigma_{22} = \text{const}$. This expression can be simplified. For abridging the recording, hereafter we shall not indicate that $y = y(x)$:

$$\begin{aligned} N(x,\,y(x)) = &\frac{1}{4g^2\mathbf{D}(x)^2E(x,\,y)^2}\left[4g^2\mathbf{D}(x)^2D_{[1]}E\,(x,\,y\,(x))^2\right.\\ &-8g^2\mathbf{D}(x)^2D_{[1]}E\,(x,\,y\,(x))\,D_{[2]}E\,(x,\,y)\,y'\\ &-4g^2\mathbf{D}(x)D_{[2]}E^2(x,\,y)(y')^2 - 4g\mathbf{D}(x)\mathbf{D}(1,\,x)E\,(x,\,y)\,D_{[1]}E\,(x,\,y)\\ &+4g^2\mathbf{D}(x)\mathbf{D}(1,\,x)E(x,\,y)D_{[2]}E(x,\,y)y' \end{aligned}$$

$$+ 2g^2\mathbf{D}(x)^2 E(x, y)\, D_{[1,1]} E(x, y)$$
$$+ 4g^2\mathbf{D}(x)^2 E(x, y)\, D_{[1,2]} E(x, y)\, y'$$
$$+ 2g^2\mathbf{D}(x)^2 E(x, y) D_{[2,2]} E(x, y) y'^2 + 2g^2\mathbf{D}(x)^2 E(x, y) D_{[2]} y''$$
$$- 2g^2\mathbf{D}(x)\mathbf{D}(2, x) E(x, y)^2 - E(x, y)^2 \mathbf{D}(x) D_{[1]} E(x, y)$$
$$- E(x, y)^2 \mathbf{D}(x) D_{[2]} E(x, y)\, y' + E(x, y)^3 \mathbf{D}(1, x)\big] . \tag{4.3.35}$$

Depending on the properties of the medium on both sides of the interface, coefficient $N(x)$ has different functional forms, and solutions of the trajectory equation have different forms.

*Crack propagation across the border of media with piecewise constant properties*

Let us consider crack propagation in media with piecewise constant mechanical characteristics, $E_1$, $E_2$.

In real materials, a transition from one material to another takes place in a narrow zone of contact of grains for layers (see Fig. 4.13). The width of this transition zone depends on technological and other factors and defines the integration interval in the second term of expression (4.3.33). For an ideal material, the width of the transition zone from one material to another is zero. With account of expressions (4.3.34) and (4.3.35), we have:

$$N(x, y(x)) = \frac{1}{2}\frac{\mathbf{D}(2, x)}{\mathbf{D}(x)} + \frac{1}{4}\frac{\bar{E}\mathbf{D}(1, x)}{\sigma^2 \mathbf{D}(x)^2}, \tag{4.3.36}$$

where $\bar{E} = (E_1 + E_2)/2$. Solution of equation (4.3.20) with the coefficient of the type (4.3.36) is difficult. However, real materials do not have any zero transition zone. Then, with account of the real width of contact zone $2\varepsilon$, the delta-function can be approximated by a smooth function [253]:

$$\frac{1}{\varepsilon\sqrt{\pi}} \exp\left(-\frac{x^2}{\varepsilon^2}\right) \underset{\varepsilon \to 0}{\longrightarrow} \mathbf{D}(x).$$

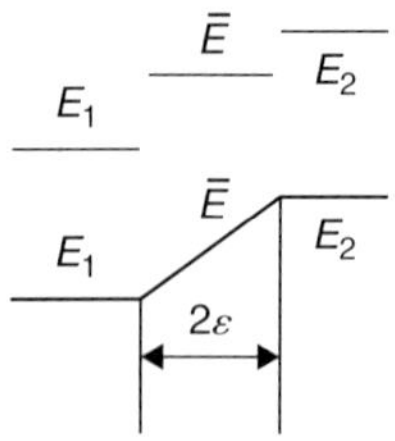


Fig. 4.13. Schematic behavior of the properties of the material in a transition zone.

In this case, for the coefficient in the trajectory equation we have the following expression:

$$N(x, y) = -\frac{1}{4E(x, y)^2\varepsilon^4\sigma^2}\Big[-4\sigma^2\varepsilon^4 D_1(E(x, y))^2$$

$$-8\sigma^2\varepsilon^4 D_1(E)(x, y(x))\, D_2(E)(x, y)\, D(y)(x)$$

$$-4\sigma^2\varepsilon^4 D_2(E)(x, y)^2 D(y)(x)^2 - 8\sigma^2 x\, E(x, y)\,\varepsilon^2 D_1(E)(x, y)$$

$$-8\sigma^2 x\, E(x, y)\,\varepsilon^2 D_2(E)(x, y)\, D(y)(x) + 2\sigma^2 E(x, y)\,\varepsilon^4 D_{1,1}(E)(x, y)$$

$$+4\sigma^2 E(x, y)\varepsilon^4 D_{1,2}(E)(x, y) D(y)(x)$$

$$+2\sigma^2 E(x, y)\varepsilon^4 D_{2,2}(E)(x, y) D(y)(x)^2$$

$$+2\sigma^2 E(x, y)\,\varepsilon^4 D_2(E)(x, y)\,(D^{(2)})(y)(x)$$

$$+4\sigma^2 E(x, y)^2\varepsilon^2 - 8\sigma^2 x^2 E(x, y)^2$$

$$+3\sqrt{\pi}\exp\left(x^2/\varepsilon^2\right)E(x, y)^2\varepsilon^5 D_1(E)(x, y)$$

$$+3\sqrt{\pi}\exp\left(x^2/\varepsilon^2\right)E(x, y)^2\varepsilon^5 D_2(E)(x, y)\, D(y)$$

$$+6\sqrt{\pi}\exp\left(x^2/\varepsilon^2\right)E(x, y^3\varepsilon^3 x\Big]. \tag{4.3.37}$$

*Crack propagation in linear medium*

Same as earlier (see § 4.3.3), let us assume that the properties of the medium change under a linear law and depend only on the $x$-coordinate, $E(x, y(x)) = kx$, where $k$ is the coefficient characterizing the medium. In this case, out of expression (4.3.37), it is easy to obtain:

$$N(x, y(x)) = -\frac{1}{4x^2\varepsilon^4\sigma^2\sqrt{\pi}}\Big(-4\sqrt{\pi}\,\sigma^2\varepsilon^4 - 4\sqrt{\pi}\sigma^2 x^2\varepsilon^2$$

$$-8\sqrt{\pi}\sigma^2 x^4 + k\,x^2\,\varepsilon^5\pi\,\exp\left(\frac{x^2}{\varepsilon^2}\right) + 2\,k\,x^4\,\varepsilon^3\,\pi\,\exp\left(\frac{x^2}{\varepsilon^2}\right)\Big). \tag{4.3.38}$$

By decomposing in standard manner exponent $\exp(x^2/\varepsilon^2)$ into a series, for coefficient $N(x, y(x))$, we have:

$$N(x, y(x)) = N(x) = x^{-2} - \left(\varepsilon^{-2} - \frac{1}{4}\frac{\varepsilon k\sqrt{\pi}}{\sigma^2}\right)$$

$$+\left(-\frac{2}{\varepsilon^4} - \frac{3}{4}\frac{k\sqrt{\pi}}{\varepsilon\sigma^2}\right)x^2 - \frac{5}{8}\frac{k\sqrt{\pi}}{\varepsilon^3\sigma^2}x^4 + O(x^6)$$

$$\equiv x^{-2} - A - Bx^2 + O(x^4). \tag{4.3.39}$$

When decomposing the exponent, we took into account the terms not higher than the fourth order by $x$. For the trajectory equation, in this case we have:

$$y'' - \left[x^{-2} - A - B\,x^2\right] y = 0. \tag{4.3.40}$$

Solution of this equation represents the crack trajectory in the transition zone. Unfortunately, because of the complexity of the coefficient, we managed to obtain a precise analytical solution of equation (4.3.40) only for the case $A = 0$:

$$y(x) = C_1 \sqrt{x}\, J_{\frac{1}{4}i\sqrt{3}}\Big(\frac{1}{2}\sqrt{-B}\,x^2\Big) + C_2 \sqrt{x}\, I_{\frac{1}{4}i\sqrt{3}}\Big(\frac{1}{2}\sqrt{-B}\,x^2\Big), \tag{4.3.41}$$

where $J_\nu$ is the Bessel function of the complex argument (McDonald function) and $I_\nu$ is the modified Bessel function of the second order (Weber function). Values $C_1$, $C_2$ are integration constants, defined by initial and boundary conditions.

## 4.4 Crack trajectory in media with random structure. Trajectory stochastization

Materials with a random structure are becoming more and more widely applied in industry. Examples of this sort of materials are composition materials with unregulated reinforcement [58], and natural biological composites. However, many issues of mechanics of composites are insufficiently studied. Among these issues are, for example, studies of crack propagation in layered and periodical structures, regulations of destruction of composites, and the onset of the modes of chaotic fracture. As a rule, the mechanical behavior of composites is simulated by introduction of efficient medium (that is, the averaging of the matrix properties and reinforcing fibers by this or that method [95, 114, 327, 362]). However, introduction of averaged characteristics prevents studying delicate effects, connected exactly with the *presence of the structure* of the material. In reality, even traditional structural materials, for example steels, have inhomogeneous internal structure (grains, dislocations, various phases of the main material, etc.), and at this sort of averaging, the information about development of inhomogeneities can be lost.

Most of the works on stochastization of the crack trajectory relate it with random characteristics of the medium, in which the crack propagates. As a rule, authors consider Markovian processes [347, 381]. In this case, since characteristics of the medium are random values, the stressed-deformed state is automatically taken to be random. This sort of 'stochastization' is not caused by physical reasons, connected with the process of crack propagation, but is induced by external reasons. Unfortunately, there are very few works on the topic of stochastization of the crack trajectory in a homogeneous medium. Thus, for sapphire monocrystals [326] it was defined that the length of decorrelation makes 65 reference points. This corresponds to approximately $90\mu$m of the physical dimension of monocrystal. In this case, the profile of the real crack is nicely simulated by a determined dynamic equation, and is essentially different from the random distribution generated both by the phase and serial randomization. In this case, the crack propagation as a whole takes place along the rectilinear axis. However, in the implicit form some presentation about stochastization can be found in many models of fracture. Thus, A. A. Shanyavskiy [319–322] speaks about microtunnelling at a crack development, and other authors [182, 183] observed a dissociation of the mainline cracks into the field of randomly scattered microcracks with subsequent coagulation into a more branched macrocrack.

### 4.4.1 Correlation function and stochastization length

Let us consider the onset conditions of stochastic modes at crack propagation. Let us decompose the components of the displacements vector of the crack edges $u_i(x, y(x))$ in equation (4.2.5) into the Taylor series by $y$ and restrict ourselves to the linear term by $y$. By expressing the coefficient at the linear term, $\partial u_i/\partial y$, as elastic deformations through stresses, we obtain:

$$\begin{aligned} N = \sigma_{ij} n_i u_j &= \sigma_{12} n_1 s_{22kl} \sigma_{kl} \tau_2 + 1/2 \sigma_{22} n_2 s_{22kl} \sigma_{kl} \tau_2 \\ &\quad + 1/2 \sigma_{11} n_1 s_{12ik} \sigma_{kl} \tau_2 + \sigma_{21} n_2 s_{12ik} \sigma_{kl} \tau_2, \end{aligned} \tag{4.4.1}$$

where $\tau$ is the unit tangent vector and $s_{ijkl}$ is the tensor of elasticity moduli. In the general case we cannot take components of the tensor of elasticity moduli as constant. Assuming that such stressed-deformed state is executed and supported that only $\sigma_{22} \neq 0$, (4.4.1) is transformed into the type:

$$N = \frac{1}{4} \sigma_{22}^2 s_{2222} \frac{2 \tan \pi}{1 + \tan^2 \pi},$$

where $\pi$ is the angle of the tangent line to the crack with axis $OX$. Let us express $f_{\{1,2\}}$ through $N$:

$$f_1 = \frac{\partial N \sqrt{1 + y'^2}}{\partial x}, \quad f_2 = \frac{\partial N \sqrt{1 + y'^2}}{\partial y}. \tag{4.4.2}$$

Depending on the type of function $s_{2222}(x, y)$, we shall have different variants of crack propagation. Let us consider a model of a composite material, for which:

$$f^0_{2,y} = -\omega^2, \quad f^0_{2,y^2} \equiv 0, \quad f^0_{2,y^3} = -\alpha\omega^2, \quad f_1 = 0. \tag{4.4.3}$$

This corresponds to representation of $\ln Q(x, y)$ in the form as:

$$\ln Q\,(x, y) = \ln Q(y) + \nu(x, y), \quad \ln Q(y) = -\omega^2 y^2 - \alpha\omega^2 y^4. \tag{4.4.4}$$

At $f_1 = 0$, there is no continuous change of inhomogeneity in direction of the normal to the line of crack front in direction of the propagation. Let us assume that along axis $OX$ the properties of the medium are piecewise-constant (crack propagation perpendicular to the border of the layers), then:

$$\nu(x, y) = \sum_{n=-\infty}^{\infty} \Theta\,(x - nX), \quad f_1 = \varepsilon \sum_{n=-\infty}^{\infty} \delta\,(x - nX), \tag{4.4.5}$$

where $X = 2\pi/\Omega$ and $\Theta$ is the Heaviside function. By assuming $\Omega << \omega$, $\varepsilon << 1$, we obtain that between points $x = nX$ $(n = -\infty \ldots \infty)$, the deviation of the crack $y$ away from axis $x$ satisfies equation of Duffing type [58, 310]:

$$\ddot{y} + \omega^2 \left(1 + \alpha y^2\right) y = 0. \tag{4.4.6}$$

In all the points of axis $x$, the Duffing equation has the form:

$$\ddot{y}_1 + \omega^2 \left(1 + \alpha y_1^2\right) y_1 = \varepsilon\omega \sum \delta\left(x - kX\right). \tag{4.4.7}$$

With the precision down to the term of order $\alpha$, we write down the solution of equation (4.4.7) in the form of periodic function [310]:

$$y = A \cos\left[(\omega + \Delta\omega) x + \varphi\right], \quad \Delta\omega = 3/8\alpha A^2\omega. \tag{4.4.8}$$

By denoting $K_n = \varepsilon\Delta\omega_n T = 3/8\alpha A_n^2 \varepsilon\omega/\Omega$ and substituting (4.4.8) into (4.4.7), for $A$ and $\varphi$, it is easy to obtain equations in final differences. They have the form: $\varphi_{n+1} = \{\varphi_n + K_n \sin 2\varphi_n\}$, $A_{n+1} = A_n\left(1 + 1/2\varepsilon \sin 2\varphi_n\right)$. Then we calculate the correlation function $R_m$ for $\varphi$ in the ordinary way as:

$$R_m = \frac{\int_0^1 \left(\varphi_{n+m} - \langle\varphi_{n+m}\rangle\right)\left(\varphi_n - \langle\varphi_n\rangle\right) d\varphi_n}{\int_0^1 \left(\varphi_n - \langle\varphi_n\rangle\right)^2 d\varphi_n}. \tag{4.4.9}$$

With the help of correlation function $R_m$, we find the condition of stochastization of the crack trajectory [229]:

$$\lim_{m\to\infty} R_m = 0. \tag{4.4.10}$$

At $K \gg 1$, we can obtain the estimate:

$$R_1 \approx \frac{1}{K_1}. \tag{4.4.11}$$

The condition of stochastization is equal to the condition of phase splitting [60, 310]; it means that the phase behavior is chaotic, and we need some statistical description for the crack. The condition (4.4.10), with account of (4.4.11), can be written in the form as:

$$K_1 = \varepsilon\frac{\Delta\omega}{\Omega} \gg 1. \tag{4.4.12}$$

The estimate of the length of trajectory, starting from which the behavior of the crack becomes stochastic, has the form:

$$x_* = \frac{X}{2\ln K_1} = \frac{1}{2\Omega \ln K_1}. \tag{4.4.13}$$

For large $K$, the probability for the trajectory to get in the phase plane into the consistency area is small, and in the case of getting into the area, the time of stay of the phase point in it decreases.

Thus, we have: at $K < 1$ $(\varepsilon \ll 1)$, the crack propagation is stable in relation to $y = 0$; at $K \gg 1$, the propagation is stochastic; at $K \sim 1$, the propagation is in a transition area. We have obtained in the explicit form an essential parameter that defines the mode of crack propagation. Usually, this parameter was defined on the basis of experimental data [347].

### 4.4.2 *Fokker–Planck–Kolmogorov equation and probability description of crack trajectory*

Equation (4.4.7) is determinate, and consequently we should pass over to the probabilistic description of trajectory $y(x)$ for the $x$ satisfying condition (4.4.13).

In the deterministic problem for (4.4.6), this small parameter represents a ratio of the effective thickness of local inhomogeneity to the distance at which an essential change of $y(x)$ takes place. Let us consider a possibility of probabilistic description of the crack behavior. From the problem statement (movement of a quasi-particle or a beam in the field of random sources), the Fokker–Planck–Kolmogorov (FPK) equation seems to be the most suitable mechanism for description of this sort of process.

*Fokker–Planck–Kolmogorov equation and energy dissipation direction*

Let us reduce equation (4.4.7) to the system of two equations in relation to variables of action $I$ and phase $\theta$ [58, 59, 386]. We have:

$$\dot{I} = \varepsilon\Psi(I, \theta, x), \quad \dot{\theta} = \omega(I), \quad \Psi = I \sum_{k=-\infty}^{\infty} \delta(x - kX)\sin(2\theta). \tag{4.4.14}$$

Let us now introduce the probability density function $p(x, I, \theta)$ and present $p(x, I, \theta)$, $\Psi(x, I, \theta)$ in the form of decomposition into series:

$$p(x, I, \theta) = \sum_{k=0}^{\infty} p_k(x, I)\, e^{ik\theta} + p_{-k}(x, I)\, e^{-ik\theta},$$

$$\Psi(x, I, \theta) = I \sum_{k=-\infty}^{\infty} \delta(x - kX)\sin 2\theta$$

$$= \sum_{k=0}^{\infty} \Psi_k(x, I)\, e^{ik\theta} + \Psi_{-k}(x, I)\, e^{-ik\theta},$$

where $p_{-k} = p_k$, $\Psi_{-k} = \Psi_k$, and function $p(x, I, \theta)$ satisfies the Liouville equation, from which for $p_0(x, I, \theta)$ we obtain the equation of the FPK type:

$$\frac{\partial p_0}{\partial x} = -\frac{\partial}{\partial I}(\alpha_1 I p_0) + \frac{1}{2}\frac{\partial^2}{\partial I^2}\left(2\alpha_1 I^2 p_0\right), \tag{4.4.15}$$

where $\alpha_1 = \varepsilon^2 8\pi\Omega$, $K_1 = \alpha_1 I$ are deflection coefficients, $K_2 = 2\alpha_1 I^2$ are diffusion coefficients.

The boundary condition for (4.4.15) has the form:

$$p_n(x_*, I, \theta) = p_0(x_*, I)\,\delta_{n0}. \tag{4.4.16}$$

Here, $x_*$ is defined by condition (4.4.13). Equation (4.4.15) at condition (4.4.16) can be correlated with the stochastic differential equation:

$$\dot{I} = \sqrt{\frac{\alpha_1}{N}} I(x)\eta(x). \tag{4.4.17}$$

Here, $\eta(x)$ is a purely random function:

$$< \eta >= 0, < \eta(x)\,\eta(x_1) >= R_0\delta(x - x_1).$$

By passing over to variable $y$, let us transform (4.4.17) to the form of:

$$\dot{y} = y\left[\sqrt{\frac{\alpha_1}{R_0}}\eta - \frac{1}{2}(\ln Q)_{,x}\right]. \tag{4.4.18}$$

Equations (4.4.18) and (4.4.7) define one and the same value $y$.

The FPK equation, corresponding to (4.4.18), has the form:

$$4\frac{\partial f(y,x)}{\partial x} = -\frac{\partial}{\partial y}\left[(\alpha_1 - 2(\ln Q)_{,x}\,y)\,f\right] + \frac{\partial^2}{\partial y^2}\left[\alpha_1 y^2 f\right]. \tag{4.4.19}$$

Let us write down the solution of equation (4.4.19) in the form as:

$$f(y,x) = \frac{1}{y\sqrt{\pi\alpha_1 x}}\exp\left\{-\frac{\ln^2 y\, Q^{1/2}}{\alpha_1 x}\right\}. \tag{4.4.20}$$

We have a stationary solution of equation (4.4.19) under the natural condition $\partial f/\partial x = 0$:

$$f(y) = \frac{2c}{\alpha_1 y^2}\exp 2\left\{2\frac{|y| - |y_*|}{|yy_*|} - \frac{(\ln Q)_{,x}}{\alpha_1}\ln\left|\frac{y}{y_*}\right|\right\}, \tag{4.4.21}$$

$$c^{-1} = \frac{4}{\alpha_1}\int_0^\infty \exp 2\left\{2\frac{y - y_*}{yy_*} - \frac{(\ln Q)_{,x}}{\alpha_1}\ln\frac{y}{y_*}\right\}\frac{dy}{y^2}.$$

Let us calculate the beam entropy, the probability density of which has the form of (4.4.20), and the stationary distribution of (4.4.21). From the general ideology of building the crack trajectory as the minimum of the functional of the fracture energy, these beams will correspond to the directions of propagation of the fracture energy.

The beam entropy density is calculated according to the formula [344]

$$h(x) = \frac{1}{2}\left\langle\frac{a^2(y(x), x)}{b(y(x), x)}\right\rangle, \tag{4.4.22}$$

$$a = \alpha_1 - 2y(\ln Q)_{,x}, \quad b = 2\alpha_1 y^2.$$

It is seen from (4.4.22) with account of type (4.4.19), that the entropy will sharply increase in points $x = nX$. In this way, the preferred energy dissipation directions are identified.

Let us consider a crack trajectory in periodically inhomogeneous medium, defined by the correlation of the type:

$$(\ln Q\,(x,\,y))_{,x} = \delta + \gamma \cos \omega x. \tag{4.4.23}$$

By integrating (4.4.23), we can obtain a clear expression for function $Q(x) = E(x)^{-1}$:

$$Q(x) = C_1 \exp\left(-\frac{\delta\, x\, \omega + \gamma \sin(\omega x)}{\omega}\right), \tag{4.4.24}$$

where $C_1$ is the integration constant. At $C_1 = 1$, the behavior of function $Q(x)$ is presented in Fig. 4.14.

The behavior of the inhomogeneity is in general exponential (the slope degree of the exponent is regulated by index $\delta$). A deviation of the function from smoothness (scatter of the properties of the composite by layers) is regulated by parameter $\gamma$ (see Fig. 4.14).

In general, behavior $Q(x)$ is similar to the earlier considered expression for an exponential medium (4.3.26).

By substituting (4.4.23) into (4.4.4), we obtain the modified Duffing equation, containing $\dot{y}$:

$$\ddot{y} - y + y^3 + \varepsilon\delta\dot{y} = \varepsilon\gamma\dot{y}\cos\omega x, \tag{4.4.25}$$

$$-f^0_{2,y} = +1, \quad -\frac{1}{2}f^0_{2,y^2} = 0, \quad -\frac{1}{3}f^0_{2,y^3} = -1.$$

It has been shown in [188] that in this case the system is not the Hamiltonian one, and it is necessary to find the conditions for the onset of deterministic chaos. In a similar

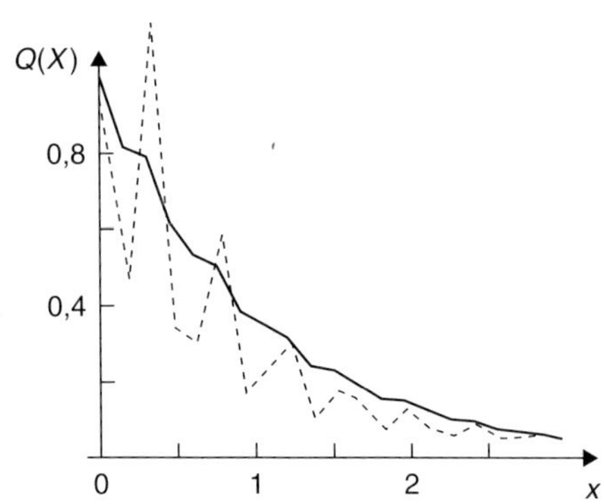


Fig. 4.14. Dependence of reciprocal fracture energy from coordinate $x$. $\omega = 15$, $\delta = 1$. Solid curve $\gamma = 1$; dotted curve $\gamma = 15$.

Table 4.1
Behavior modes of crack trajectory in the phase plane for a continuously periodical medium.

| $\gamma$ | Motion mode |
|---|---|
| 0.76–0.83 | Intersection of separatrix, chaotic motion |
| 0.84–0.94 | Attraction to one of the foci |
| 0.95–1.08 | Bifurcations cascade of doubling of the period of both foci |
| 1.08–1.15 | Limit period cycle. Appearance of a strange attractor |
| 1.2–2.45 | A strange attractor appears |

way [386], we obtain the conditions for transition from the deterministic motion of the beam trajectory to the chaotic one[3].

*Analysis of chaotic behavior of crack trajectory*

The general behavior of solutions of the standard Duffing equation is rather well known [127, 188], Table 4.1.

Let us write equation (4.4.25) in the form of a system:

$$\dot{y} = v, \quad \dot{v} = y - y^3 + \varepsilon \dot{y} \left[ \gamma \cos \omega x - \delta v \right]. \tag{4.4.26}$$

The analysis shows that for equation (4.4.26) there is one hyperbolic point $y = v = 0$ with one separatrix $H_0 = 0$. For the trajectory equation on separatrix, we have the equation:

$$\frac{dy}{dx} = -y\sqrt{1 - \frac{y^2}{2}}. \tag{4.4.27}$$

From (4.4.27), we find:

$$y(x) = \frac{\sqrt{2}}{\cosh x}, \quad v(x) = -\sqrt{2}\,\frac{\sinh x}{\cosh^2 x}. \tag{4.4.28}$$

It follows from the analysis of an unperturbed system that the incoming and outgoing separatrices in hyperbolic point $y = v = 0$ coincide. We can show [188] that in a perturbed Hamiltonian system the separatrix 'splits', that is, the incoming and outgoing separatrices split apart, do not coincide already and, generally speaking, intersect each other, resulting in an endless number of homoclinic points and chaotic motion.

In the case when the Duffing equation has $\dot{y}$, there exist two possibilities in the behavior of separatrices:

1) the separatrices never intersect, and any of them can completely embrace the other;
2) the separatrices intersect in an endless number of points; in this case, a chaotic motion of trajectories appears.

[3] It may correspond to the stochastic generation of dislocations of the moving crack.

According to Melnikov's method [204, 205], to find the conditions of intersection of separatrices, it is necessary to calculate the distance $d$ between the separatrices in a certain point $x_0$. If $d$ changes its sign at any $x_0$, a chaotic movement appears.

In the considered case:

$$d = -\int_{-\infty}^{\infty} dx \left[\gamma v_0^2 (x - x_0) \cos \omega x - \delta v_0^2 (x - x_0)\right]. \tag{4.4.29}$$

By calculating (4.4.29) similar to [205, 233], we have:

$$d(x_0, x_0) = \frac{\sqrt{2}}{3} \pi \gamma \omega \frac{\sin \omega x_0}{\coth\left(\frac{\pi \omega}{2}\right)} + \frac{4\delta}{3}. \tag{4.4.30}$$

A chaotic movement in the proximity of the separatrix appears at the condition of intersection of separatrices, that is, when $d(x_0)$ changes its sign. It follows from formula (4.4.30) that it happens at:

$$\delta < \delta_c = \frac{3\sqrt{2}\pi \gamma \omega}{4 \cosh\left(\frac{\pi \omega}{2}\right)}. \tag{4.4.31}$$

The Duffing equation (4.4.25) at $\delta = 0$ always has chaotic solutions in the proximity of the separatrix. The chaotic movement in this case is observed in a narrow layer and is limited by invariant curves. At $\delta > 0$, equation (4.4.25) is an analogue of the Duffing equation in time with account of dissipation, and, consequently, the existence of stochastic attractors is possible. Indeed, at $\delta > 0$, destruction of invariant lines takes place, which restrict the stochastic behavior in the proximity of the separatrix, and phase trajectories can get away from it rather far and find themselves in the attraction area of the sustainable focus or cycle. Thus, as has been shown with the help of analogue simulation [233], at fulfilment of conditions (4.4.31) and $\delta > 0$, the trajectory wanders in the vicinity of the separatrix until it gets on some attractor, either simple or strange (stochastic).

The results of building phase portraits and the trajectory of crack propagation for different initial conditions are presented in Figs. 4.15–4.22. Figures 4.15–4.18 have been built for parameter $\gamma = 2.19$. The integration step was defined by the capacities of the used computing machinery and in any case did not exceed 0.5[4].

It follows from the curves (see Figs. 4.15 and 4.16), that the system has a weak chaos. In conformity with the canonical Duffing equation solution properties [188], (Table 4.1), for our modified Duffing equation in Fig. 4.16 we can observe a weak strange attractor.

Figures 4.17 and 4.18 illustrate the dependence of the behavior of the trajectory on parameter $\omega$. It is seen that the trajectories are similar in principle. The secondary weak build-up of the trajectory in the proximity of $x = 80$ looks interesting. In Fig. 4.18 we can see the sensitivity of the solution to the initial conditions—the trajectories run to different foci. Along with that, it is necessary to remark that the obtained solutions stabilize. It means that at propagation of the mainline crack, only the initial stage of its

---

[4] After making a series of trial calculations, no essential dependence on the integration step in this range has been revealed.

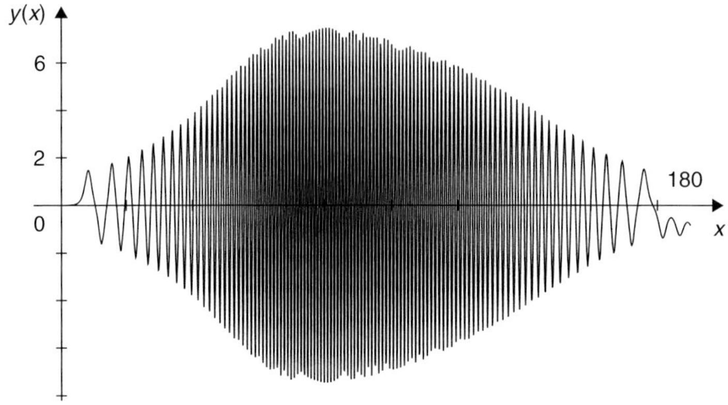


Fig. 4.15. Crack trajectory behavior in a composite material. Values $\varepsilon = 0.1$; $\gamma = 2.19$; $\omega = 0.01$; $\delta = 1$.

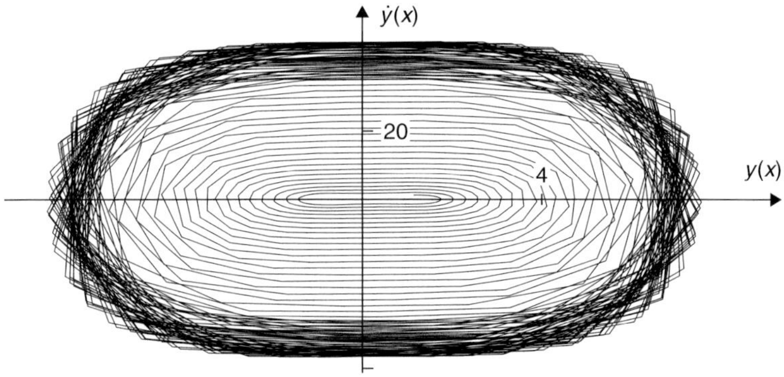


Fig. 4.16. Phase portrait of the crack. Values $\varepsilon = 0.1; \gamma = 2.19; \omega = 0.01; \delta = 1$. Initial conditions $y(0) = 0$; $\dot{y}(0) = 0.001$

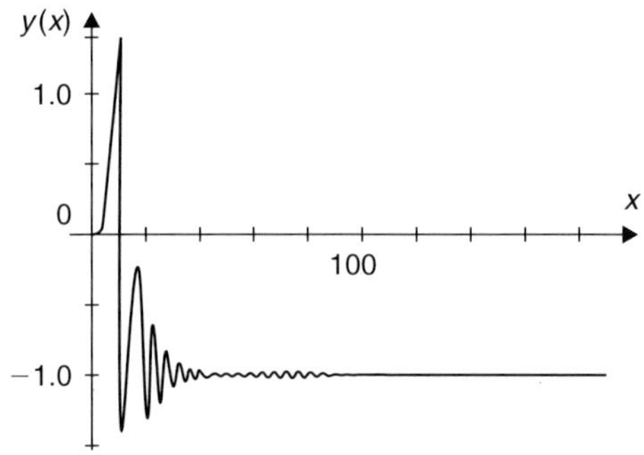


Fig. 4.17. Crack trajectory. Values $\varepsilon = 0.1$; $\gamma = 2.19$; $\omega = 0.1$; $\delta = 1$. Initial conditions $y(0) = 0$; $\dot{y}(0) = 0.001$.

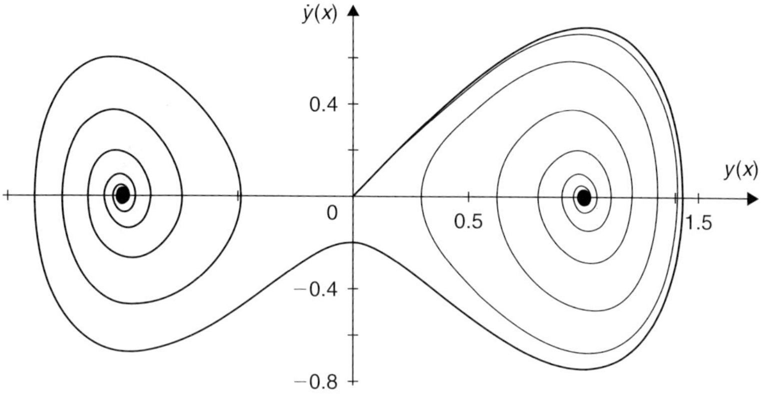


Fig. 4.18. Phase portrait. Values $\varepsilon = 0.1$; $\gamma = 2.19$; $\omega = 0.1$; $\delta = 1$. Initial conditions $y(0) = 0$; $\dot{y}(0) = 0.00001$ dotted line; $y(0) = 0.0001$; $\dot{y}(0) = 0.00005$ solid line.

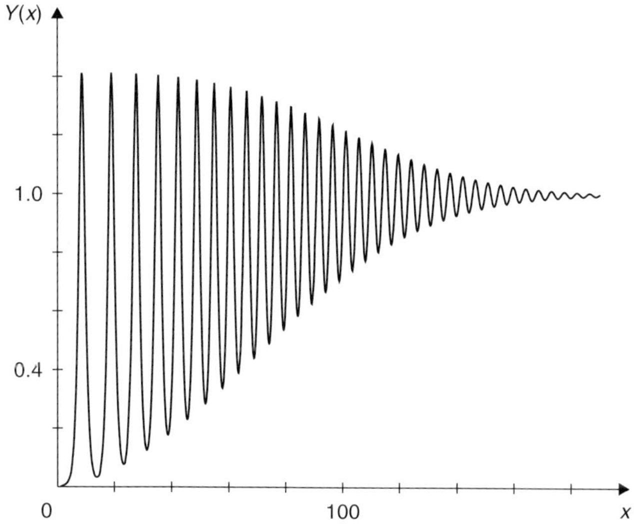


Fig. 4.19. Crack trajectory. Values $\varepsilon = 0.1$; $\gamma = 1$; $\omega = 0.01$; $\delta = 1$. Initial conditions $y(0) = 0$; $\dot{y}(0) = 0.001$.

motion is unstable, and after the crack acquires a certain speed of propagation, it becomes stable. At the initial stage, a transition is possible of a crack from one layer into another. After stabilization, the crack propagates within the boundaries of one material. This behavior can be explained as a 'search' by the crack of an optimum propagation trajectory. Figures 4.19–4.22 are built for parameter $\gamma = 1$. Unlike the previous case, phase portraits

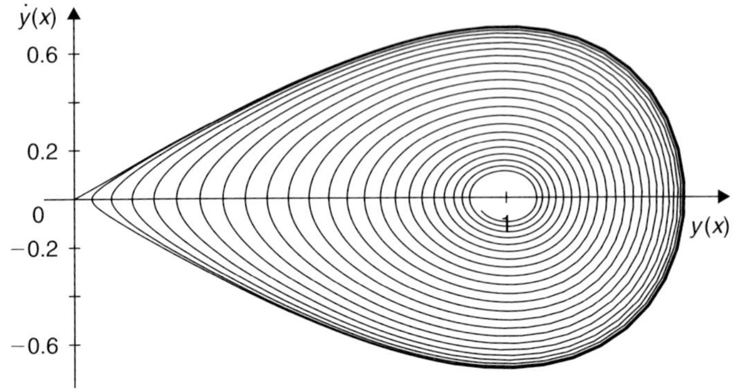


Fig. 4.20. Phase portrait. Values $\varepsilon = 0.1$; $\gamma = 1$; $\omega = 0.01$; $\delta = 1$. Initial conditions $y(0) = 0.0001$, $\dot{y}(0) = 0.00005$.

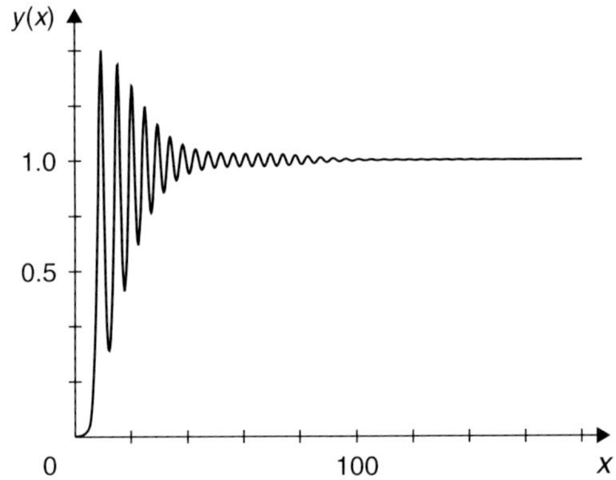


Fig. 4.21. Crack trajectory. Values $\varepsilon = 0.1$; $\gamma = 1$; $\omega = 0.1$; $\delta = 1$. Initial conditions $y(0) = 0.0001$, $\dot{y}(0) = 0.00005$.

for the trajectories with close initial values in Fig. 4.20 are indistinguishable. It has been noted that the behavior of the trajectory depends on the initial conditions. Thus, in case $\varepsilon = 0.1$; $\gamma = 1.9$; $\omega = 6$; $\delta = 1$, if $0.965 < \dot{y}(x) < 1.285$, the trajectory focus is +1; if $\dot{y}(x)$ does not get into this range, point 1 is the focus. This corresponds to propagation of the crack 'as a whole' in this or other material.

**Remark 4.4.1.** Comparison of behavior of the smooth exponential medium (4.3.26) with a medium with point sources of the type (4.4.23) shows that the averaging procedure can essentially influence the results. For example, in the latter case, no exponential deviation from the initial direction of propagation was obtained.

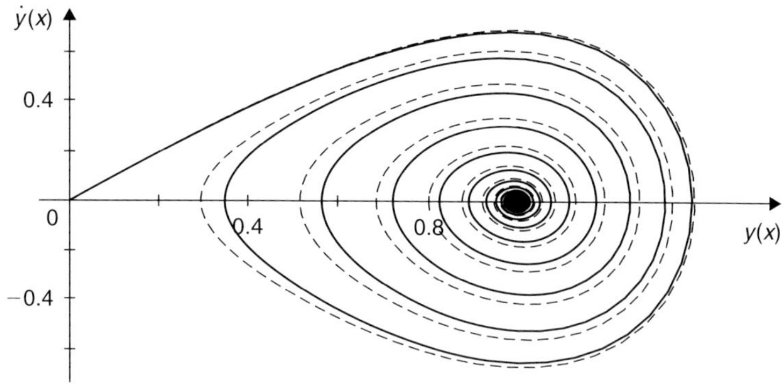


Fig. 4.22. Phase portrait. Values $\varepsilon = 0.1$; $\gamma = 1$; $\omega = 0.1$; $\delta = 1$. Initial conditions $y(0) = 0.0001$, $\dot{y}(0) = 0.00005$ solid line; $y(0) = 0.00005$, $\dot{y}(0) = 0.00005$ dotted line.

The considered models of inhomogeneity correspond to the conclusion that a change of inhomogeneity along axis $Y$ ensures a waveguide character of crack propagation along axis $Y$. In particular, a cut-out may play the role of such a waveguide. However, even a rigid determination of the initial and final points is insufficient to have the crack trajectory $y(x)$ to be a deterministically forecasting function. In the considered models of piecewise-constant and periodic media, the change of inhomogeneity along the direction of propagation leads to stochastization of the trajectory; consequently, the forecast is possible only in the probabilistic sense [59].

Certain discrepancies between the results of the analogue (Table 4.1) and discrete simulation (attractors in Figs. 4.15–4.22) have been caused, probably, by the round-off mistakes in calculation models of discrete simulation.

### *4.4.3 Physical reasons of crack trajectory stochastization*

Equation (4.4.25) gives a mathematic justification of a possibility of chaotization of the crack trajectory. Let us consider *material* reasons for onset of chaotization. From the general sense of the FPK equation (4.4.19), expression $2(\ln Q)_{,x}$ is the forcing coherent (microscopic) force acting in the system; term $\alpha_1$ is the fluctuation of the force acting in the preset time interval. Consequently, stochastization of the trajectory can arise only in case of real existence in the system of fluctuating forces acting on the crack. Such forces can in principle arise in two cases: they can be generated by randomly distributed power sources, or caused by random properties of the material in which the fracture propagates. It is obvious that in real systems both mechanisms are realized simultaneously. Random sources are stochastically distributed defects (for example, dislocations [119]); random properties of the material are inevitable consequences of technological processes. Since the medium is taken not random, but piecewise-continuous according to (4.4.5), parameter $\Omega$ is responsible for the intensity of the pulse force acting in the system.

It is known that there exists a weak dependence of the direction of propagation of the mainline crack on the distribution of microcracks in the proximity of the crack tip [242, 267, 299]. If we take the criterion of maximum normal stresses [275] as the criterion of the growth direction of the mainline crack, this dependence within the classic elasticity theory can be described as:

$$\Theta^M = 2 \arctan \frac{2(K_{II}/K_I)}{1 + \sqrt{1 + 8\,(K_{II}/K_I)^2}}, \tag{4.4.32}$$

where $K_I$, $K_{II}$ is the SIF of the body with a field of microcracks in the proximity of the tip of the mainline crack. For composite materials, in the volume of which there exists a field of randomly distributed microcracks, interacting with the mainline crack, the expression for SIF can be taken in the second approximation as:

$$\begin{aligned} K = K^{\pm}_{I02} - iK^{\pm}_{II02} = -\frac{p\lambda^2\sqrt{l_0}}{4} \\ \times \sum_{k=1}^{N} \left\{ \frac{1}{2} \left[ 1 - e^{-2i\alpha_k} - \frac{u_k}{\sqrt{u_k^2 - 1}} - \frac{\overline{u_k}}{\sqrt{\bar{u}_k^2 - 1}} - \frac{(\bar{u}_k - u_k)e^{-2i\alpha_k}}{\left(\bar{u}_k^2 - 1\right)^{3/2}} \right] \right. \\ \times \left[ \frac{1}{(\bar{u}_k \mp 1)\sqrt{\bar{u}_k^2 - 1}} - \frac{e^{2i\alpha_k}}{(\bar{u}_k \mp 1)\sqrt{u_k^2 - 1}} \right] \\ + \left[ 1 - e^{2i\alpha_k} - \frac{u_k}{\sqrt{u_k^2 - 1}} - \frac{\overline{u_k}}{\sqrt{\bar{u}_k^2 - 1}} - \frac{(u_k - \bar{u}_k)e^{2i\alpha_k}}{\left(u_k^2 - 1\right)^{3/2}} \right] \\ \left. \times \left[ \frac{(\bar{u}_k - u_k)(2\bar{u}_k \pm 1)e^{-2i\alpha_k}}{2(\bar{u}_k^2 - 1)^{3/2}(\bar{u}_k \mp 1)} + \frac{1 - e^{-2i\alpha_k}}{2(\bar{u}_k \mp 1)\sqrt{\bar{u}_k^2 - 1}} \right] \right\}, \end{aligned} \tag{4.4.33}$$

where the overline means a complex conjugation, $u_k = z_k/l_0$, $2l_0$ is the length of the macroscopic crack, $k$ is the number of the microcrack, $k = 1 \ldots N$, $z_k$ is the coordinate of the centre of the microcrack, $\alpha_k$ is the inclination angle of the $k$-th crack to the abscissa, $u_k$ is the normal displacement, $\lambda = l/l_0$ is the small parameter, $l$ is the length of the microcrack, taken for simplicity same for all the microcracks, and $p = \sigma_{22}^{\infty}$.

Expression (4.4.33) does not include the terms describing the interaction of microcracks, since the second approximation describes only the interaction between macrocracks and each of the microcracks. With account of (4.4.33), we have:

$$\frac{K_I}{K_{II}} = \frac{\text{Re}\,K}{\text{Im}\,K}. \tag{4.4.34}$$

Strictly speaking, all the existing methods of calculation of the SIF for a real medium do not yield satisfactory results from the viewpoint of mathematical rigor. This is connected, among other things, also with the fractality of the fracture surface. The integral evaluation of the influence of the relief roughness on the equivalent SIF should include simultaneous account of misorientation at different scale levels, the relief height and peculiarities of the relief profile in the direction of the crack growth and perpendicular to it [320]. Such integral assessment can be in the form of the fractal dimensionality of the break, that is, the efficient SIF should be the function of $D$. For further investigation, it is important for us that the general expression (4.4.33) can be represented in the form as:

$$K_{I02}^{\pm} = pN\sqrt{l_0}l^2\xi(\alpha, v, s),$$

$$K_{II02}^{\pm} = pN\sqrt{l_0}l^2\zeta(\alpha, v, s),$$

where $N = rs/l_0^2$ is the density of microcracks. In such presentation, values $r$ and $s$ are connected with the coordinates of microcracks, as $x_n = l_0 n/r$, $y_n = l_0 m/s$, where $n, m = 1, 2, \ldots$ are natural numerals, indicating the number of microcracks in the length section $l_0$ along axes $X$, $Y$, accordingly. Then, expression (4.4.34) takes the form:

$$\frac{K_I}{K_{II}} = \frac{\xi(\alpha, v, s)}{\zeta(\alpha, v, s)} \equiv \Upsilon(\alpha, v, s). \tag{4.4.35}$$

Use of expression (4.4.33) for definition of clear form of functions $\xi$, $\zeta$ causes essential mathematic difficulties. However, since we are in fact interested in the type of function $\Upsilon(\alpha, v, s)$, the analytical form of $K$ seems not so important.

We have already mentioned that equation (4.4.32) has been derived in supposition of interaction of the crack with the field of randomly distributed microcracks. The development of the theory in this direction requires a consecutive consideration hereafter of the interaction of the macrocrack with the field of randomly distributed dislocations. However, this complication is not always represented to be justified, since the main part of function $\Upsilon(\alpha, v, s)$ is defined by rather simple physically justified modification of the classic solution of the ideal problem (4.4.32). Thus, reference [343] considers, on the basis of ideas of Nowozhilov [261, 262] about the existence of structural units in the fracturing body, propagation of a fragile crack as a process of serial microbreaks of structural bonds. Unlike reference [346], it was accounted that the speed of microfracture *is not the speed of crack propagation*. In general, the process of crack propagation was considered as a three-stage one: break of structural bond; development of deformation until the net cell; and then break. Similar assumptions were put into the basis of computer simulation of fracture in references [182, 183]. In conformity with the fact that processes generating a fracture develop within the limits of one cell, it was assumed that only the front itself affects the structure of peculiarities in the vicinity of the considered front (self-consistent consideration). Important in this case is the fact that even at single-axis loading, the front is really realizing a complicated stressed-deformed state. In this case, an analytical expression was obtained [343] for deviations of dedicated surface area in the breakage front as:

$$\Theta = \frac{\pi}{2} - \arctan\left(\frac{\sigma_{yy}}{4\tau_{xy}} \pm \sqrt{\frac{\sigma_{yy}}{4\tau_{xy}} + \frac{1}{2}}\right). \tag{4.4.36}$$

Analysis of this expression shows that only in case of a uniform three-axis stretching does the fracture propagate in the direction of the initial defect. These results coincide in principle with the results of numeric simulation [346]. Appearance of stresses in inclined pads corresponds to refusal from the principle of local symmetry [1]:

$$K_{II} = 0 \Leftrightarrow \text{ smooth crack propagation.}$$

Besides, equation (4.4.36) allows studying the onset of bifurcation mode at crack propagation.

Earlier, a similar study was conducted for the field of microscopic cracks [366] on the basis of supposition about the fatigue fracture as the process of aggregation, limited by diffusion. The microscopic progress of the crack was written in the form of iteration of discreet mapping:

$$\phi_{m+1} = \phi_m + \Theta_m^M, \quad l_{m+1} = l_m + s_m \cos \phi_m, \tag{4.4.37}$$

where $\phi_m$ is the angle of crack deviation away from axis $OX$ in the $m$-th step and $\Theta_m^M$ is the angle responding to the maximum of $r(\Theta)$.

The value of the deviation angle can be chosen from various assumptions, for example, in conformity with reference [385]. However, in view of mathematical complications, function $\Upsilon$ can be selected from the physically justified suppositions and limitations. Thus, in reference [366] the function was selected in the form as:

$$\Upsilon = \lambda(l) \sin(-\phi_m). \tag{4.4.38}$$

In the physical sense, $\lambda(l)$ is the scale factor (similarity factor between the first and second loading modes with the crack growth and change of $\phi_m$). G. V. Vstovskiy assumed that $\lambda(l) = \chi\sqrt{s/l}$, where $\chi$ accounts for strengthening of the second mode on inclusions, grain borders and other inhomogeneities (this can be simulated, for example, by the Heaviside function), $s$ is the step (pitch) of the macrocrack progress (length of an element increment). This representation with account of (4.4.37) corresponds to the independence of $K_I$ from $\phi_m$. Function $r(\Theta)$ describes the border of the local zone of fatigue fracture. Inside this zone, the equations of the elastic fracture mechanics are not true. It is logical to suppose that the cluster of defects in the next step develops in the direction $\max r(\Theta)$. An analytical expression for $r(\Theta)$ can be obtained by various methods.

The chaotic properties of the mapping $\phi \to \phi$ depend on the choice of the type of function $\Theta$. The behavior of $\phi$ for $\Upsilon = \lambda \sin(-\phi_m)$ is presented in Fig. 4.23.

From analysis of Fig. 4.23 in logarithmic coordinates, we see that bifurcations arise in the proximity of points $\log \lambda = 0.5$. We note here that for a complete analysis of the behavior of the mapping, it would be necessary to use expression for $\Upsilon$ from (4.4.34). However, the existing computing possibilities so far do not allow this, since the obtained expressions are too complicated. An essential influence of the type of the used function $\Upsilon$ can be illustrated in the simplest way. Thus, by using expression for $\Theta$ from (4.4.36) in equation (4.4.37), we can study the chaotic properties of mapping $\phi \to \phi$ in the classic statement. The analysis shows that accounting in (4.4.32) of the expression for $\Upsilon$ the form of $\Upsilon = \lambda \sin(-\phi_m)$ results in qualitatively different results. In the simplest case of accounting for microdamage for the angle according to expression (4.4.32) and scale dependence in the form of (4.4.38), the mapping has the form as represented in Fig. 4.24. A practical proximity to zero of the deviation angle for small $\lambda$ indicates that at small distances the fracture propagates stably and gets 'stochastized' only at the lengths longer than the critical length of phase uncoupling $x_*$ according to (4.4.13). We find important the presence of only two sustainable states. A precise start of bifurcation (value $\lambda_{cr}$) is difficult to define because of the inconsistency of the calculation scheme. In general, the process of changing the direction of the crack can be represented as a change of sustainable propagation in the direction of the previous growth into the sustainable growth in the direction, defined by angle $\phi$. This coincides in principle with the results of studying the stability of crack propagation and responds to the transition between two stable foci according to the results in §4.4. In general the appearance of the bifurcation and the existence of two stable branches in the bifurcation diagram correspond to a possibility of the crack deviation by $\pm\phi$ after achieving the critical length.

The appearance of bifurcations in the process of crack propagation can be related also with achieving by the crack of the critical speed $v_c$ [246]. It is obvious that the critical speed can be related with critical length $x_*$.

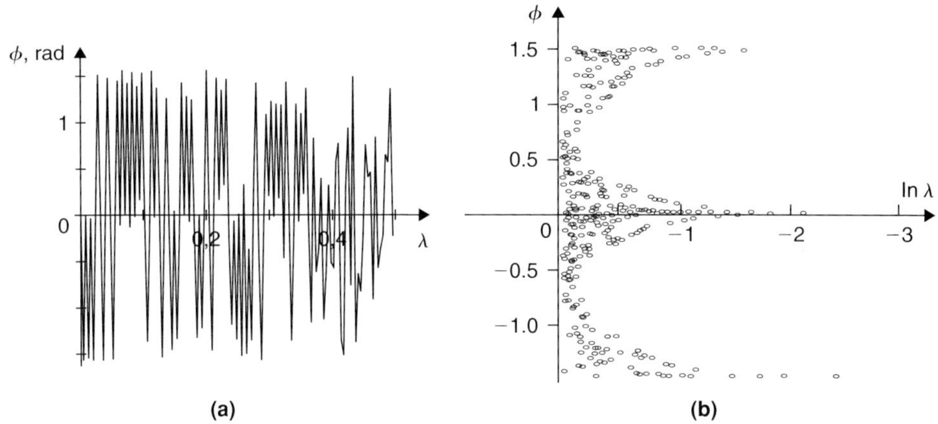


Fig. 4.23. Dependence of the deviation angle of the crack end on λ: (a) in Cartesian coordinates, (b) in logarithmic coordinates.

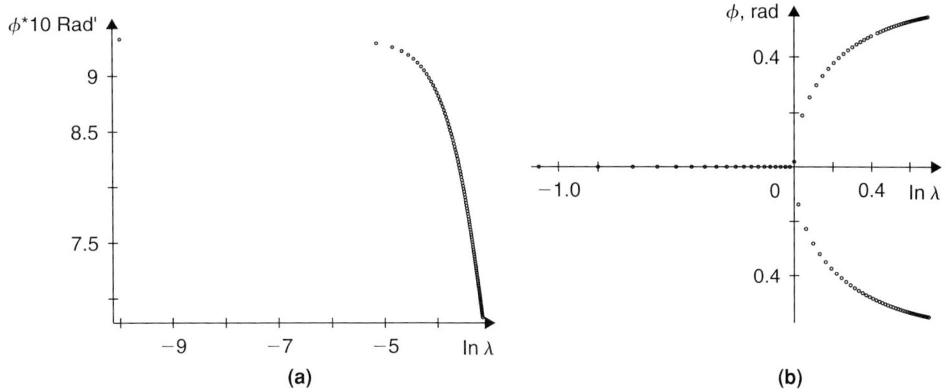


Fig. 4.24. Dependence of the deviation angle of the crack tip on λ with account of the microdamage: (a) small values of λ, (b) high values of λ.

## 4.5 Deformation and fracture with account of electromagnetic fields

It is known that the dipole moment of a dielectric filling the volume $V$ is defined by formula

$$M_j = \int_V x_j \rho_l dV,$$

where $\rho_l$ is the volumetric density of bound charges and $x_j$, $j = 1, 2, 3$ are spatial coordinates. Then, the medium polarization vector is set as:

$$P_j = \frac{dM_j}{dV}. \tag{4.5.1}$$

It means that polarization vector $P_j$ is the electric dipole moment of a dielectric volume unit.

Electric field intensity $\mathbf{E}$ in a dielectric conforms to the Maxwell equations system

$$\operatorname{div} \mathbf{D} = \rho_e, \tag{4.5.2}$$

$$\operatorname{rot} \mathbf{E} = 0, \tag{4.5.3}$$

where $\mathbf{D} = \epsilon_0 \mathbf{E} + \mathbf{P}$ is the electric displacement vector, $\epsilon_0 = 8.854 \times 10^{-5}$F/m is the dielectric constant and $\rho_e$ is the free electron density.

Equations (4.5.2), (4.5.3) are tractable in the material medium, if the functional relations $\mathbf{P} = \mathbf{P}(\mathbf{E})$, $\mathbf{D} = \mathbf{D}(\mathbf{E})$ have been preset. As a rule, these relations for the media are defined experimentally. In the simplest case, a linear dependence of the field intensity from the electric inductance vector is taken. For an anisotropic medium, the linear relations have the form of

$$D_i = \epsilon_{ij} E_j, \tag{4.5.4}$$

where $\epsilon_{ij}$ is the symmetric tensor of dielectric constants. In more complicated media, like a ferroelectric, a more complex relation of field intensity with the electric induction vector may be implemented. For them, the polarization vector may differ from zero at $E_j = 0$.

In the case of a direct piezoelectric effect, electric charges appear on the surface of the crystals of a certain symmetry class under the effect of mechanical loading. It has been experimentally proved for piezoelectrics that in the absence of the electric field, the piezoeletric's polarization vector is linearly related, similar to the law (4.5.4), with the components of mechanical stress tensor $\sigma_{kl}$, $k, l = 1, 2, 3$:

$$P_i = d_{ijk} \sigma_{jk}, \tag{4.5.5}$$

where $d_{ijk}$ is the third-rank tensor characterizing the piezoelectric body modules (tensor of piezoelectric modules).

In the case of piezoelectric media, the electric induction vector $D_i$ takes the form of:

$$D_i = d_{ikl} \varepsilon_{kl} + \epsilon_{ij} E_j. \tag{4.5.6}$$

The reverse piezoelectric effect assumes a change of shape and dimensions of piezocrystals under the effect of the external electromagnetic field. For the reverse piezoeffect, there exists a linear relation between the electric field intensity $E_i$ and deformation tensor $\varepsilon_{ij}$ of a mechanically free ($\sigma_{ij} = 0$) crystal.

$$\varepsilon_{ij} = d_{kij} E_k. \tag{4.5.7}$$

In case of a joint action of electric and mechanical fields, the defining relations in a piezocrystal ('Hooke's law') take the form of

$$\sigma_{ij} = c_{ijkl}\varepsilon_{kl} - d_{kij} E_k, \tag{4.5.8}$$

where $c_{ijkl}$ is the tensor of elastic constants. Equations (4.5.6) and (4.5.7) form, together with equilibrium equations, a full system of equations for solving the problem of crack propagation in the presence of an electromagnetic field [202, 283].

In real piezoelectric crystals, mechanical, electric and thermal properties are interrelated and need to be studied in combination. Thermodynamic methods may be used for this purpose.

The energy balance equation (4.1.26) for the piezoelectric exposed to external loads, temperature and external electromagnetic field takes the form of:

$$\frac{d}{dt}\int_V \left(\frac{1}{2}\rho\dot{u}_i\dot{u}_j + U\right) dV = \int_V \left(X_i\dot{u}_i + E_i\dot{D}_i + W\right) dV + \int_{\partial V} (t_i\dot{u}_i - q_i n_i)\, dS, \tag{4.5.9}$$

where $U$ is the internal energy density, $u_i$ is the mechanical displacement vector, $X_i$ are body forces, $t_i = \sigma_{ij} n_j$ are surface loads, $q_i$ is the heat flow vector, $W$ is the intensity of heat sources, $\rho$ is the density of materials, $n_i$ is the unit normal vector to surface $\partial V$, and the overdot means the time derivative.

In accordance with the second law of thermodynamics (1.2.6), in equation (4.5.9) the term:

$$dW_{ext} = \int_V X_i\dot{u}_i dV + \int_{\partial V} t_i\dot{u}_i dS \tag{4.5.10}$$

means the power of external forces.

In the general case, the flow of electromagnetic energy through a body surface makes

$$U_{el} = \int_{\partial V} S_i n_i dS = -\int_{\partial V} \operatorname{div} \mathbf{S} dV = \int_V (\mathbf{E} \operatorname{rot} \mathbf{H} - \mathrm{H} \operatorname{rot} \mathbf{E})\, dV \tag{4.5.11}$$

where

$$\mathbf{S} = \mathbf{E} \times \mathbf{H} \tag{4.5.12}$$

is the Pointing vector, which means the energy flow vector. In the case of electrostatics, the electric field is potential, rot $\mathbf{E} = 0$ and with account of Maxwell equations rot $\mathbf{H} = \dot{\mathbf{D}}$, the term

$$U_{el} = \int_V E_i \dot{D}_i dV \tag{4.5.13}$$

in equation (4.5.9) is the flow of electromagnetic energy in electrostatic approximation.

By using standard methods [283], we may transform the surface integrals included in equation (4.5.9) to the type of:

$$\int_V \dot{U} dV = \int_V \left( \sigma_{ij}\dot{\varepsilon}_{ij} + W + E_i \dot{D}_i - \frac{\partial q_i}{\partial x_i} \right) dV, \tag{4.5.14}$$

where deformations are taken small and the deformation tensor has the form of (2.3.3). The contribution of external heat sources $(W, q_i)$ may be converted into the differential form on the basis of thermodynamic relations and the Clausius–Duhem inequality:

$$T\dot{S} = T\frac{\partial}{\partial x_i}\left(\frac{q_i}{T}\right),$$

where $S$ is the entropy density and $T$ is the temperature. Finally, the full differential of the internal energy function $U$ takes the form of

$$dU = \sigma_{ij} d\varepsilon_{ij} + E_m dD_m + TdS. \tag{4.5.15}$$

Respectively,

$$\sigma_{ij} = \left(\frac{\partial U}{\partial \varepsilon_{ij}}\right)_{D,S}, \quad E_m = \left(\frac{\partial U}{\partial D_m}\right)_{\varepsilon,S}, \quad T = \left(\frac{\partial U}{\partial S}\right)_{\varepsilon,S}.$$

### *4.5.1 Destruction of piezoelectric materials*

Relatively new investigation objects for the fracture mechanics are the media and systems which cannot be satisfactorily described within the linear fracture mechanics. Among others, such media are, for instance, piezoelectric and magnetostriction substances broadly used in the manufacture of sensors, actuators, transformers and for military purposes. One of the first fundamental works in the field of piezomaterial fracture [283] formulated the basic problems, various aspects of which, especially mathematic ones, are intensely studied [117, 139, 186, 189, 305, 337, 369].

*Statement of the problem*

We assume that a piezoelectric has no volume forces and free charges. The values in their initial states have index (0); the values in their final states have index (1). We are considering adiabatic processes, in this case $dS = 0$ and the expression for the internal energy (4.5.15) acquires the form of:

$$U = \frac{1}{2}\int_V \left(\sigma_{ij}^{(k)}\varepsilon_{ij}^{(k)} + E_m^{(k)}D_m^{(k)}\right)dV, \quad (k = 1, 2). \tag{4.5.16}$$

The increment of internal energy as a result of transition from state (0) into state (1) has the form of [283]:

$$\begin{aligned}\Delta U &= \frac{1}{2}\int_V \left[\left(\sigma_{ij}^{(1)} + \sigma_{ij}^{(0)}\right)\left(\varepsilon_{ij}^{(1)} - \varepsilon_{ij}^{(0)}\right) + \left(E_j^{(1)} - E_j^{(0)}\right)\left(D_j^{(1)} - D_j^{(0)}\right)\right]dV \\ &= \frac{1}{2}\int_{\Sigma+\Delta\Sigma} \left[\left(\sigma_{ij}^{(1)} + \sigma_{ij}^{(0)}\right)\left(u_i^{(1)} - u_i^{(0)}\right) - \left(D_j^{(1)} - D_j^{(0)}\right)\left(\phi^{(1)} - \phi^{(0)}\right)\right]n_j dS.\end{aligned} \tag{4.5.16}$$

Equation (4.5.16) takes into account that

$$\frac{\partial \sigma_{ij}^{(k)}}{\partial x_j} = 0, \qquad \frac{\partial D_i^{(k)}}{\partial x_i} = 0, \qquad E_j^{(k)} = \frac{\partial \phi^{(k)}}{\partial x_j}.$$

When considering full work as the work on the fracture surface and the work of fracture increment, we may express the change of internal energy as:

$$\Delta U = \Delta A_\Sigma + \Delta A_{\Delta\Sigma} \tag{4.5.17}$$

where

$$\Delta A_\Sigma = \frac{1}{2}\int_\Sigma \left(\sigma_{ij}^{(1)}u_i^{(1)} - D_j^{(1)}\phi^{(1)}\right)n_j dS - \frac{1}{2}\int_\Sigma \left(\sigma_{ij}^{(0)}u_i^{(0)} - D_j^{(0)}\phi^{(0)}\right)n_j dS, \tag{4.5.18}$$

$$\Delta A_{\Delta\Sigma} = \frac{1}{2}\int_{\Delta\Sigma} \left(\sigma_{ij}^{(0)}u_i^{(1)} + D_j^{(0)}\phi^{(1)}\right)n_j dS - \frac{1}{2}\int_{\Delta\Sigma} D_j^{(1)}\left(\phi^{(0)} + \phi^{(1)}\right)n_j dS. \tag{4.5.19}$$

By its physical sense, the value $\Delta A_{\Delta\Sigma}$, which makes part of expression (4.5.17), defines the energy flow at fracture formation $\Delta\Sigma$, and integration, like in the usual fracture mechanics [157], is made by two surfaces $\Delta\Sigma_1$, $\Delta\Sigma_2$ of the additional fracture, while the normals to the surface are directed inside the crack. With account of expression (4.5.19), the fracture energy criterion (Griffith criterion) of start of crack growth takes the form of:

$$\gamma(\Delta\Sigma_1 + \Delta\Sigma_2) = -\Delta A_{\Delta\Sigma}, \tag{4.5.20}$$

where $\gamma$ is the surface energy density.

*Conditions of electric contact*

With account of expressions (4.5.18) and (4.5.19), the crack propagation depends on electric boundary conditions. Two types of electric boundary conditions are studied:

- **Conditions of imperfect contact (impermeable crack)**. Dielectric permeability of the material between the crack edges is smaller than the dielectric permeability of the basic material, and in the limit case is equal to zero and $D_j^+ = D_j^- = 0$, [117, 369].
- **Conditions of perfect contact (permeable crack)**. Electric potentials $\psi^{(1)}$ and $D_j^{(1)} n_j$ are continuous in the surface $\Delta\Sigma$ [283].

The case of imperfect contact with an intermediate value of dielectric permeability is physically most justified. We may specifically consider complex media, where the dielectric permeability of impurity is higher than that of the basic material (for example, inclusion of a massive of nanotubes into piezoceramics volume). For the model of perfect contact, the local fracture condition for the piezoelectric medium coincides with the known fracture condition for a resilient body, and in the case of imperfect contact, the stress intensity coefficient depends on both the mechanical and electrical components [139, 283].

## 4.5.2 *Crack propagation in piezoelectric material*

By using the well-known Sedov decomposition of the full crack problem into subproblems, (Fig. 4.25), we may present the full crack problem in piezoelectric medium as superposition of mechanical and electrical problems [63, 202, 391]

$$\bar{\sigma}_{2i}^{+} = \bar{\sigma}_{2i}^{-} = \bar{\sigma}_{2i}, \quad \bar{\sigma}_{2i}\sigma_{2i}^{\infty} = 0$$
$$\bar{D}_2^{+} = \bar{D}_2^{-} = \bar{D}_2, \quad \bar{D}_2 + D_2^{\infty} = \epsilon_0\left(\bar{E}_2 + \bar{E}_2^{*}\right).$$

For brevity we take into account case stressed state that only $\sigma_2 \neq 0$. This decomposition allows using the earlier developed methods for consideration of the crack propagation problem in the presence of electromagnetic field on the basis of a system of analogues

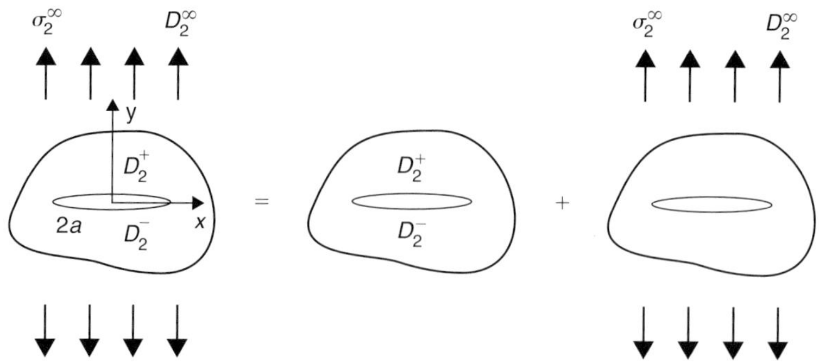


Fig. 4.25. Decomposition of full problem into elementary subproblems.

between the respective laws and values [91]. Thus, for stress intensity coefficients in the case of an elliptical crack with half-axes $a$, $b$, we have [202]:

$$\begin{pmatrix} K_{II} \\ K_I \\ K_{III} \\ K_D \end{pmatrix} = \left(\{\sigma\}_2^{\infty} - \{\sigma\}_2^{*}\right)\sqrt{\pi a}. \tag{4.5.21}$$

In the equation $\left(\{\sigma\}_2^{*}\right)^T = (\sigma_{21}, \sigma_{22}, \sigma_{23}, D_2)$ is a matrix column, the asterisk refers the value to the internal cavity, and $K_D$ is the crack tip electric displacement factor. The intensity factors are defined to give $\{\sigma_2\} = \{K\}/\sqrt{2\pi r}$ on the plane ahead of the crack tip where $r$ is the distance from the crack tip.

To study the crack propagation with account of the electromagnetic field in the material, let us express the reverse energy $Q(x, y)$, (4.4.24), as a superposition of mechanical and electrical components,

$$Q(x, y) = Q_m + Q_e, \tag{4.5.22}$$

$$Q_m = C_1 \exp\left(\frac{-\delta x\omega + \gamma \sin(\omega x)}{\omega}\right), \quad Q_e = C_2 \cos(\Omega x).$$

Then expression (4.4.23) takes the form of

$$(\ln Q(x, y) =)_{,x} = -\frac{\delta Q_m + \gamma Q_m \cos(\omega x) + Q_e \Omega}{Q_n + Q_e},$$

while the trajectory equation (4.4.25) may be written in the form of:

$$\ddot{y} - y + y^3 = \varepsilon \dot{y}\left(-\frac{\delta Q_m + \gamma Q_m \cos(\omega x) + Q_e \Omega}{Q_n + Q_e}\right). \tag{4.5.23}$$

The crack trajectory and phase images may be studied with the help of the Maple package. In the presence of the electromagnetic component, the crack behavior is essentially different. Figures 4.26–4.31 present crack trajectories and phase images at identical

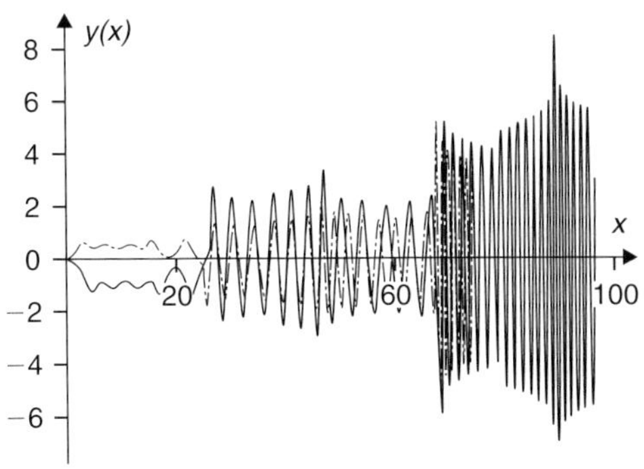


Fig. 4.26. Crack under effect of electric field. $\Omega = 3\omega$, $\epsilon = 0.1, \gamma = 1, \omega = 0.1, \delta = 1$. Initial condition $y(0) = -0.0001, \dot{y}(0) = -0.01$ solid line; $y(0) = -0.0001, \dot{y}(0) = 0.01$ dash-dotline.

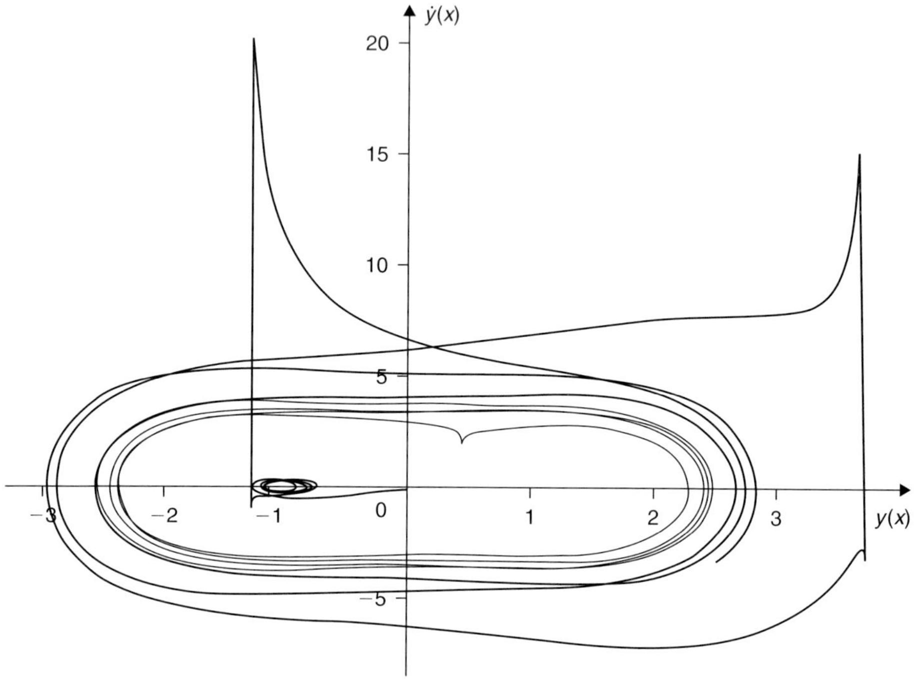


Fig. 4.27. Phase image, $\Omega = 3\omega$. Initial condition $y(0) = -0.0001$, $\dot{y}(0) = -0.01$.

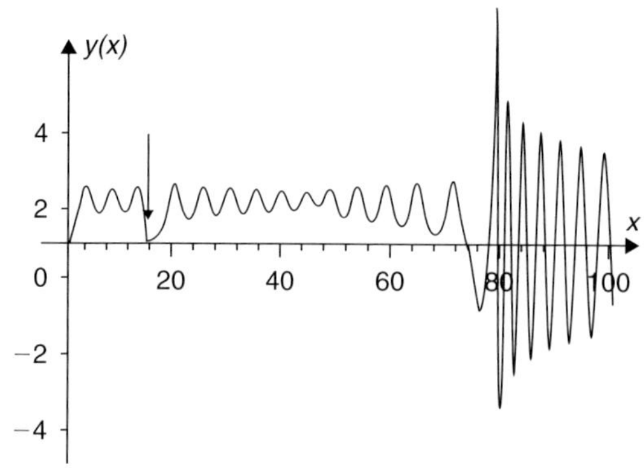


Fig. 4.28. Crack under effect of electric field, $\Omega = 3\omega$. Initial condition $y(0) = -0.0001$, $\dot{y}(0) = -0.01$. The bifurcation near $x = 15$ is marked by the arrow.

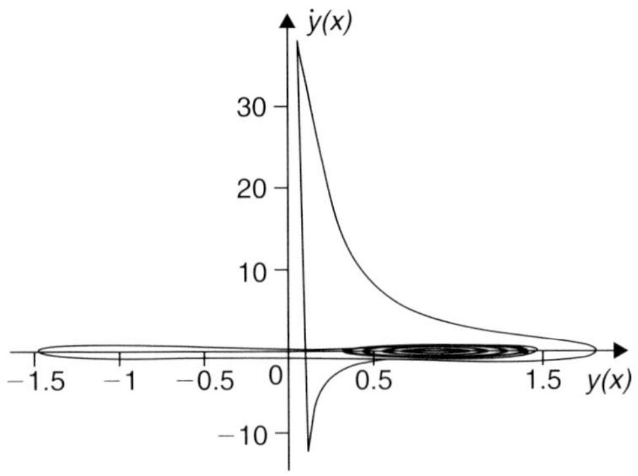


Fig. 4.29. Phase image, $\Omega = \omega$.

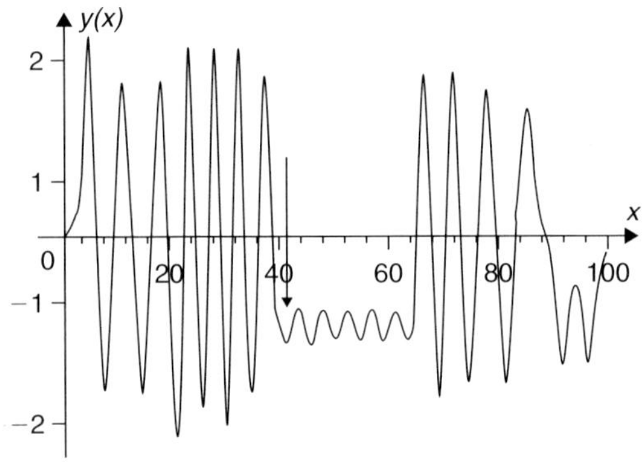


Fig. 4.30. Crack under effect of electric field, $\Omega = 10\omega$. Two bifurcation points near $x = 40, 60$.

initial conditions in the presence of electric field and different ratios of mechanical and electrical periodicity $\omega$, $\Omega$.

The general analysis of the crack behavior indicates that the presence of the electric field essentially increases the chaotic behavior of the trajectory. It follows from Figs. 4.28–4.31 that even in case of high-frequency electromagnetic field, no 'suppression' of mechanical oscillations takes place, and the trajectory has multiple bifurcation points (sharp breaks of phase trajectory).

In Fig. 4.28 the bifurcation of the trajectory occurs near point $x = 16$.

In a special ideal case, when a crack propagates only under the effect of the sinusoidal electromagnetic field, Figs. 4.32 and 4.33, at the background of the general growth of

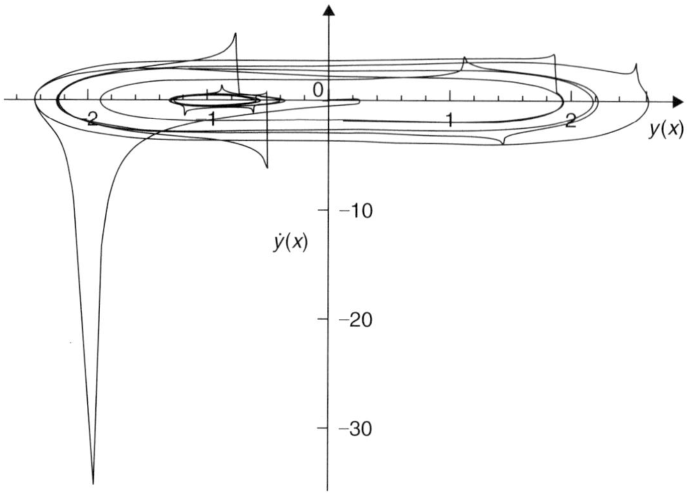


Fig. 4.31. Phase image, $\Omega = 10\omega$.

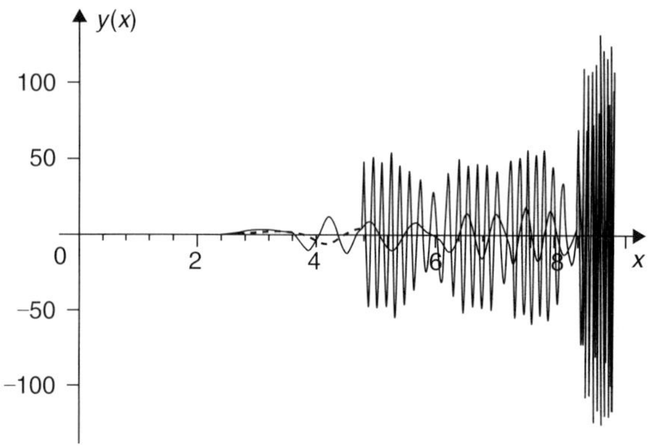


Fig. 4.32. Crack without mechanical loading, $\Omega = 22\omega$.

oscillations of the crack trajectory local oscillations exist (a shorter mode). Thus, the Landau mechanism should be implemented for transition to the chaotic state. As a whole, this behavior conforms with the experimental data [252].

**Remark 4.5.1.** Since the trajectory behaves essentially in the linear and singular way, the calculated schemes are unstable; therefore, the reliability of the obtained curves for high values of argument $x$ requires a more detailed examination.

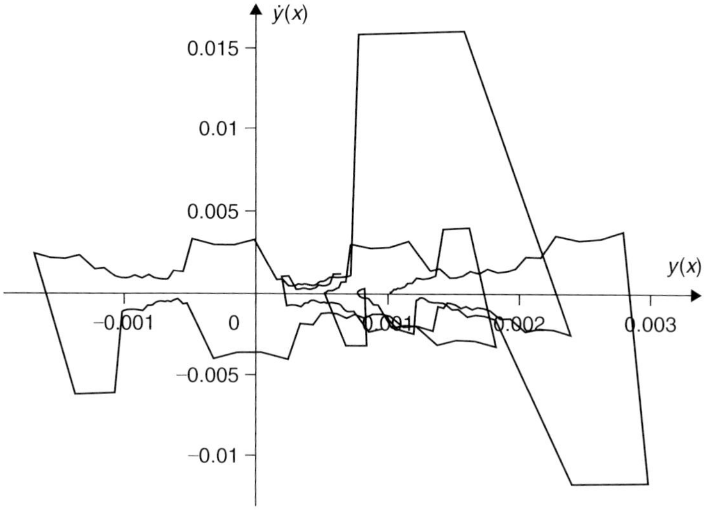


Fig. 4.33. Crack without mechanical loading, $\Omega = 22\omega$.

## Certain outcomes

- At crack propagation in a non-uniform medium, in the conditions of a non-uniform stressed state in the material, there are closure or non-passage crack areas in the material for particular media. Design of the media, in which closure areas have been created artificially, allows improving the operation properties of materials.
- The effect of the structural boundary may be accounted for by introduction of a smooth approximation of fracture characteristics of the material. Whereby, the real physical thickness of the transition zone between the interacting layers is an approximation parameter.
- The crack propagation across the interface depends on the ratio of the stresses versus mechanical properties. This allows one, in principle, to relate the Dundurs parameters [79, 80] (4.1.37) with the destruction work of the connection zone. Whereby, since the ratio 'stress/properties' enters both the linear and cubic by $x$ members, it will be essential both for the linear crack propagation (within the fracture mechanics of ideal bodies), and for evolvement of nonlinear crack propagation modes.

  According to equations (4.3.40), and (4.3.41), the deviation of the crack trajectory from a straight line when passing across the structural boundary depends not only on the properties of the medium, but also on the stressed-deformed state and width of the transition zone. In the known Dundurs parameters (4.1.37), the crack deviation angle $\alpha$ depends on the Young modulus $E(x)$ and Poisson ratio $\nu$. In our case, for several layers of identical materials with a sharp boundary according to (4.3.40), we have $\tan(\alpha) = y' = f(E, \sigma, \varepsilon)$. This dependence provides broader opportunities for the crack propagation modes.

Definition of the rotation angle at the border allows putting a principal question on managing the direction of crack propagation in layered composites by varying the width of the transition zone or its properties.

- The proposed approach allows analyzing the dependence of the crack behavior from the properties of the medium, through which the crack propagates, finding the areas of unstable propagation, and assessing the stochastization (randomization) length. Forecasting of the crack trajectory in the case of meeting the stochastization condition is possible only through use of probability methods, for example, on the basis of Markov process theory.
- The forecasted crack behavior is logically explained in the model of composites with adhesion links [53]. The initial instability (chaotic state) of propagation of a mainline crack may be physically realized only through formation of lateral microcracks in the boundaries of the structural elements of the composite. Whereby, the fracture energy flow relaxes in the process of destruction of adhesion links in conformity with the Kelly theory [144, 145]. In development of the destruction process, the damage will accumulate in the layers passed by the crack, and a weak relaxation on destruction of adhesion links will be insufficient, the energy will start to liberate along the trajectory of the mainline crack and stabilize it.
- The considered problem may be an example of layered medium synthesis, in which the boundary is a barrier in the crack route, since in the considered example, the boundary may break the condition $y'^2 \ll 1$, the crack propagates along the boundary, and then along the layer again.

  The work [52] considers a model of crack propagation without a possibility of fibering. The accounting of the possibility of fibering enables us to demonstrate that a crack of normal fracture at layer fibering transforms into the lateral crack shear, the propagation of which takes more energy, and, hence, the crack is decelerated.

  Thus, stochastization of the crack trajectory in a layered medium leads to an impossibility to forecast its propagation and, therefore, the impossibility of taking measures, at the stage of designing and manufacturing of an article, to improve crack resistance. On the other hand, the crack stochastization that enables its propagation along the fibering ensures the crack resistance of the article, since the shear fracture viscosity is higher than the fracture destruction.
- The results of studying the trajectory stability at crack propagation in composites coincide with the results obtained during computer simulation [103] of homogeneous media.

  For uniform media and small crack speeds, the roughness decreases with the growth of crack speed and increase of applied shear; the bifurcation point has been found, related with the crack speed $\sim 0.6$ of the sound speed.
- In the presence of an electric field, the crack behavior is much less stable than in the absence thereof. At crack development, bifurcation transitions are possible (sharp outbreaks at phase diagrams). In general, this sort of behavior coincides with experimental

observations [252]. These bifurcations are responding, in accordance with the problem put forward, to the stepwise change of crack propagation (transition from propagation across the layers to the intralayer growth or bifurcation mode). Unlike earlier studied cases [229], propagation is not realized in the waveguide mode. Along with that, in the case of $\Omega >> \omega$ it seems impossible to stabilize the crack trajectory by means of 'suppressing' oscillations.

# 5 Surface, Fractals and Scaling in Mechanics of Fracture

## Contents



In metal sciences, when designing articles and mechanisms, specialists have always paid attention to obtaining homogeneous articles with ideal surfaces (for example, maximum flat and smooth surfaces). However, the development of principally new technologies did not bring about principally new results, which has caused certain bewilderment. Only with development of the fractal theory does it seem possible to offer an explanation. In this connection, a problem appears to formulate general ***exclusion principles*** in metal sciences and metrology, which are similar in their idea, for example, to the indeterminacy principle of quantum mechanics or Hölder conditions. The need to formulate these principles is also related with the fact that, despite the seeming simplicity and well-mastered methods of running experiments, there exist rather contradictory evaluations in relation to interrelations of mechanical and fractal properties of materials [49]. This requires a more detailed consideration of physical grounds of fractal behavior of real materials.

A similar situation exists in fracture theory. The propagation dynamics of a macroscopic crack, as has been experimentally and theoretically demonstrated in recent works, has a number of peculiarities, connected with the presence of inconsistency, dynamic stochasticity and branching [61, 222, 246]. Among these peculiarities is the presence of bifurcation points (a transition from rectilinear propagation to the mode of branching with stochastic dynamics [229, 246, 366]). A complete description of the transition process from disperse fracture (accumulation of damages) to macroscopic fracture (crack propagation) seems problematic so far because of the complexity of the processes. In this respect, of certain interest are the attempts to describe the fracture processes on the basis of mathematic analysis of the main regularities of fracture, for example, the catastrophe theory [199]. Thus, the classic objects of the fracture theory can be presented as the objects of the catastrophe theory: the Griffith problem—as a folding catastrophe, deformation through a spring – as an assembly catastrophe; the crack length – with a cusp catastrophe. Then, a transition from a fuzzy fracture to a mainline crack, or appearance of bifurcation points, corresponds to a change of the catastrophe type.

The basic fractal characteristics of fracture (for example, the fractal dimension of the surface fracture, its dependence on the loading parameters and on the properties of the material) are a subject of much debate and studies of recent years [16, 56]. Nevertheless, the physical mechanisms which cause just the fractal character of fracture are paid insufficient attention. However, certain individual aspects can be considered with a sufficient degree of detailing.

One such aspect is the fractal properties of macroscopic fracture. Strictly speaking, the fractal properties are connected with the dynamics of the motion of the fracture crack [282]. For a quasi-static crack propagation, the effects are levelled, the roughness factor $\zeta = 0.5$. The basic problem—***what causes the fractal character of fracture***—is not practically considered today.

## 5.1 Some conceptions of theory of fractals

It is well known that fractal objects are not quite adequately described as objects pertaining to the Euclidean space [49]. The typical example is the snowflake of H. von Koch. It is endowed with finite area but infinite perimeter. Furthermore, it is a self-similar object: any part of the object is similar to the whole object entirely and at any scale level. The dimension of the fractal objects, which is determined by a standard algorithm, also exceeds the respective Euclidean dimension. The discrepancy lies in that the snowflake belongs to the Euclidean plane featuring dimension two, it is constructed from Euclidean segments featuring dimension one and, what is more, it features its own dimensions of 1.2619.

The anomalous mathematical properties of Euclidean space were stated many years ago but were less publicized. Recall the area paradox by Schwartz [197]. In the late nineteenth century Herman Amandus Schwartz showed that the common triangulation for the cylinder featuring unit radius and unit height for the surface area depends on the method of fragmenting the surface into triangles. Thus for the area of the surface we may obtain a random value from true $2\pi$ to infinity.

More reasonably, the 'Schwartz area paradox' stimulated Hermann Minkowski to devise his safe definition of length and area via the volumes of increasingly thin Minkowski 'sausages' of curves and 'comforters' of surfaces. These are the $\varepsilon$-neighborhoods made of all points within $\varepsilon$ of a point on the curve or surface. Minkowski defines the area of an ordinary surface as $\lim_{\varepsilon\to\infty}(1/2\varepsilon)$ (volume of the $\varepsilon$-comforter) and for the unit cylinder he obtains as unambiguous $2\pi$. In detail this description can be found in Appendix D.

Sometimes the conception of the fractal applied to the fracture is limited to the conception of 'self-similar' and thus only the development of self-similar structures [302] is investigated. Mathematically self-similar objects are the objects invariant in relation to the affine transformation:

$$x_i \to \lambda_i x_i, \quad i = 1, \ldots, d.$$

The group structure of the transformation is additionally required.

Considerable confusion exists concerning the relationship between self-affine fractals that can be rescaled by transformations that require different change in length scale in different directions and self-similar fractals that can be rescaled via the same change of length scales in all directions [203]. The structure must effect all of the values $\lambda_i$ such that $\lambda_i$ must be the homogeneous function from one of the set values, for example $\lambda_1$:

$$\lambda_i = \lambda_1^{\zeta_i}. \tag{5.1.1}$$

The set of the indices $\zeta_i$ characterizes the scaling properties of a self-affine *object* and depending on the problem is called *homogeneity exponent* or *roughness exponent*.

Taking into account resultant § 1.1, fractal behavior is only a special limited case of general hierarchical evolution. In hierarchical methodology, the fractal processes are described by the help of two hierarchical variables the level $\Lambda$ and unit object $\Omega$. $\Lambda$ is compared to the length scale and $\Omega$ is compared to $\zeta_i$. Other hierarchical variables are not defined and the system is hierarchical simplified.

### *5.1.1 Roughness of fracture surface*

In general, according to the current standards the surface roughness is defined as maximum amplitude of the crack profile:

$$R_t = \max_{0<x<L} z(x) - \min_{0<x<L} z(x).$$

One of the most promising approaches towards the characterization of rough surfaces is via the height difference correlation functions

$$C_q(\mathbf{r}) = \langle |h(\mathbf{x}) - h(\mathbf{x}+\mathbf{r})|^q \rangle^{1/q}. \tag{5.1.2}$$

In equation (5.1.2) $h(\mathbf{x})$ is the height of the surface above position $\mathbf{x}$ on a smooth reference surface. In many cases it has been found that the correlation functions have the form:

$$C_q(r) \sim r^{H_q}, \quad C_q(r) = \langle |h(\mathbf{x}) - h(\mathbf{x}+\mathbf{r})|^q \rangle_{|\mathbf{r}|=r} \tag{5.1.3}$$

over a substantial range of length scales. $C_q(r)$ is the average of $C_q(\mathbf{r})$ over all directions in the smooth reference surface.

The roughness of the profile is a complex function of the measure base [74]. Experimentally, the roughness can be measured by a variety of methods (by a profiler, optical methods, sound waves dissipation method and others). Theoretically roughness of the surface can be viewed on the basis of different assumptions. For example, J. W. Morrisey and J. R. Rice [240, 241] considered the equation of the crack surface on the basis of the classical elasticity theory; in their works [36, 37, 246] they investigated the percolating growth model of the fracture clusters.

The analysis of the experimental data showed that the height of the crack profile $z$ was proportional to the current measuring position $R$: $z \sim \tau^{\zeta}$. Thus the exponent $\zeta$ is physically equal to the roughness exponent for the self-affine processes.

If the set of indices $\zeta_i$ consist of only one index, we obtain the Hurst exponent and the self-affine scaling properties of a surface $z(\mathbf{x})$ with $h(\mathbf{0})$ can be represented as:

$$z(\mathbf{x}) \equiv \lambda^{-H} z(\lambda \mathbf{x}) \tag{5.1.4}$$

where $\equiv$ implies equivalent statistical properties.

In physical systems, the self-affine scaling regime is bounded by upper and lower correlation length [203] $x^+$ and $x^-$ in both horizontal ($||$) and vertical ($\perp$) directions, i.e. self-affine scaling is found over the range

$$x_{||}^- < \delta r < x_{||}^+ \quad x_{\perp}^- < \delta h < x_{\perp}^+$$

where $\delta h$ and $\delta r$ are the deviation in height and length and these correlation lengths are related by

$$\left(x_{||}^-/x_{||}^+\right) = \left(x_{\perp}^-/x_{\perp}^+\right)^{-\zeta}.$$

It is common knowledge that a relationship between the self-affine Cantor set structure and a self-affine surface profile $z(x)$ can be presented through the use of the structure function [370]

$$S(\tau) = \langle [z(x+\tau) - z(x)]^2 \rangle,$$

where $S(\tau)$ physically represents the mean square of the difference in height expected over any spatial distance $\tau$, and $\langle * \rangle$ implies averaging over the statistical ensemble of $z(x)$. It has been shown that the structure function for a fractal profile can be expressed as:

$$S(\tau) = \Lambda^{2D-2} \tau^{4-2D}. \tag{5.1.5}$$

In equation (5.1.5) $D$ is the self-affine fractal dimension and $\Lambda$ is the characteristic parameter of the fractal function referred to as the topothesy [93]. The self-affine fractal dimension $D$ of the surface profile is dimensionless and falls in the range $1 < D < 2$, while the topothesy $\Lambda$ can take any positive value and has the dimension of length.

Topothesy is defined in two ways: as a measure of anisotropy of 3-D surface $z(x, y)$ [334], and as a length scale over which the profile has a mean slope of 45° [77, 333]. The smaller is $\Lambda$, the flatter the profile appears in the macroscopic scale.

**Remark 5.1.1.** From the viewpoint of the atomic theory, the representation of fractality of real objects is nonsense. For material bodies, there exists the minimal dimension (size of the atom). Theoretical physics has a concept of the Planck length. Therefore, the endless divisibility of a fractal object is just a mathematic idealization, and its correlation with physical reality is idealization and supposition.

If we ***assume*** that the fracture surface has a self-affine character, some general conditions must be met. If we introduce parameterization for the fracture surface as $z(x)$, for $x = (x_1, \ldots, x_{\tau-1})$, the normal displacement of the edges of the cut can be presented as

$$h(\tau) = \max_x \left(z(x) - z(x+\tau)\right). \tag{5.1.6}$$

The self-affinity condition demands that $h(\tau) \propto |\tau|^{\zeta}$. The description of the real profile of fracturing surface is quite problematic if we take into account its self-affinity. The complete and consistent solution to this problem so far is impossible. However, the condition of self-affinity allows us to narrow the range of possible solutions to the problem.

We can investigate the profile of the real fracture using an analogue model of the fracture. It is known that to fracture ideal elastic–plastic continua an equivalent electric model can be constructed [121, 387]. In this model the width of the fracture zone, which is determined as a minimum of fracture energy applied along the chosen path, equals the width of the zone (number of electrical links), which reached the threshold value (peak current). In this case the characteristics of the element $P(t)$ distribution function, the elements which achieved the threshold of conductivity, correspond to the asymptotical behavior of the fracturing process. The crack will include the first continuous line of elements which reached the saturation threshold and pass through the whole net. The distribution function $P(t)$ can be chosen in one way or another but the index of the system homogeneity lies in the range of $\zeta = 0.78 \pm 0.03$.

### *5.1.2 Fractal shape of fracture surface*

**Weibull and Weierstrass functions.** With the development of the fractal theory, many mathematic functions long known in various other sciences have found their application, sometimes unexpectedly, in the fractal geometry. The Weibull function is one of them. This function was initially introduced by V. Weibull for approximation of experimental data related to steel breaking strength [374] and has the form:

$$F_w(t, p, \sigma, \mu) = \begin{cases} 1 - \exp\left[-\left(\frac{t-m}{\sigma}\right)^p\right] & \text{at} \quad t > \mu, \\ 0 & \text{at} \quad t \leq \mu. \end{cases} \tag{5.1.7}$$

The formula for the probability density function of the general Weibull distribution is:

$$f(x) = \frac{\gamma}{\alpha}\left(\frac{x-\mu}{\alpha}\right)^{p-1} \exp\left[-\left(\frac{x-\mu}{\alpha}\right)^p\right] \quad x \leq \mu; \quad p, \alpha > 0, \tag{5.1.8}$$

where $\gamma$ is the shape parameter, $\mu$ is the location parameter, and $p$ is the scale parameter. The case where $\mu = 0$ and $\alpha = 1$ is called the standard Weibull distribution. The case where $\mu = 0$ is called the two-parameter Weibull distribution. The equation for the standard Weibull distribution reduces to:

$$f(x) = px^{p-1}\exp\left(-(x^p)\right) \quad x \leq \mu; \; p > 0.$$

The Weibull distribution is used in statistics for different process description, for example, for description of failure regularities of ball bearings, vacuum or electronic devices.

As mentioned in previous paragraphs, the self-affine character of fracture demands the functional dependence of the roughness exponent. Because the Weibull function was introduced for the surface description, the roughness exponent may be linked with the Weibull distribution. And the mechanical behavior of the Weibull function is connected with physical properties of fracture.

The fractal objects are interesting by their continuity but non-differentiable character (naturally, on the scale of more than atom the ideal solid is continuous). The mathematical formalism may satisfy this demand. K. Weierstrass in the nineteenth century introduced this kind of function [372]. Applying to the surface shape description we can express it as:

$$z(x) = Z_0 \left(1 - \gamma^{D-2}\right) \sum_{n=0}^{\infty} \frac{\cos(2\pi \gamma^n x/L + n\Phi)}{\gamma^{(2-D)n}}, \tag{5.1.9}$$

where $D$ is the fractal dimension of the curve, $1 < D < 2$, $Z_0$ is the scaling parameter for the scale matching the profile with experimental data, with dimension of length, and $L$ is the maximal wavelength of the curve.

The sum in equation (5.1.9) is the superposition of the sinusoidal waves with the wavelengths and amplitudes decreasing as a geometric series with the exponent $\gamma^n$ and $\gamma^{(2-D)n}$ respectively. The exponent $\gamma$, which is determined by the decreasing of harmonics, is constant, $\approx 1.5$ [24, 25]. Phase $\Phi$ can be taken as arbitrary for more close concordance with the experimental data of profile of fracture surface.

## 5.2 Physical way of fractal crack behavior

Taking into account a possibility of stochastic behavior of the crack according to §4.4, justifiable is the provision that fractal properties of the surface are caused by generation of random affections ('noise of fracture') at breakage of interatomic bonds. In this case, the crack front should be described by a stochastic equation of the type of the Langevin equation [87–89]. The characteristics of random sources in the Langevin equation should be selected in some physically justified way.

Let us consider a simple hierarchical model of crack propagation in a medium under additional noise.

### *5.2.1 Problem statement*

Let us consider crack propagation along axis $X$. The crack front is a curve, sufficiently smooth and unlimited in $Z$ direction. In real systems, three different zones arise: a pre-fracture zone, a zone of undisturbed material $y < 0$, and a zone of disturbed material $y > 0$. In the case of propagation in a homogeneous material, the properties of the material in undisturbed areas are taken to be identical. In the case of crack propagation along the border of a composite (an exfoliation crack) the properties are different. Borders can be grains, different phase zones, or other structural irregularities. A border (a break of physical properties) can also be formed by energy flows in front of the crack tip, waves and other dynamic processes.

In the strict sense, the influence zone (zone of pre-fracture of Barenblatt–Dugdale, p. 119) of the crack is an inhomogeneous thin wedge, which is clutched between thick elastic (or elastic–plastic) foundations without slippage on the boundary [101, 191].

A crack growth can be represented as indentation of the influence zone into undisturbed material under the action of the facial load. For the sake of simplicity, we approximate the shape of the influence zone not as a wedge, but as a thin equivalent plate. The edge effects are not present in the system because the plate is unlimited along $Z$ axis, and the full problem reduces to a 2-D problem of beam deformation of the Winkler (elastic) foundation. We can investigate the loss of stability of the beam with thickness $h$ and width $b$ clutched between elastic thick foundations under the effect of facial load $P(x,t)$ and additional 'noise of fracture'.

In the ideal case, when building a microscopic description this noise of fracture should be connected with the microscopic parameters of the fracture. We shall not consider the physical nature of transverse perturbations and assume that the frequency of the given perturbation is the generation frequency of transverse oscillations caused by breaking of a fracture cell, $\Omega = 1/t_{char}$. In the investigation, we should differentiate between two stages of fracture. The first one is the stage of the elementary cell fracture with characteristic time $t_{char}$, and the second one is the fracture propagation between elementary cells with characteristic time $t* \neq t_{char}$. According to definition [363], the additional perturbation is a shock one. The principal difference of the considered processes is in the fact that the shock acts not along the beam axis and the system loses its stability not as a result of a shock load but as a result of a quasi-static load in the condition of parametrical perturbation. The inhomogeneity of this model is represented by additional terms (change of parameters of noise of fracture).

In this case we should observe energy transfer of the energy of longitudinal compression into the energy of lateral oscillations [360, 363]. The proposed model of parametric excitation of a previously non-bent (perfect) composite is essentially different from the well-known model of fibering of a non-perfect (previously supplied with a defect) composite [129]. In our model, no limitation exists in the principle related to the minimal size of the initial imperfection, because the energy transfer is made in the resonance mode.

The system of equations, accounting for shift, rotation inertia, and the influence of longitudinal oscillations on the lateral beam motion, has the form [361, 363]:

$$kFG(w_x - \psi)_x + EF\left[u_x\left(w_x + w_x^0\right)\right]_x + p(x,t) = \rho F w_{tt}; \tag{5.2.1}$$

$$EI\psi_{xx} + kFG\,(w_x - \psi) = \rho I \psi_{tt}; \tag{5.2.2}$$

$$EFu_{xx} = \rho F u_{tt}, \tag{5.2.3}$$

where $u(x,t)$, $w(x,t)$ are longitudinal and lateral displacements, respectively, $\psi$ is the angle of tangent to the curve of bend, $x, t$ are the longitudinal coordinate and time, $E, G$ are elasticity and shear moduli, $F, I$ are area and moment of inertia, $k$ is the coefficient of the shape of section, $\rho$ is the density of the material, $p(x,t)$ is the unit load which is local orthogonal to the beam axis, and $w^0$ is the initial bend.

The loading is realized by load $-P$, essentially exceeding the Euler critical load, from the free undisturbed by $t = 0$ end of the beam.

Thus, the initial and boundary problem can be stated in the form of

$$w = 0; \quad \psi_x = 0 \text{ by } x = 0,\ l_0,\ t \leq 0;$$

$$w(x, 0) = 0; \quad w_t(x, 0) = 0; \quad \psi(x, 0) = 0; \quad \psi_t(x, 0) = 0;$$

$$EFu_z(0, t) = -P, \quad u(l_0, t) = 0 \quad \text{or} \quad u_x(l_0, t) = 0;$$

$$u(x, 0) = 0; \quad u_t(x, 0) = 0.$$

The moving lateral load in the system in each given point is determined by a lateral wave, initiated on the free end of the beam. The physical source of this excitation is the periodical break of interatomic or interblock bonds. Since the frequency $\Omega$ of such oscillations is very high, despite the small excitation amplitude, high stress gradients are observed. In this case, the dynamical investigation looks justified. Since in real systems we always have excitation decay caused by dissipation, we assume that in the coordinate frame that is moving together with the crack tip, we can represent the load in the form of:

$$p(x, t) \equiv p(x) = A\exp(-\lambda x)\sin(\Omega x), \tag{5.2.4}$$

where $A$ is the normalizing constant and $\lambda$ is the logarithmic decrement of energy dissipation by the bend of an elastic–plastic beam.

In this statement of the problem, the system is considered as a 1-D system with slowly changing dimensions, in which oscillations are parametrically excited [360]. In this sort of system, resonance phenomena are possible. In the case of inter-grain boundary fracture or stripping, the excitation can localize near the interface, producing additional energy gradients and adding to the destruction caused by fracture [340].

### *5.2.2 Equation of beam bending*

The set of equations (5.2.1)–(5.2.3) can be written in the linearized form in the usual way [166]. In the result, we obtain the equations of the bend for a beam on an elastic foundation in the form of [361]:

$$TJw_{,xxxx} + cw + Pw_{,xx} + \rho Fw_{,tt} = p(x, t) - Pw_{1,xx}, \tag{5.2.5}$$

where $TJ$ is the beam rigidity, $T = EG/(E + G)$ is the reduced module (von Karman coefficient), where $E$ is the material Young modulus and $G$ is the shear modulus,

$$J = \frac{bh^3}{12(1 - \mu_0^2)}$$

is the moment of inertia of the cross-section with account of the restricted spread of the beam, $c$ is the coefficient of elastic foundation [361, 363]

$$c = \frac{E_0}{1-\mu_0^2}\frac{b}{3},$$

where $\mu$ is the Poisson ratio,

$$E_0 = \frac{E}{1-\mu^2}, \qquad \mu_{i0} = \frac{\mu_i}{\mu_i - 1}$$

where $\mu_i$ is the Poisson ratio of the foundation, indices $i$ refer to the bottom and top plate, respectively, $\rho$ is the density of the material, $F$ is the cross-section area, and $w_1$ is the initial imperfection of the axes. In as much we investigate the forced oscillations under applied transversal force, we assume further $w_1 = 0$ for simplification.

If we investigate an exfoliation crack of laminated composites, the real influence zone includes the areas, located in both materials, and the beam is two-layered. In the microscopic scale, the same mechanism is realized by the grain boundary cracking. Since we consider a beam clutched between two elastic foundations, equation (5.2.5) with account of (5.2.4) can be modified as follows:

$$\begin{aligned} &TJw_{,xxxx} + (c_1 + c_2)\exp(-\lambda x - Vt)\sin(\Omega x - Vt)\, w + Pw_{,xx} \\ &\quad = (\rho_1 F_1 + \rho_2 F_2)\, w_{,tt} = 0, \end{aligned} \tag{5.2.6}$$

where $c_1$, $c_2$ are the foundation coefficients of the bottom and top elastic foundations, respectively, $\rho_i$, $F_i$ are the density and the cross-section area of the first and second materials of the beam, respectively, and $V$ is the velocity of crack propagation (crack growth). In the quasi-static case $V = 0$. Generally, in the derivation of (5.2.6), we assume that the bend is small [363]. In the case of high gradients of beam deformation, we can take into account the nonlinear terms in the series of expansion of the bend [226], and (5.2.6) is only the first approximation.

### *5.2.3 Solution analysis*

We shall seek solution of equation (5.2.6) in the following form:

$$w(x,t) = a(x)\exp(-i(kx - \phi t)) \tag{5.2.7}$$

where $a(x)$ is the amplitude, generally complex, $k$ is the wave number, and $\phi$ is the frequency. Substituting (5.2.7) into (5.2.6) and separating the real and imaginary parts, we obtain the set of equations:

$$TJ\frac{d^3}{dx^3}c(x) - \left(\frac{1}{2}P - TJk^2\right)\frac{d}{dx}c(x) = 0 \tag{5.2.8}$$

$$TJ\frac{d^4}{dx^4}b(x) - Q\frac{d^2}{dx^2}b(x) + b(x)\left[R + \exp(Zx)\right] = 0, \tag{5.2.9}$$

where $a(x)$, $b(x)$ are imaginary and real parts of the amplitude, respectively,

$$Z = (-\lambda + \Omega - 2Vt), \quad Q = P - 6TJk^2,$$

$$R = -Pk^2 + TJk^4 - (\rho_1 F_1 + \rho_2 F_2)\phi^2.$$

Strictly speaking, we need an additional condition for dispersion relations but in the first approximation it is possible to study the medium without dispersion [360]. The formal solution of the first equation (5.2.8) has the form:

$$c = C_1 + C_2 \exp\left(\frac{\sqrt{-P + 2TJk^2}x}{\sqrt{2TJ}}\right) + C_3 \exp\left(-\frac{\sqrt{-P + 2TJk^2}x}{\sqrt{2TJ}}\right) \tag{5.2.10}$$

where $C_1$, $C_2$, $C_3$ are integration constants defined from the initial conditions. The general behavior of solution (5.2.10) depends on the sign under the root and we can obtain both periodical oscillations and the exponential increase. The formal solution of the second equation of the system can be found with the help of the Maple package in the form:

$$\begin{aligned} b(x) = {} & C_4 e^{-\frac{\delta x}{2Z}} {}_0F_3\left([], \{1\}, -\frac{e^{(Zx)}}{TJZ^4}\right) + C_5 e^{-\frac{\delta x}{2Z}} {}_0F_3\left([], \{2\}, -\frac{e^{(Zx)}}{TJZ^4}\right) \\ & + C_6 e^{-\frac{\delta x}{2Z}} {}_0F_3\left([], \{3\}, -\frac{e^{(Zx)}}{TJZ^4}\right) + C_7 e^{-\frac{\delta x}{2Z}} {}_0F_3\left([], \{4\}, -\frac{e^{(Zx)}}{TJZ^4}\right) \end{aligned} \tag{5.2.11}$$

where $C_4$, $C_5$, $C_6$ are the integration constants, defined from the initial conditions, $d = -Q^2 + 4TJR$, $\beta_1 = Z^2 Q$,

$$\beta_2 = \sqrt{-\alpha^4 d}, \quad \delta_1 = -2\beta_1 - 2\beta_2, \quad \delta = \sqrt{\frac{\delta_1}{TJ}}, \quad \delta_2 = \sqrt{\frac{(4\beta_2 + \delta_1)}{TJ}},$$

and expressions ${}_0F_3(0; \{i\}; -z)$ are the generalized hypergeometric functions of the power (0,3) (this function is also known as Barnes's extended hypergeometric function).

In general, the hypergeometric function is given by

$$F(a, b, c; z) = \sum_{s=0}^{\infty} \frac{a(a+1)\ldots(a+s-1)b(b+1)\ldots(b+s-1)}{c(c+1)\ldots(c+s-1)} \frac{z^s}{s!}$$

by the additional limitation $c \neq 0$ or $c$ not equal to negative integer number. The hypergeometric function is convergent series by $|z| < 1$.

The generalized hypergeometric functions of the power (0,3) have the form [269]

$$F(a_0; c_1, c_2, c_3; z) = \sum_{s=0}^{\infty} \frac{(a_0)_s}{(c_1)_s (c_2)_s (c_3)_s} \frac{z^s}{s!}$$

where $a_i$ is the Pochhammer symbol which is defined for a positive integer $s$ and complex number $a$ as

$$a_s = a(a+1)\dots(a+s-1).$$

The parameters of the hypergeometric functions {1}, {2}, {3}, {4} are the functions of the material properties and condition of loading:

$$\{1\} = \left[-\frac{-Z^2+\delta}{Z^2}, -\frac{1}{2}\frac{-2\,Z^2+\delta+\delta_2}{Z^2}, -\frac{1}{2}\frac{-2\,Z^2+\delta-\delta_2}{Z^2}\right]$$

$$\{2\} = \left[\frac{Z^2+\delta}{Z^2}, \frac{1}{2}\frac{2\,Z^2+\delta-\delta_2}{Z^2}, \frac{1}{2}\frac{2\,Z^2+\delta+\delta_2}{Z^2}\right]$$

$$\{3\} = \left[\frac{Z^2+\delta_2}{Z^2}, -\frac{1}{2}\frac{-2\,Z^2+\delta-\delta_2}{Z^2}, \frac{1}{2}\frac{2\,Z^2+\delta+\delta_2}{Z^2}\right]$$

$$\{4\} = \left[\frac{Z^2-\delta_2}{Z^2}, -\frac{1}{2}\frac{-2\,Z^2+\delta+\delta_2}{Z^2}, \frac{1}{2}\frac{2\,Z^2+\delta-\delta_2}{Z^2}\right].$$

Additional conditions on parameters arise from the necessity to have physical solutions (for example, the Mandelstam radiation principle) and mathematical properties of generalized hypergeometric functions [269]. Since for the order of the hypergeometric function, the condition $0 < 3$ is satisfied, the function is an entire one and is converging at all $x$.

In the general case, we can use the asymptotic representation of the generalized hypergeometric function [380]:

$$_pF_q(z) = \sum_{n=0}^{\infty} \frac{f(n)}{\Gamma(n+1)} z^n \tag{5.2.12}$$

where $f(n)$ is the predetermined function.

Thus, the general solution of equation (5.2.11) is the sum of exponents with some coefficients. The behavior of the solution depends on functions $Z$, $R$ and in principle both periodical oscillations and the resonance regime with exponential increment are possible [225]. This behavior is consistent with the behavior of the beam at the dynamical columnar deflection [166]. For the rod it was also shown that the exponential growth of deflection is present with a superimposed fast sinusoidal component.

Since a physically unlimited growth of oscillations is impossible, the system should reach its bifurcation point, which switches the system over into a qualitatively different state. Such bifurcation is the start of fracture, after the onset of which the model ceases to be valid. From the viewpoint of the MHST, the system changes the hierarchical level, and operator $\rho^+$ is a crackon.

## 5.3 Crackon

### *5.3.1 Movement of dynamical system*

It is a well-known practice to represent the motion of a random dynamic system with the help of geodesic lines [9, 377]. *In this presentation, the equations of the trajectory of the given dynamic system in the space of states coincide with the equation of geodesics in a certain surface.* The general principle was originally applied to the crack theory [236]. The crack tip was presented as a quasi-particle called a crackon. According to the ray approximation, a crackon, like any other free particle, propagates along the geodesics [237, 284]. Many works are known, in which the crackon mass is assessed or the kinetic crack energy is considered [13, 98, 96, 198]. Only recently, have data become available [323, 324] on experimental detection of crack inertia in glass by the method of localization of wave fronts. However, these experimental data contradict theoretical models of a dynamic crack [1].

According to general formalism, let us investigate a crackon with the Hamiltonian $H(q_1, \ldots, q_n, p_1, \ldots, p_n) = h$, where $q_i$ are the generalized coordinates, $p_i$ are the generalized impulses and $h$ is the initial constant. Introduce the function $\Upsilon$ such that $p_1 = \Upsilon(q_1, \ldots, q_n, p_2, \ldots, p_n, h)$ if in the present region $\partial H/\partial p_1 \neq 0$. The Legendre function transformation leads to the canonical conjugate function [9, 377]:

$$P = P(q_2, \ldots, q_n, q'_2, \ldots, q'_n, h) = \sum_{j=1}^{n} q'_j p_j - \Upsilon$$

where $q'_j = dq_j/dp_j$. For this function, $\Pi = 2\sqrt{h - H}$ results, where $\Pi$ is the potential energy of the particle and $H = T + \Pi$. Let us introduce the distance between two neighboring points as

$$ds^2 = (h - H)\sum_{i,k=1}^{n} a_{ik}dq_i dq_k. \tag{5.3.1}$$

From the Jacobi principle, it is found that $\delta s = 0$. Hence, the problem of quasi-particle trajectory determination is reduced to the geometrical problem of geodesic line building in space with the metric given in equation (5.3.1) [9, 377]. From a more general point of view, this optomechanical analogy is associated with the Jacobi theorem for the generating function.

**Remark 5.3.1.** With account of the concepts of the thermofluctuational strength theory (§3.1.2), a crackon is nothing else but a dynamically stable macroscopic complex of dilatons. This complex appears as a result of a cooperative interaction of individual dilatons. Obviously, from the mathematical viewpoint, we can consider a crackon as a solution of the equation of dilaton dynamics.

### 5.3.2 Mass of crackon

To define the distance, corresponding to (5.3.1), we have to know the kinetic and potential crack energy, which requires to introduce the mass of the crackon. A certain problem lies in the fact that according to the physical sense, the crack mass, similar to the mass of a point defect or a vacancy, is the negative effective mass [143]. At the same time, the negative mass leads to specific characteristics of the crackon. Sometimes, as a crack mass one considers the full mass of the volume of the material that moves at the crack growth or spread of its edges [96]. However, this approach seems not quite correct—in this case, for motion of any macroscopic body in a medium, the mass of the body should be not only the mass of the body itself, but also the mass of the medium disturbed at motion. This definition of mass is usually used in electrodynamics (mass of an electron is the mass at rest plus effects related to interaction with the medium) [178]. It was already noted that from the microscopic viewpoint, fracture is a process in the local area near the crack tip, where the state of the material changes from a non-deformed lattice to breakage of bonds. In this local zone, the dilaton density changes from zero to the critical one. Traditionally, it is assumed that these processes are realized in the pre-fracture zone; therefore, the crackon mass is connected with the characteristics of this area[1]. In this case, the motion of the volume of the material is the effect of post-action and not connected with the crack mass. Let us consider first the 'mass at rest' of the crackon.

*Mass at rest*

We take for simplicity, within the framework of the linear fracture mechanics, that the pre-fracture zone is deformed elastically. In a more general case, the full deformation can be represented as a sum of its elastic and plastic components. With account of additivity of deformations, the macroscopic elastic deformation of the pre-fracture zone can be represented as a sum of elementary elastic deformation of the lattice, while the total plastic deformation in the pre-fracture zone is caused by a combination of elementary breaks of interatomic bonds (appearance of point defects, dislocations and disclinations). Then, we can think that:

$$m_{cr} = m_{el} + m_{pl}, \tag{5.3.2}$$

where $m_{cr}$ is the mass of crackon, $m_{el}$ is the elastic mass, and $m_{pl}$ is the plastic mass. It is known that for elastic deforming:

$$\frac{\rho}{\rho_0} = \sqrt{g}, \tag{5.3.3}$$

where $\rho$, $\rho_0$ is the density of the material before and after deforming, $g = |g_{ij}| = 1 + 2I_1$ is the covariant metric tensor, $I_1 = \varepsilon_{11} + \varepsilon_{22} + \varepsilon_{33}$, and $\varepsilon_{ij}$ is the tensor of elastic

[1] In this case we have a bare mass [143].

deformations. We assume that Hooke's law is observed in the elastic area, and we have a possibility to restore the displacements by the stress tensor. Then, we have:

$$m_{el} = \int_{V_1} (\rho - \rho_0)\, dV = \rho_0 \int \left(\sqrt{1 + 2I_1} - 1\right) dV, \tag{5.3.4}$$

$$m_{pl} = \int_{V_2} m_* n dV, \tag{5.3.5}$$

where $V_1$, $V_2$ are the volumes of the zone of the elastic and plastic deforming near the crack tip, respectively, $m_*$ is the mass of unit dislocation, and $n$ is the density of dislocations.

To integrate in (5.3.4), (5.3.5) we need to know the shape of the pre-fracture zone and the mass of unit dislocation. To integrate by volume, we take the crack to be an ellipsoidal inclusion, and the pre-fracture zone of total length $h$ as a superposition of a circular cone with height of $h_1$ with the base on one of the diameters of the ellipse (elastic pre-fracture, zone 1) and a sphere-shaped area of diameter $a$ near the crack tip (plastic deformation, zone 2), Fig. 5.1.

The estimates show that approximation of the real zone of plastic deformation by a sphere-shaped area brings in an error of the order of 25%, which is quite satisfactory for a coarse model. For the mass of unit dislocation in the plastic area we can use the well-known expression [126]

$$m_* = \frac{W_0}{Lc_t^2}, \tag{5.3.6}$$

$$W_0 = L\frac{\mu b}{4\pi} \ln \frac{\alpha R}{b},$$

where $W_0$ is the energy of the stress field of a stationary rectilinear dislocation, $\mu$ is the shear modulus, $b$ is the Burgers vector, $R$ is half of the average distance between dislocations, $\alpha \cong 4$ is the physically justified parameter [126], $L$ is the length of the dislocation line, and $c_t^2$ is the velocity of lateral waves. With account of the fact that $2R^3 = n$, where $n$ is the dislocation density, we have:

$$m_{pl} = m_*\, n V_2 = \frac{a^3 n \mu b^2}{3c_t^2} \ln \frac{\alpha}{2bn^{1/3}}. \tag{5.3.7}$$

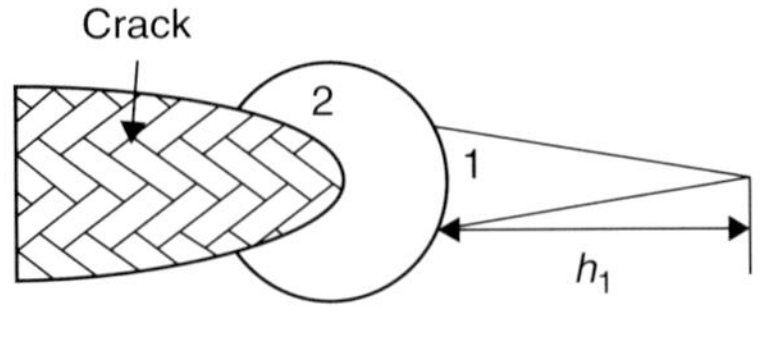


Fig. 5.1. Pre-fracture zone.

The density of dislocations emitted by the crack can be defined by the experimental data or calculated by well-known formulas. With account of the form of zone 1, we have for the elastic mass of the crack the expression:

$$m_{el} = \frac{1}{3}\rho_0 \pi a^2 h_1. \tag{5.3.8}$$

*Catastrophic fracture*

One of the practically important fracture cases is catastrophic fracture, when under unchanged externally applied stress, the crack propagates to considerable distances without halting. In the terms of quasi-particle propagation, it means that a quasi-particle moves in the medium, feeling no resistance. Since the interaction with the medium is connected with the mass of the crackon, it is possible only in the case when $m_{cr} = 0$. Since stresses in the material depend on the crack velocity [1, 96, 98], we should have made in (5.3.4) integration with account of $\varepsilon = \varepsilon(v)$, which leads to an essential complication of expressions. Therefore, taking for simplicity that deformation is permanent, we can write down:

$$\begin{aligned} m_{cr} &= \frac{a^3 n\mu b^2}{3c_t^2}\ln\frac{\alpha}{2bn^{1/3}} + \left(\rho_0 \int \left(\sqrt{1+2I_1}\right) dV - \frac{\pi}{3}\rho_0 a^2 h\right) \\ &= \frac{a^3 n\mu b^2}{3c_t^2}\ln\frac{\alpha}{2bn^{1/3}} + \left(\rho_0 \left(\sqrt{1+2I_1}\right) \int dV - \frac{\pi}{3}\rho_0 a^2 h\right) \\ &= \frac{a^3 n\mu b^2}{3c_t^2}\ln\frac{\alpha}{2bn^{1/3}} + V_1\rho_0\left(\sqrt{1+2I_1} - 1\right) = 0. \end{aligned} \tag{5.3.9}$$

For motion, the considered physical volumes and energy of dislocations vary in conformity with Lorentz transformations. The condition of equality to zero of the dynamic mass takes the form:

$$m_{cr} = h_2 a^2 n \frac{\mu b^2 \ln \dfrac{\alpha}{2bn^{1/3}}}{12c_t^2\sqrt{1 - \dfrac{v^2}{c_t^2}}} + \frac{\pi}{3}a^2 h_1 \rho_0 \left(\sqrt{1+2I_1}\right) = 0. \tag{5.3.10}$$

After elementary transformations, we have for the velocity of the catastrophic crack:

$$v = \sqrt{c_t^2\left[1 - \left(\frac{D}{12c_t^2 B}^2\right)\right]}, \tag{5.3.11}$$

$$D = h_2 n\mu b^2 \ln\frac{\alpha}{2bn^{1/3}}, \quad B = \frac{\pi}{3}h_1\rho_0\left(\sqrt{1+2I_1}\right).$$

It follows from (5.3.11) that the crack velocity of catastrophic fracture is *a fortiori* smaller than the velocity of lateral waves, since in real materials $n \neq 0$, respectively,

$D \neq 0$. The limit case $\alpha = bn^{1/3}$ (in this case $D = 0$) is of no real interest. In ideally elastic materials, dislocations are absent, $m_{pl} = 0$, and the crack velocity is equal to the velocity of lateral waves. On the other hand, in the case of $D = 12c_t^2 B$, the velocity of the catastrophic crack is equal to zero, the crack is arrested. Thus, by technologically regulating the density of dislocations, we can achieve the absence of catastrophic fracture. It is possible that these conditions will correspond to the mode of accumulation of damage and stochastically distributed microcracks.

## 5.4 Crackon in medium

In fracture mechanics, the pre-fractured zone is described by various methods [101, 284]. In principle, the formation mechanism of this zone is clear—it is weakening and fracture of interatomic bonds (excitation of dilaton states). However, it seems that until now no attention has been paid to the mechanisms of bond weakening. Let us show that irradiation of energy by a propagating crack can be a mechanism of this sort.

### *5.4.1 Acoustic approximation*

It is well known that propagation of a dynamic crack can be described by means of wave equations for the wave potential $\varphi$ [101]. In a Lagrangian coordinate system $(x, y, z, t)$, the wave equation has the form:

$$\Delta\varphi - \frac{1}{c^2}\frac{\partial^2\varphi}{\partial t^2} = 0, \tag{5.4.1}$$

where potential $\varphi = \psi \exp(i\omega t)$ is a harmonic time function.

Depending on the processes and cracks under consideration—dynamic or quasi-static—equation (5.4.1) describes different waves existing in the system. For quasi-static fracture, these may be, for example, Danilov–Zuev deformation waves, while for dynamic fracture these may be longitudinal and lateral elastic. Depending on the problem considered, the respective equations should be studied.

Let the motion trajectory equation of a crackon have the form:

$$x = X(t), \quad y = Y(t), \quad z = Z(t). \tag{5.4.2}$$

In this case, we can introduce the Euler coordinate system $(\xi, \eta, \zeta, \tau)$, which is connected with the crackon:

$$\xi = x - X(t), \quad \eta = y - Y(t), \quad \zeta = z - Z(t). \tag{5.4.3}$$

Then, we can consider a flow-around of the crackon by a flow at the velocity of:

$$V_{0x} = -\frac{dX}{dt} = -v_x, \quad V_{0y} = -\frac{dY}{dt} = -v_y, \quad V_{0z} = -\frac{dZ}{dt} = -v_x. \tag{5.4.4}$$

Let us transform equation (5.4.2) into the coordinate system $(\xi, \eta, \zeta, \tau)$ [30]. Note that:

$$\varphi(x, y, z, t) = \varphi\left(\xi + X(t), \eta + Y(t), \zeta + Z(t), \tau\right),$$

such that:

$$\frac{\partial \varphi}{\partial x} = \frac{\partial \varphi}{\partial \xi}, \quad \frac{\partial \varphi}{\partial y} = \frac{\partial \varphi}{\partial \eta}, \quad \frac{\partial \varphi}{\partial z} = \frac{\partial \varphi}{\partial \zeta},$$

consequently:

$$\nabla_{xyz}\varphi = \nabla_{\xi\eta\zeta}\varphi = \nabla\varphi;$$

$$\frac{\partial \varphi}{\partial t} = \frac{\partial \varphi}{\partial \tau} - (\mathbf{v}, \nabla)\varphi;$$

$$\frac{\partial^2 \varphi}{\partial t^2} = \frac{\partial^2 \varphi}{\partial \tau^2} - 2(\mathbf{v}, \nabla)\frac{\partial \varphi}{\partial \tau} + (\mathbf{v}, \nabla)(\mathbf{v}, \nabla)\varphi - \left(\frac{d\mathbf{v}}{dt}, \nabla\right)\varphi.$$

With account of these expressions, wave equation (5.4.2) acquires a familiar appearance:

$$\Delta\varphi - \frac{1}{c^2}\frac{\partial^2 \varphi}{\partial \tau^2} + \frac{2}{c^2}(\mathbf{v}, \nabla)\frac{\partial \varphi}{\partial \tau} - \frac{1}{c^2}(\mathbf{v}, \nabla)(\mathbf{v}, \nabla)\varphi + \frac{1}{c^2}\left(\frac{d\mathbf{v}}{dt}, \nabla\right)\varphi = 0. \qquad (5.4.5)$$

Let us choose a coordinate system in such a way as to have axis $x$ directed locally along the crack motion. In this case expression (5.4.5) acquires the form:

$$\Delta\varphi - \frac{1}{c^2}\frac{\partial^2 \varphi}{\partial \tau^2} + \frac{2v}{c^2}\frac{\partial^2 \varphi}{\partial \tau \partial \xi} - \frac{v^2}{c^2}\frac{\partial^2 \varphi}{\partial \xi^2} = 0. \qquad (5.4.6)$$

If we introduce a compressed coordinate system:

$$\xi^* = \frac{x - vt}{\sqrt{1 - \beta^2}}, \quad \eta = y, \quad \zeta = z, \quad \tau = t,$$

where $\beta = v/c$, we obtain (5.4.6) in the form [30]:

$$\Delta\varphi - \frac{1}{c^2}\frac{\partial^2 \varphi}{\partial \tau^2} - \frac{2\beta}{\sqrt{1 - \beta^2}}\frac{1}{c}\frac{\partial^2 \varphi}{\partial \tau \partial \xi^*} = 0. \qquad (5.4.7)$$

Equation (5.4.7) permits a double interpretation. It can be interpreted as the equation of flow-around by a flow of a motionless particle, and as an equation for propagation of a perturbation in the medium that moves at the velocity of $v$.

### 5.4.2 Crack-Generated energy flow

In the general case, the issue of energy irradiation by the crack was studied by Freund [101] on the basis of the work by Kostrov and Nikitin [168]. Let us consider a possibility of energy localization in the pre-fracture zone.

Let us suppose that a source of wave fields (crackon) is moving at a random velocity along a random trajectory

$$x = X(t), \quad y = Y(t), \quad z = Z(t).$$

We are not interested in the real nature of the source, but in the behavior of the energy flows only. Let us suppose that an oscillation is created by a volume force concentrated in the localization place of the crackon. Then, in wave equation (5.4.1), a source appears:

$$\Delta\varphi - \frac{1}{c^2}\frac{\partial^2\varphi}{\partial t^2} = -4\pi\, Q(x, y, z, t), \tag{5.4.8}$$

where $Q$ is the force of the volume source.

In the case of dynamic fracture, the growing crack is a source of such perturbation. According to classic solutions of the elasticity theory, in the crack tip stresses have a peculiarity. This peculiarity is a point source $Q$, localized in the tip. Since fracture cells have identical dimensions in a homogeneous material (in the simplest case it is the period of the crystal lattice), and an actual crack propagation is a succession of processes, bond break–accumulation of energy–new break, the point source in the first approximation has a periodic character. Then, force $Q$ can be represented as:

$$Q(x, y, z, t) = F(t)\delta(\xi)\delta(\eta)\delta(\zeta), \tag{5.4.9}$$

where $\delta$ is the Dirac delta-function.

Value $F(t)$ gives a force–time dependence in the system connected with the source, and delta–functions ensure the point character.

The solution of equation (5.4.8) in the Lagrangian coordinate system with account of representation (5.4.9) has the form of the Lienar–Wichert potentials:

$$\varphi(x, y, z, t) = \frac{F\,(t - R/c)}{R^*\sqrt{1-\beta^2}}, \tag{5.4.10}$$

where $R$ is the only positive root of the equation:

$$f(R) = [x - X\,(t - R/c)]^2 + [y - Y\,(t - R/c)]^2 + [z - Z\,(t - R/c)]^2 - R^2 = 0. \tag{5.4.11}$$

Value $R$ has a clear physical sense—this is a distance, perturbation from which in the current moment reaches the given point (effective distance). Value $R^*$ has the sense of the distance in the Euler contracted coordinate system from the observation point $P$ to the point of wave surface:

$$R^* = \sqrt{\xi^i\xi_i}, \tag{5.4.12}$$

and $\xi^i$ is transformed to new coordinates in accordance with the Lorentz transformations [178, 294].

The obtained solution (5.4.11) represents the field of zero source (i.e., of such source which has decomposition by spherical harmonics of zero order only). By combining such forces with the respective duly located phases, we can represent any wave field. With account of the expression for the potential, the wave amplitude is expressed as:

$$\psi(x, y, z, t) = \frac{F(t - R/c)}{R^* \sqrt{1-\beta^2}} e^{-i\omega t}. \tag{5.4.13}$$

Let us consider the energy carried over by the wave. In the general case, the power, averaged by period $T_0$, irradiated into the media or arriving through surface $\Omega$ is [30, 94]:

$$\langle E \rangle = -\frac{\omega}{2} \iint_\Omega \mathbf{uq} dx dy, \tag{5.4.14}$$

where $\mathbf{u}$ is the vector of medium displacement and $Q = \mathbf{q}\exp(-i\omega t)$ is the harmonic load.

For greater clearness, expression (5.4.14) is written down in the normalized dimensionless form. For fast processes we can neglect the energy delivered from the outside of surface $\Omega$ at the account of kinetic losses (for example, diffusion or heat exchange). In case of a non-super strong dynamic loading, generation of secondary waves from the volume also looks unlikely. Then, the maximum amount of energy that can be generated within the volume makes:

$$\mathcal{E} = -\frac{\omega}{2} \int_0^T \langle E \rangle dt, \tag{5.4.15}$$

where $T = l/v$ is the travel time of the crack along the trajectory, $l$ is the distance from the start of the crack motion to the given point (the trajectory length). Value $\mathcal{E}$ is the full energy, carried out of the volume with the waves. With account of the general balance condition (4.1.32), we can elementarily obtain the condition of crack halting

$$\int R da > \mathcal{E}.$$

Strictly speaking, in expression (5.4.15) we need to consider the sum of all the wave types that exist in the body. If we consider the wave propagation along the interface (border of the pre-fracture zone), equation (5.4.1) should be written down for the Rayleigh surface waves. In this case, the velocity of Rayleigh waves $v_R$ should be considered as the characteristic velocity of excitation transmission. In the first approximation, for small distances, the track trajectory may be taken to be rectilinear, and axis $X$ can be matched with the crack growth direction. For the volume deformation, in the simplest case, we can assume that the media displacements coincide with the amplitude of the incoming wave, and the source frequency coincides with the wave frequency. Then, for amplitude (5.4.13) we have:

$$\psi(x, y, z, t) = \frac{F(t - R/c)}{R^* \sqrt{1-\beta^2}} e^{-i\omega t} e^{-i\omega t} = \frac{F(t - R/c)}{R^* \sqrt{1-\beta^2}} e^{-2i\omega t}. \tag{5.4.16}$$

If a medium has only Rayleigh waves, in the first approximation, term $(\mathbf{uq}) \approx 0$—displacements are normal to the crack propagation direction. The same follows from the principle of local symmetry of the crack theory. In the case of existence of several wave types in matter, the displacement function of this point takes a more complicated character. With account of the fact that the average value of the harmonic function during the whole time of propagation has the form of $\langle \sin^2 \omega \rangle = (1/2)N$, where $N = l\omega/v$, at transition to dimensional variable for (5.4.16), we have [30]:

$$\mathcal{E} = -\frac{4\rho l\omega^3}{T_0} \int_0^T \frac{\mathbf{q}^2 (t - R/c)}{(x - vt)^2 + y^2 (1 - \beta^2) + z^2 (1 - \beta^2)} \frac{1}{v} dt. \tag{5.4.17}$$

Since the source periodicity was accounted for in expression (5.4.16), the amplitude of the harmonic load can be taken as constant. In this case: $\mathbf{q} = \text{const} \neq \mathbf{q}(t)$. The crack growth rate is also dependent on external conditions, and provided they do not change it can be taken as constant. Then we have:

$$\mathcal{E} = -\frac{4\rho l\omega^3}{T_0 v} \mathbf{q}^2 \int_0^T \frac{dt}{(x - vt)^2 + y^2 (1 - \beta^2) + z^2 (1 - \beta^2)}. \tag{5.4.18}$$

Should the crackon move at the sound speed $\beta = 1$, equation (5.4.11) has two coinciding roots, since $df/dR = 0$, and function $f(R)$ has an extremum. From the physical viewpoint, the existence of two effective distances indicates an appearance of bifurcation. For the case $\beta = 1$, equation (5.4.18) is elementarily integrated and yields infinity:

$$\mathcal{E} = -\frac{4\rho l\omega^3}{T_0 c^2} \mathbf{q}^2 \left[ \frac{1}{x - cT} - \frac{1}{x} \right] = -\infty. \tag{5.4.19}$$

In this case, the surface of equal phase degenerates into an ellipse, degenerated along axes $Z, Y$ [218]. Equation (5.4.19) takes into account that the length of the crack $l$ coincides with the current coordinate $x$. The equation gives us only a scalar value of energy. If the direction of energy propagation is essential in our problem, we can use the formalism of Pointing vector (§5.7.1).

Thus, the existence of the limit crack velocity—the velocity of the Rayleigh waves—is connected with the development of the pre-fracture zone. In the case of an impact loading, when the loading rate exceeds the limit velocity, the energy that arrives into the pre-fracture zone is insufficient to excite the sufficient number of dilatons, and no break of atomic bonds take place. For this reason the high-velocity dynamical elasticity constants essentially exceed the quasi-static one. For the high-speed dynamical fracture, post-loading effects are essential, for example unloading by shock waves.

## 5.5 Fractal dimensionality

The fractal dimension is one of the basic characteristics connected with fractal objects. There are some definitions of fractal objects which are used for analysis of different characteristics of the objects. The problem of different definitions follows from the fact that we obtain different values of fractal dimension not only by using different measures but by using the same definition for global and local measures [197].[2]

**Experimental measure of fractal objects.** The most 'geometrical' method of measurement is the method of vertical sections (MVS). The MVS consists in investigating the relation between the length of the profile of the crack surface and the scale of the measurement. To enhance the statistical significance of the result obtained, the measurement must be conducted for several different orientations of the surface profile obtained by polishing the vertical section of the crack surface. It has been established that the dependence of the profile length of the measurement scale $\varepsilon$, obeys the asymptotic law,

$$L(\varepsilon) = L_0 \varepsilon^{-(d_x - d_T)}, \tag{5.5.1}$$

where $d_x$ is the first Rényi dimension according to definition and $L_0$ is a constant. Taking into account equation (D.5) for $d_x$ we obtain

$$d_x = -\lim_{\varepsilon \to 0} \frac{\ln M(\varepsilon)}{\ln \varepsilon}. \tag{5.5.2}$$

Here $M(\varepsilon)$ is the minimum number of $d$-dimensional cubes of face $\varepsilon$, required to cover all the structure elements for the multi fractal structures the dimension Rényi is congruent to coarse Hölder exponent $\alpha P$.

One of variants of the MVS is to measure the ratio, $R_L(\varepsilon) = L(\varepsilon)/L'$, where $L'$ is the length of the profile projection on the plane parallel to the crack surface:

$$R_L(\varepsilon) = C \varepsilon^{-(d_x - d_T)}. \tag{5.5.3}$$

## 5.6 Internal geometry of the media and fractal properties of the fracture

After publication of Bentoit Mandelbrot's articles about fractals [196], theory of fractal structures came to be actively applied to the theory of fracture. The different aspects of the fracture processes are under investigation—there are self-similarity of the cracks, scaling rule and other parameters. In [4] the area $\Omega$ with internal volume $V$ was investigated. *A priori* it is believed that the full density of the microcracks $\rho$ in area has a fractal character with hyperbolic distribution:

$$\rho(a_c) = \frac{N_r(a > a_c)}{V} = \lambda a_c^{-D}, \tag{5.6.1}$$

where $a_c$ is the critical size of the crack, $\lambda$ is constant, and $D$ is a usual fractal dimension.

The value $N_r(\cdot)$ is the number of microcracks, which satisfy the given condition. In our case it is the number of microcracks with size exceeding the critical length. In general $N_r(\cdot)$ is the discontinuous stepwise function because the number of the microcracks is natural. But the number of the cracks is large enough and behavior of the function $N_r(\cdot)$ is smooth by a small modification of the number of the microcracks.

[2] The definition of different fractal dimension is given in Appendix D.

The density of the microcrack changing by $\Delta a \ll a_c$ can be approximated as:

$$\Delta\rho(a_c) = \frac{N_r(a_c < a < a_c + \Delta a)}{V} = D\lambda a_c^{-(D+1)}\Delta a. \tag{5.6.2}$$

In this case both $\rho$ and $\Delta\rho$ have hyperbolic distribution. The index $D$ is limited $0 < D < 3$. If $D > 3$ we obtain the unlimited value of total volume of the microcracks (divergence of solution). It is impossible from a physical point of view. Values $D < 3$ have a physical meaning; so, if $D > 1$ total diameter (it means the sum of diameters of all microcrack) tends to infinity by $a_c \to 0$; if $D > 2$ the total area of surface of all microcrack tends to infinity by $a_c \to 0$. The number of microcracks $N_r > 0$ is the asymptotic to $a_c = 0$.

The appearance of divergence (an endless number of microcracks) should not cause any special concern. The situation is similar to the appearance of endless stresses in the proximity of the crack tip in the classic elasticity theory. This evidences an exit outside the applicability of this or that equation (for example, appearance of the plasticity zone in the proximity of the crack tip).[3]

### *5.6.1 Damage mechanics and stable growth of microcrack*

The growth of any crack is the complex process of different causes of interaction. These causes tend to the development and stopping of the crack growth simultaneously. In the simplest case according to the results of § 3.3 and taking uniform propagation of the crack in time we believe that applied stress supports the growth and relaxation hinders the development. In this case, for the stable growth of the microcrack we have the energetic balance condition (4.1.32).

The crack propagates unstably if $dG > dR$. Since $\sigma$, $a$ are the independent parameters equation (4.1.32) takes a form:

$$\frac{\partial G}{\partial \sigma} d\sigma + \frac{\partial G}{\partial a} da = \frac{dR}{da} da. \tag{5.6.3}$$

Selection of function $G$ determines the solution of equation (5.6.3). In the simplest case of brittle fracture, taking into account the principle of local symmetry, for the stress applied normal to crack propagation direction we have:

$$G = \frac{4\sigma^2 a}{\pi E}. \tag{5.6.4}$$

Substituting expression (5.6.4) in (5.6.3) we obtain:

$$\frac{da}{a} = \frac{8\sigma d\sigma}{\pi E}\left(\frac{dR}{da} - \frac{4\sigma^2}{\pi E}\right)^{-1}. \tag{5.6.5}$$

[3] Origin of divergence of solution (unlimited number of microcracks) is not a cause of problems for interpretation. The situation is analogous to the endless stresses near the crack tip in the classical elasticity. This situation is evidence of overcoming the limits of the adaptability equation. In classic theory of fracture stress singularity near the tip is reduced through plasticity flow.

It is evident for homogeneous materials by the small movement of the crack tip the value $dR/da$ is constant which is independent of crack length. In this case the value $da/a$ is constant also. In the strict sense $dR/da = \text{const}$ is an additional postulate which demands additional verification. For this reason talk about scaling is given in terms of sizes and velocities. As an example, in [313] was investigated the crack propagation in granite blocks. From experiments it follows that a crack propagates as a self-similar object but the roughness exponent depends on specimen size and position of measure. Generally it was shown that roughness increase by crack growth can be approximated by a two-parametric function of size of the investigated area and distance of the initial line. The self-similarity demands the corrections in equation (4.1.32). These corrections result from fractal nature of energy of fracture generation [301]. Usually chosen is:

$$W = kD^{(n-2)}, \tag{5.6.6}$$

where $k$ is the constant of proportionality which depends on material, $D$ is the fractal dimension on the observation scale, and $n$ is the exponent of fractal dimension.

In most cases, the fractal dimension of different objects is determined from experiments. For many materials (both fragile and plastic), a universal value of roughness factor was found experimentally: $\zeta = 0.8$ [246]. The roughness factor for spatial structures is connected with the fractal dimension as $D = 3 - \zeta$.

For minor spatial scales, it is considered that the universal nature is breached, and the factor takes the value of $\zeta = 0.3$–$0.5$ [246, 314]. In this case, a transition is possibly made from logarithmic correlations to self-similar ones.

The necessity to make corrections in equation (4.1.32) is also related to the fact that, according to fractographic studies, the crack propagation is not a continuous process, but rather a discrete, step-by-step one. Thus, for a fatigue crack [366], progress of the tip takes place after reaching, in the process of cycling in plastic zone, a definite damage level. In this case, equation (5.6.6), according to experimental data [108], takes the form:

$$W_f = kN_f^{\alpha} + c,$$

where $N_f$ is the number of cycles to fracture, and $\alpha, c$ are the constants of the material, which may be found from experiments.

### *5.6.2 Distribution of microcracks in a sample*

The Griffith criterion (4.1.25) for a brittle fracture can be written down in the form as:

$$\sigma_f = \sqrt{\left(\frac{\pi EW}{2a_c}\right)}, \tag{5.6.7}$$

where $E$ is the Young modulus.

In the first approximation, it is adopted that a brittle crack starts its motion at such a stress, normal to the crack propagation line, that $\sigma > \sigma_f$. The simplest method to account for the quasi-brittle mechanism of fracture and plasticity is to consider that the value of the fracture energy $W$ comprises both elastic and non-elastic parts.

It is necessary to note that for fracture, a propagation (growth) of one crack only is sufficient. Therefore, the strength limit of a damaged material is defined by one 'luckily' located crack. However, as shown in [387], the roughness factor for inhomogeneous anisotropic media (for example, mountain rocks) *does not depend* on the orientation of applied stress. This allows supposing that in combination, the process of microfracture is homogeneous. Then, we can state the directions of distribution of microcracks to be equally possible, and for small samplings from a great volume of sample $V$, containing $\rho$ critical cracks in the unit of volume, the probability of fracture is:

$$F = 1 - \exp(-\rho V). \tag{5.6.8}$$

Equation (5.6.8) represents a special case of the Poisson distribution, where $F$ is the probability to find at least one critical crack in the volume of sample. By substituting equation (5.6.2) into (5.6.8), with account of (5.6.7) we have:

$$F = 1 - \exp\left[-V\left(\frac{\sigma_f}{\Theta}\right)^{2D}\right], \tag{5.6.9}$$

where $\Theta = \sqrt{\left[\left(2\lambda^{\frac{1}{D}}\right)/(\pi E W)\right]}$.

Equation (5.6.8) has the form of the Weibull distribution for fracture stresses. This distribution, introduced earlier [374] in an empirical way for the description of these experiments on brittle fracture, now acquires a theoretical justification.

If the existence of the crack with length longer than $a_u$ is forbidden, then we arrive at a three-parametric Weibull distribution [4]:

$$F = 1 - \exp\left[-V\left[\left(\frac{\sigma_f}{\Theta}\right)^{2D} - \left(\frac{\sigma_u}{\Theta}\right)^{2D}\right]\right], \tag{5.6.10}$$

where $\sigma_u$ is the destructive stress corresponding to the maximum length of crack $a_u$.

Equations (5.6.10) and (5.6.9) place restrictions on the power index, at which the distribution can have a fractal character ($2D \leq 6$). Drawing of experimental curves $\sigma - N_r$ allows one to verify these restrictions.

### *5.6.3 Fractal characteristics of a macroscopic crack*

Fractal properties of fracture can reveal themselves in two ways. The properties of fractals may belong to directly measurable values, connected with the fracture (for example, the crack length and surface profile). This can also be a fractal distribution of characteristics, connected with fracture (for example, the fissuring tensor). The difference in the physical essence of these values is connected with the fact that the former are independent values that are not dependent on the observer (with certain reservations, characteristic for the whole physical problem of observed parameters and interaction of the object and observer). The latter are the *interpretation* of the object by the observer, and reflect both the properties of the object and the properties of the observer. Therefore, the interpretation of fractal characteristics of the second order in the terms of observed values requires certain care.

Most experimental works are dedicated to definition of fractal characteristics of the first order for the fracture surface. This can be fractography in metallurgy, measuring of roughness, or polar-recording charts. Then, in most cases, to describe the surface characteristics, they use random distributions, for example, trigonometric ones [81] or the Weierstrass function (5.1.9) [74]. The influence of the neighboring regions on the characteristics, measured in the local point, is accounted for through introduction of correlation functions.

Most theoretical conceptions for the problem of fractal crack measures arise from a simple assumption that the geometry of fracture surface is the 'heritage' of the geometry of crack front at fracture development or elementary geometrical considerations [185]. Earlier, it was shown that the front moves along equation (3.4.7). In this case, the component of the front line in the fracture plane is the plane (*in plane*) roughness, and the component perpendicular to the plane is out-the-plane roughness (*out plane*) [312]. Only out-the-plane roughness defines the roughness of the fracture surface. The issue of independent plane and out-the-plane roughness is currently a topic of debate.

*Multifractals and fracture*

As mentioned above, it has been shown experimentally that depending on the scale level and considered fracture model, the fractal dimension index takes different values. For heterogeneous media, at a quasi-static fracture, a flat crack has $D \approx 2.8$; for a 3-D case $D \approx 2.6$, for fracture by mode I of Plexiglas $D \approx 2.5$ [246, 314]. The dependencies of the fracture energy $W$ (or of the surface energy $\gamma$ [185]) on the fractal dimension also look different. For a brittle ceramic $E$ increases with the growth of $D$; for metals $E$ decreases with the growth of $D$ [378]. If we assume that fracture mechanisms of various subsystems and hierarchical levels are acting rather independently (the principle of independence, not superimposition, of effects holds), this behavior can be explained on the basis of the 2-D Cantor mapping (Fig. 5.2).

This 2-D mapping can be considered as a special case of multifractal at $n = 2$. Multifractal should not be perceived as a real geometrical object, for example a fracture surface. This looks more like the *distribution of measurements*, arising from studying objects. An example can be measurement of the fracture energy. The spectrum of multifractal dimensions dependent on the structural level is connected with the structural elements of the considered material. The highest dimension ($D_0$) characterizes a full fractal; the following dimension corresponds to the next highest substructure involved in fracture. As a rule, the value of fractal dimension for a higher level is bigger than for the lower one [320]. This corresponds to the supposition that in the hierarchical sequence of fracture processes, a direct cause of fracture in the upper level is the processes in the preceding level [107] (the correlation of structure and process or, in another system of variables, of the cause and measure). Thus, in consideration of damage accumulation as a percolation process [246], the second dimension is connected with the core of the fracture cluster. As one of possible pairs of characteristics of the fracture process, we can choose a spectrum of dimensions of the fractal, connected with the reasons (damage) $f$ and value $\alpha$. This

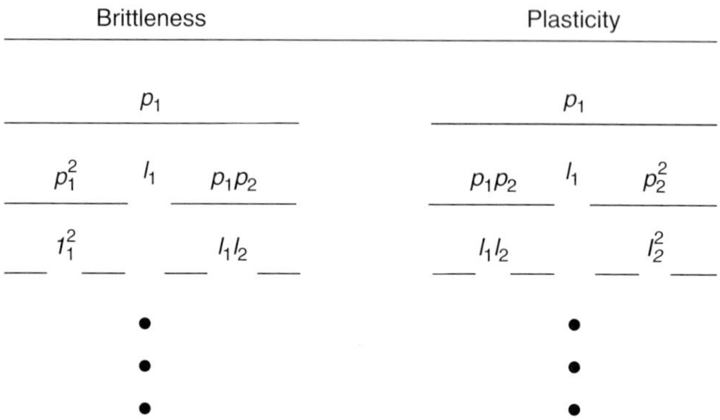


Fig. 5.2. Model of 2-D Cantor fractal.

value $\alpha$ is the dimension of the fracture process, $E \sim L^{\alpha}$, where $L$ is the characteristic length of fracture ($L^2$ is the area in case of 2-D fracture) and $E$ is the volume of fracture, connected with the fissuring tensor (1.2.1) or a free volume of the damaged body.

**Stability of multifractal structures.** For multifractal dimensions we can introduce restrictions connected with stability of fractal structure [136]. A multifractal in conditions of dynamic self-organization preserves its capability to self-organization until the connection between dimension of the manifold $D_q$ and index of fractal weight $d = \tau(q)$ is defined only by one variable. In this case:

$$D_q = \frac{\tau_q}{q-1}.$$

The loss by the multifractal of the capacity to topological self-organization can occur either upon reaching by the fractal of the percolation threshold, or of the elasticity threshold. In this case, for flat fractal manifolds at $q = q^*$ and $D_q = D_{q^*}^{min}$, we achieve the minimal, and at $q = -q^*$ and $D_q = D_{-q^*}^{max}$ the maximal density of the fractal medium [136].

Definition of threshold values $q = q^*$ and $D_q = D_{q^*}$ represents essential methodological difficulties. They are overcome by an accurate choice of the scale of observations and parameters of the discrete structure. For example, for a 2-D Cantor manifold, it has been shown that $q = q^* = 40$ [120].

In the general case, the multifractality characterizes the progress of simultaneously unfolded fracture processes in different scale levels, and also a successive change of fracture mechanisms [319, 320]. With account of the definition of §1.1, the multifractal medium is not a completely stratified one, and is a system of decision-making.

*Control parameters and hierarchy of fractal structures*

The hierarchical approach to fracture allows representing the defining equations of the fracture process as a system of equations in relation to the stress distribution law for the structural elements according to their rigidity and kinetic equation of damage accumulation [332]. The development of fracture process corresponds to the motion of the system along the hierarchical tree. The boundary conditions, connected with the fracture by different subsystems, are introduced by means of superposition of the conditions of the transition between the levels of the hierarchical system. This can be interpreted as information transfer in critical points, when a multifractal structure reaches its consistency threshold. In inconsistency points of the system, the demonstration by multifractals of information properties is caused by a strong excitation of the medium, accompanied by appearance of nonlinear waves and vortex at transition from the old structure that has lost its capacity to stable self-organization to a new one, in the other spatial-temporal level [215, 278].

For the 2-D Cantor mapping, the elements are associated with fracture processes of different character. Segments of length $l_1$ correspond to a brittle fracture, $l_2$ to the plastic one, $l_1 l_2$ to the fracture of mixed character. In the capacity of a physical value, 'cut-out' by the fractal, we can take the area of the fracture surface. The fracture surface itself can be chosen in conformity with any of the adopted strength theories. Each of the segments is connected with the measured values by means of probabilities, that is, $p_1$ is the probability of the fact that the fracture arises in $l_1$; $p_1 p_2$ is the probability of the fact that the fracture arises in $l_1 l_2$. Then, we have [120]:

$$\begin{aligned} &\ln(n/m)\ln(l_1/l_2) - \ln(n/m-1)/ln(l_1) \\ &\quad = q\left[\ln(p_1)\ln(l_2) - \ln(p_2)\ln(l_1)\right], \end{aligned} \tag{5.6.11}$$

$$\alpha = \frac{\ln(p_1) + (n/m-1)\ln(p_2)}{\ln(l_1) + (n/m-1)\ln(l_2)}, \tag{5.6.12}$$

$$f = \frac{(n/m-1)\ln(n/m-1) - (n/m)\ln(n/m)}{\ln l_1 + (n/m-1)\ln l_2}, \tag{5.6.13}$$

$$(q-1)D_q = q_\alpha(q) - f(q), \tag{5.6.14}$$

where $n, m$ is the number of segments of each type accordingly, $q$ is the index, and $D_q$ is the dimension, connecting values $f$ and $\alpha$. For the preset values $l_1$, $l_2$, $p_1$, $p_2$, $q$, we can use equations (5.6.11)–(5.6.13) to build the dependence of $f$ on $\alpha$ for fracture. For this purpose, we express, from (5.6.11) by successive approximations, $n/m$, and then, by substituting into (5.6.12), (5.6.13) we find other parameters. Equation (5.6.14) allows us to make some elementary analysis. At $q = 0$, $f = D_0$, the dimension of the process coincides with the dimension of a complete fractal fracture. At $q = \infty$, the dimension of the process $\alpha = \ln p_1/\ln l_1$, at $q = -\infty$, the dimension of the process $\alpha = \ln p_2/\ln l_2$.

For a system comprising 4096 elements, a number analysis of multifractal distribution was made in [332]. It has been shown that the fractal dimension of this cluster tends to the natural limit $D = 2$ at increased homogeneity of the material of the sample. In this case, the stochastic behavior of the fracture process decreases, and all the elements are destroyed simultaneously.

A fractal analysis of the hierarchical structure of deformed body in the scale of sizes $\delta \le l \le L$, where $\delta$ are the grain dimensions and $L$ is the size of the problem, the inhomogeneity scale of external fields, was performed in the work by R. V. Goldstein and A. B. Mosolov [107]. They considered a cascade transfer of fracture energy from the upper layers of the hierarchical structure to lower ones (from a full fractal to partial ones). The energy is transferred until it dissipates in the microlevel to form a new surface. The levels of the fracture structure are defined by the change of the fracture mechanism. If in scale $l$, the crack tip has moved by $\Delta l$, then in $l_\lambda \to \Delta l_\lambda = \Delta l / R^\lambda$. Here, $R$ is the scaling parameter which characterizes the change of scales at transition between the levels. If we considered the crack propagation as a one-dimensional process, the change of energy in the respective level can be written as:

$$\Delta U = W_0 \Delta l = W_1 G_1 \delta l_1 = \ldots = W_n G_n \Delta l_n = \ldots, \tag{5.6.15}$$

and the law of energy conservation at the non-dissipative transition between the levels has the form:

$$W_n G_n \delta l_n = W_{n+1} G_{n+1} \Delta l_{n+1}. \tag{5.6.16}$$

By assuming that $W_n = W(l_n)$, for the energy density we have a typical renormalization group equation:

$$W(l_n) = \frac{1}{R} \frac{N_{n+1}}{N_n} G\left(\frac{l_n}{R}\right). \tag{5.6.17}$$

In the presence of dissipation in the material (5.6.16), the law of energy conservation has the form:

$$q W_n G_n \delta l_n = W_{n+1} G_{n+1} \Delta l_{n+1}, \quad q < 1. \tag{5.6.18}$$

On analyzing the crack fractal dimension, one should take into account that the fractal dimension depends on the stage of crack propagation. At the initial development of the crack $D \approx 1.9$; in the process of stationary propagation $D \approx 1.3$–$1.5$. This is connected with realization of different fracture mechanisms in different modes. Another factor that needs accounting for is the spatial structure of the crack front. Since the shape of the fracture surface is caused by a range of factors (as indicated in §3.4.1), when calculating the general fractality factor, it is necessary to study fractality in different directions. The following value [320] can be taken as the average dimension:

$$D_\Sigma = \left(D_1^2 + D_2^2\right)^{1/2},$$

or, according to Mandelbrot:

$$D_\Sigma = \frac{\left(D_1^2 + D_2^2\right)}{2}.$$

Here, subscripts denote different directions, respectively.

The split of the full fractal structure into a number of hierarchical fractal substructures allows us to justify the independence of energy absorption by defects of different fractal dimensions and interaction of defects with each other [304]. Thus, let us consider a defect with fractal dimension $\nu_D$, which is in the macroscopically observed volume $V$. Let $F$

represent a respective set of defects and $F_\delta$ be a $\delta$-parallel body from $F$ [92, 304]. Volume $vol(D_\delta)$ characterizes the method of filling in with defects of macroscopic volume $V$. Value $\delta$, for which $F_\delta = V$, we can treat as a certain value, characterizing the macroscopic distribution of defects. For minor defects, contained in a huge (macroscopic) volume, value $\delta$ will be high; on increasing of the volume and the number of defects, $\delta$ drops to zero. In this case, the methods of measuring volume are different for small and big $\delta$. For small $\delta$, $F_\delta = \nu_D \delta^{3-D}$, where $D$ is the Minkowski dimension. In the general case, this dimension exceeds the Hausdorff dimension and coincides with the box-counting dimension. For high $\delta$, the fractal structure of the defect is practically not seen in the macroscopic scale of sizes and $vol(F_\delta) \propto \delta^3$.

Any defect in the continuum is disturbing a closed domain around itself. This perturbation can be characterized by the defect energy $\mathcal{E}$. If the perturbed areas are far from each other, they do not interact. At overlapping the perturbed areas, the energy of interaction is essential. Therefore, $\mathcal{E} = \mathcal{E}(D, \delta)$ and can be written down as [304][4]:

$$\mathcal{E} = a(D, \delta)\nu_D, \quad \nu_D \delta^{3-D} = V = \text{const}. \tag{5.6.19}$$

### 5.6.4 *Exclusion principles and fractal dimension of crack trajectory*

Since the macroscopic fracture surface is formed as a result of coagulation of and formation of the mainline crack, the problem of definition of fractal characteristics of fracture surface can be considered as a problem of definition of the fractal dimension of the crack trajectory. Most of the works dedicated to forecasting crack growth consider the *global* crack propagation and *global* trajectories, but this sort of averaged approach does not allow speaking about the fractal characteristics, which are local in their essence. Besides, at the global calculation, one should account for the influence of the sample dimensions on the characteristics of crack development [133, 320]. The *local* approach to definition of the crack trajectory is possible on the basis of supposition about a random change of the direction of crack growth, which is connected with determined parameters of the medium [80, 181]. However, in this approach the microscopic characteristics (for example, the stress field in the local zone in the proximity of the crack tip) are defined from the macroscopic approach. This sort of inconsistency cannot arouse any internal satisfaction. Besides, account of the effect of microscopic fields of stresses and deformations, caused by defects of the material, seems rather problematic with this approach.

The variation principle of the crack theory is more consistent and free from the above contradictions [284]. It allows accounting that the stress fields, used for calculation of trajectories in the 'local' approach of the crack theory, ideally should be defined from the microstructure of the fractured material. In this case, the fracture fractality can be accounted for.

[4] It would be interesting to somehow bind the geometrical characteristics of the defect with its energy, similar to § 3.2.3. This could make it possible to introduce connection of the geometry and fractal dimension.

Simulation of the character of stress distribution in the elements in the proximity of the crack tip by the method of multifractal analysis [332] allows obtaining dependence $f(\alpha)$. In this case, function $f(\alpha)$ has all the characteristic properties of the multifractal dimension $f(\alpha) \leq \alpha$ and has only one common point with bisector $f_4 = \alpha$. Analysis shows that an increase of the homogeneity of the material results in a decrease of the fractal properties of the distribution. This coincides with experimental results for the composite material polyhydroxyether–graphite [169, 260]. A critical structural defect in this composite is born in the matrix–filler interface, while the interphase adhesion is the correlation function of dimensions of the matrix and the filler and is maximal at the coincidence of the dimensions of the components of the composite material.

However, the physical reasons of the fractal character of the fracture in inhomogeneous materials are not quite clear. Nevertheless, the variation methods of the crack theory and the provision about the crack propagation along the geodesic (an extreme of energy liberation at fracture) give us a possibility to introduce the physical grounds of the fractal character of macroscopic fracture.

*Mathematical formalism of the surface description*

For simplicity, we consider a 2-D problem. Let us simulate a real fracture surface as a plane (a line in the 2-D case), filled in with triangular irregularities. We take triangles to be right triangles.

We consider the Euclidean space $\mathbb{E}$ with the standard scalar product. According to Fig. 5.3, we introduce the partitioning of the hypotenuse as a sum projections of the legs of the triangle

$$c = a_1 + b_1. \tag{5.6.20}$$

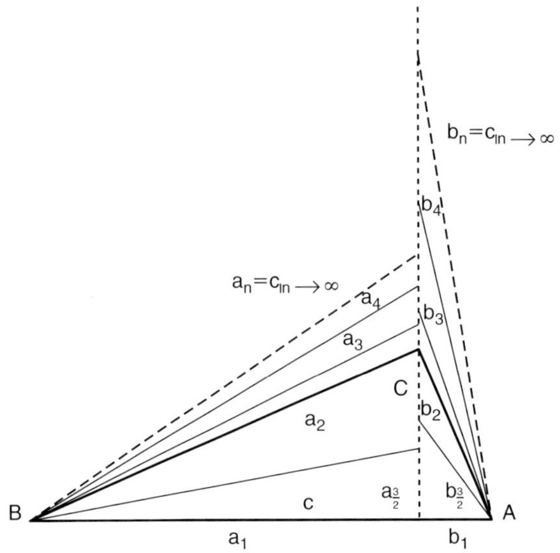


Fig. 5.3. Geometrical interpretation.

For brevity, we make calculation only for one projection, in most cases. From decomposition of (5.6.20), we have:

$$c^n = (a_1 + b_1)^n = (a_1 + b_1)\,c^{n-1} = a_1 c^{n-1} + b_1 c^{n-1} = a_n^n + b_n^n. \tag{5.6.21}$$

**Definition 5.6.1.** The $n$-th grade of projection is $a_n^n ::= a_1 c^{n-1}$ where $n \in \mathbb{N}$. Respectively, $a_n = a_1^{\frac{1}{n}} c^{\frac{n-1}{n}}$.

The expression (5.6.21) can be interpreted as a new formulation of $n$-dimensional Pythagorean theorem. By this definition, the formulations of the two-dimensional and $n$-dimensional theorems look identical. From the geometrical point of view, $a_n$ is the length of the face of an $n$-dimensional cube. The lower indices $n$ denote the dimension of the space. Thus, $a_n^n$ is the volume of the cube in $n$-dimensional space. From (5.6.21):

$$1 \equiv \left(\frac{a_n}{c}\right)^n + \left(\frac{b_n}{c}\right)^n. \tag{5.6.22}$$

It follows from (5.6.21):

$$\frac{a_1}{a_n} = \left(\frac{a_n}{c}\right)^{n-1}. \tag{5.6.23}$$

After elementary calculations we obtain:

$$\frac{a_n}{c} = \left(\frac{a_1}{a_n}\right)^{\frac{1}{n-1}} \text{ and } \left(\frac{a_n}{c}\right)^n = \left(\frac{a_1}{a_n}\right)^{\frac{n}{n-1}}. \tag{5.6.24}$$

In the case $n > 2 \in \mathbb{N}$ the set $a_i$ has a fractal nature [196] and forms the fractal as an unlimited set of $a_n$. We can make an analogous transformation with $b_n$. Now we can exaggerate the scale and stretch our space. Denote $p ::= n/(n-1)$. Taking into account (5.6.20) and (5.6.24), we obtain:

$$\left(\frac{a_1}{a_{\frac{p}{p-1}}}\right)^p + \left(\frac{b_1}{b_{\frac{p}{p-1}}}\right)^p = 1. \tag{5.6.25}$$

Taking into account that $p/(p-1) = n$ we rewrite (5.6.25) in the form of:

$$\left(\frac{a_1}{a_n}\right)^{\frac{n}{n-1}} + \left(\frac{b_1}{b_n}\right)^{\frac{n}{n-1}} \equiv 1. \tag{5.6.26}$$

From (5.6.22) and (5.6.26), the identity follows:

$$\left(\frac{a_n}{c}\right)^n + \left(\frac{b_n}{c}\right)^n = \left(\frac{a_1}{a_n}\right)^{\frac{n}{n-1}} + \left(\frac{b_1}{b_n}\right)^{\frac{n}{n-1}} \equiv 1. \tag{5.6.27}$$

Taking into account Definition 5.6.1, we can establish the dual length as follows.

**Definition 5.6.2.** $a^n_{\frac{n}{n-1}} = a_1^{n-1}c$ and thus $a_{\frac{n}{n-1}} = a_1^{\frac{n-1}{n}} c^{\frac{1}{n}}$.

This definition of dual length is in accordance with the main principles of duality:

$$a \cdot \tilde{a} = INV.$$

We examine the main principles which follow our definitions (5.6.1), (5.6.2):

$$a_n \cdot a_{\frac{n}{n-1}} ::= a_1^{\frac{1}{n}} \cdot c \cdot c^{-\frac{1}{n}} \cdot a_1^{1-\frac{1}{n}} \cdot c^{\frac{1}{n}} = a_1 c ::= a_2^2 \equiv INV. \tag{5.6.28}$$

Geometrically, this invariant is the area of a two-dimensional rectangle with legs $c$ and $a_1$. It follows from equations (5.6.28) and (5.6.21):

$$c^2 \equiv a_2^2 + b_2^2 \equiv a_n a_{\frac{n}{n-1}} + b_n b_{\frac{n}{n-1}}. \tag{5.6.29}$$

It means that we can build the invariant for the arbitrary element of the fractal consequence $a_n$.

Taking into account Definition 2.1.2, we can transform equation (5.6.27) as:

$$\left(\frac{a_1}{a_{\frac{n}{n-1}}}\right)^n + \left(\frac{b_1}{b_{\frac{n}{n-1}}}\right)^n \equiv 1. \tag{5.6.30}$$

By this transformation we are going from fractal space to space with fractal stretched unit length. In this space, the relation of two fractal values is non-fractal and our value transforms to the regular.

We can now find the limit of the fractal sum [339]. Taking into account (5.6.21) and (5.6.29), it follows that:

$$a_n + b_n = c^{\frac{n-1}{n}}\left(a_1^{\frac{1}{n}} + b_n^{\frac{1}{n}}\right) = c\left(\left(\frac{a_1}{c}\right)^{\frac{1}{n}} + \left(\frac{b_1}{c}\right)^{\frac{1}{n}}\right).$$

Thus:

$$c \lim_{n\to\infty}\left(\left(\frac{a_1}{c}\right)^{\frac{1}{n}} + \left(\frac{b_1}{c}\right)^{\frac{1}{n}}\right) = 2c. \tag{5.6.31}$$

Considering $a_1$, $b_1$ and (5.6.31), we obtain:

$$c < a_n + b_n < 2c. \tag{5.6.32}$$

It means that all the edges of the $n$-dimensional cube are limited and our fractal space is limited as well. Taking into account the evaluation (5.6.32), we can postulate the divergence of the sequence $a_n + b_n$ and obtain that the full length of fractals is unlimited.

*Topothesy of the Surface*

It is well known that multifractals, which are much used in theory of fracture, are characterized by the spectrum of Rényi dimensions (equation (D.5)) [93, 16].

Multifractal structures are characterized by the spectrum of singularity indices,

$$d_i = \frac{\ln P_i(\varepsilon)}{\ln \varepsilon}.$$

These indices are characterized by the distribution and the Sobolev spaces are appropriate apparatus for this distribution description.

All these indexes can be determined from the metallographic measure [16]. Taking into account definition (5.6.20) and condition (5.6.32) we obtain an agreement with a well-known experimental fact [16, 69] that:

$$1 < R_L(\varepsilon) < 2. \tag{5.6.33}$$

According to definition of topothesy (5.1.5) and structural function $S(\tau)$, we obtain:

$$\frac{\sum\limits_{n} (a_n - a_1)^2}{n} = \Lambda^{2D-2} c^{4-2D}. \tag{5.6.34}$$

Considering equation (5.6.29) from the expression (5.6.34), it follows that:

$$\frac{\sum\limits_{n} (a_n - a_1)^2}{n\left(a_n a_{\frac{n}{n-1}} + b_n b_{\frac{n}{n-1}}\right)^{(2-D)}} = \Lambda^{2D-2}, \tag{5.6.35}$$

and topothesy is not parameters but the determined function of scales in agrement with experimental data.

## 5.6.5 *Defect structure and fractal properties of a real crack*

In case of generalized spaces, the geodesic equation can be written down as a function of the parameter of the curve length, current coordinates and connectivity coefficients of the space, accordingly $s$, $x^\mu$, $\Gamma^\mu_{i\lambda}$:

$$\frac{d^2 x^\mu}{ds^2} + \Gamma^\mu_{\sigma\lambda} \frac{dx^\mu}{ds} \frac{dx^\lambda}{ds} = 0. \tag{5.6.36}$$

It is well-known that the processes of elastic deformation, which take place in the ideal crystal, can be viewed from the geometrical position as the processes in the Riemannian space with a positively defined metric independent from the direction. The length element in this metric can be written down as:

$$dl^{(r)} = \sqrt{{}^r g_{ij} \frac{dx^{(r)i}}{dt} \frac{dx^{(r)j}}{dt}}. \tag{5.6.37}$$

In equation (5.6.37) index $r$ at the variables defines the Riemannian space, $x^i, x^j$ are the current coordinates of the point, $t$ is the natural parameter of the curve; and the indices cover the values from 0 to the dimension of space $n$, considered in the problem. In the case of crack propagation, the crack length is usually chosen to be the natural parameter. Since each type of defect corresponds to an additional space geometrical characteristic, curvature $S$, torsion $R$, and segment curvature $K$ (the first, second and third Cartan curvature tensors; §2.4.2), then for a medium with microstructure, the metric depends not only on the position of the respective system of coordinates, but also on the vector field connected with the defects. Besides, on the basis of the physical sense of the problem, the geometrical properties depend also on the direction of motion in this space, and not only on the defining characteristics of the space, but also on their speeds [282, 294]. Then, in the general case:

$$g_{ij} = g_{ij}\left(S, R, K, x^j, \dot{x}^j\right). \tag{5.6.38}$$

For the Finsler space, we have [303]:

$$^c g^{ij} = \left(\frac{\partial \Gamma^{*i}_{k.l}}{\partial x^t} - \frac{\partial \Gamma^{*i}_{k.l}}{\partial \dot{x}^s}\frac{\partial G^s}{\partial \dot{x}^t} + \Gamma^{*i}_{q.t}\Gamma^{*q}_{k.l}\right)\left(\frac{\partial \Gamma^{*}_{kjl}}{\partial x^t} - \frac{\partial \Gamma^{*}_{kjl}}{\partial \dot{x}^s}\frac{\partial G^s}{\partial \dot{x}^t} - \Gamma^{*q}_{j.t}\Gamma^{*}_{kql}\right)^{-1}. \tag{5.6.39}$$

In equation (5.6.37) $\Gamma^*_{kjl} = \Gamma^*_{kjl}\left(S, R, K, x^i, \dot{x}^i\right)$ are symmetric connectivity coefficients of the Finsler space (in the general case, they do not coincide with the connectivity coefficients of the Riemannian space), $g^{ij}g_{ij} = \delta_{ij}$, $\delta_{ij}$ is the Kronecker symbol. Value $G^i$ appears because of the fact that the Finsler metric depends not only on the position of the respective system of coordinates, but also on the additional vector field $\xi^l$ (the field connected with defects). This field appears because of decomposition of the field derivatives by the basis of the field itself. Value $G^i$ can be found from the equation of the field derivatives [303], or obtained from the introduced connectivity. Thus, for the stationary field, for which $\xi^l_{;}(x, \xi) = 0$, where the subscript: defines a covariant $\delta$-differentiation, we have:

$$\Gamma^i_{k.j}\dot{x}^k = \frac{\partial G^i}{\partial \dot{x}^j}, \tag{5.6.40}$$

$$\Gamma^{*i}_{k.j} = \Gamma^i_{k.j} - C^i_{k.h}\Gamma^h_{r.j}\dot{x}^r. \tag{5.6.41}$$

Lifting and lowering of indices is made in the usual way with the help of the metric tensor set by equation (5.6.37). We note here that torsion (tensor (2.4.21)) can be set by different ways. Sometimes, one uses the first Cartan curvature tensor [303]:

$$S^j_{i.kl} = A^j_{k.r}A^r_{i.l} - A^j_{r.l}A^r_{i.k}, \tag{5.6.42}$$

where

$$A^j_{i.k} = F(x, \dot{x})C^j_{i.k}.$$

Since the process of length measuring takes place in the real Euclidean space, the 'excess' variables of the Finsler space are a hidden parameter, and their projection on the tangent Euclidean space defines the fractal behavior of the surface. This coincides with introduction of three curvature tensors as additional parameters of the mechanics of a deformed body [118, 317].

We note here that in the space of affine connectivity, we are interested in geodesics, and we can consider the Finsler space without torsion as the space of crack motion. This is connected with the fact that the connectivity object $\Gamma^{\mu}_{\sigma\lambda}$ defines in this manifold the same geodesics that the connectivity object without torsion $\tilde{\Gamma}^{\mu}_{\sigma\lambda}$, obtained by its symmetrization [294, p. 420]. This in principle allows restricting the consideration of the Finsler space by 1-form Finsler spaces with the Berwald–Moore metric [303]. In this case:

$$g_{ij}(x^m, \dot{x}^m) = S_i^A S_j^B g_{AB}(\dot{x}^D(x^m, \dot{x}^m)), \tag{5.6.43}$$

where $S_m^A$ is the global field of references of class $C^3$, $\cdot \overset{def}{=} \frac{d}{dt}$, and the initial metric of Minkowski space:

$$g_{AB} = \frac{1}{2}\frac{\partial^2 F^2_{Mink}}{\partial \dot{x}^A \dot{x}^B}, \tag{5.6.44}$$

depends on $x$ only through $\dot{x}^D$.

With account of (5.6.41), the equation of Finsler geodetics (2.2.3) has the form:

$$\frac{d}{dt} y_c + F^A_{BC} y_A y^B, \tag{5.6.45}$$

where $\sigma$ is the parameter of the Finsler arc length, $F^A_{BC} = S^j_B S^h_c F^A_{jh}$, and tensor $F^A_{jh}$ has the form:

$$F^A_{jh} = \partial_j S^A_h(x) - \partial_h S^A_j(x). \tag{5.6.46}$$

While defining the fractal dimension of crack $D$ as the ratio of the trajectory lengths in the real crystal to the trajectory in the ideal crystal (that is, the relation of the lengths of trajectories in the Finsler space and in the Riemannian space), we have for the case of the same parameterization of the curves:

$$D = \frac{dl}{dl^{(r)}} = \frac{\sqrt{{}^c g_{ij} \frac{dx^i}{dt}\frac{dx^j}{dt}}}{\sqrt{{}^r g_{ij} \frac{dx^{i(r)}}{dt}\frac{dx^{j(r)}}{dt}}} = \frac{\sqrt{{}^c g_{ij} dx^i dx^j}}{\sqrt{{}^r g_{ij} dx^{i(r)} dx^{j(r)}}}. \tag{5.6.47}$$

Since the symmetrical connectivity coefficients make the state function, the fractal dimension of the crack is also the state function.

With account of the fact that the valence of tensors (5.6.36), (5.6.37) can be made equal by means of lifting (lowering) of indices, and the tensors are set in one and the same point of the manifold, the deviation of the metric of the material is defined as:

$$\Delta^{ij} = {}^f g^{ij} - {}^c g^{ij}. \tag{5.6.48}$$

**Remark 5.6.1.** Earlier the idea that the real length of the crack does not coincide with the observed (measured) length was expressed from the analysis of the process of measuring the crack length [185]. However, the principal 'precise' non-measurement connected with the internal structure of the continuum of a body with microstructure has not been discussed so far.

## 5.7 Microscopic fracture of geo massifs*

Defined scaling characteristics of mountain rock fracture or earthquake processes is a popular general topic today. At the same time, a detailed theory of mechanical behavior of geological massifs in linear approximation does not bring sufficiently satisfactory results, since the real continuum of geomechanics is essentially nonlinear [298, 309]. On the other hand, the behavior of geological structures under external effects is critically important. This is related to colossal scales of possible implications—both human and material casualties—for 'geological' emergencies. Among such emergencies we can name earthquakes, mudflows and earth shell dynamics. Of considerable interest are mathematical problems of a nonlinear continuum, connected with mineral exploration activities. For example, in this respect, for description of ray propagation, which is applied for experimental geophysics (oil and gas exploration), the Finsler metric is broadly used [7, 40, 243, 354, 355].

Let us consider the process of rupture propagation in geological structures and effects related to the inhomogeneous medium structure.

### *5.7.1 Energy flux*

It is known from the vector analysis that the flux of random vector **F** through surface $S$ is defined as:

$$\mathcal{F} = \int_S \mathbf{F}(\mathbf{r}) d\mathbf{S}, \tag{5.7.1}$$

where $\mathbf{S} = S\mathbf{n}$ is the oriented surface and $\mathbf{n}$ is the normal to the surface. This definition is easily spread on tensors of random order and valence and on spaces of any mathematical structure. For simplicity, we consider rupture propagation in the medium without heat sources and neglect the traction forces. In this case, for the energy flux density (4.1.27) we have:

$$\dot{W} = \lim_{\varepsilon\to\infty} \int_{S_p} \left\{ \sigma_{ij}\dot{u}_i + \left[ \int^t \left( \sigma_{ik}\dot{\varepsilon}_{ik} \right) dt + \frac{1}{2}\rho\dot{u}_i\dot{u}_i \right] v_j \right\} n_j dS_\varepsilon \tag{5.7.2}$$

$$= \lim_{\varepsilon\to\infty} \int_{S_p} \left\{ \sigma_{ij}\dot{u}_i + \left[ \int^t \left( \sigma_{ik}\dot{\varepsilon}_{ik} \right) dt + \frac{1}{2}\rho\dot{u}_i\dot{u}_i \right] v_j \right\} d\mathbf{S} = \int_{S_p} \vec{\mathcal{E}} d\mathbf{S} \tag{5.7.3}$$

where

$$\vec{\mathcal{E}} = \sigma_{ij}\dot{u}_i + \left[ \int^{t} \left( \sigma_{ik}\dot{\varepsilon}_{ik} \right) dt + \frac{1}{2}\rho\dot{u}_i\dot{u}_i \right] v_j \tag{5.7.4}$$

is the Umnov–Pointing vector. It defines the direction of energy propagation [94, 296]. The mathematical details connected with the integration procedure of equation (5.7.2) can be found in [101]. The Umnov–Pointing vector can be presented in a slightly different form:

$$\vec{\mathcal{E}} = -\boldsymbol{\sigma}_q |\mathbf{q}|, \tag{5.7.5}$$

where $\boldsymbol{\sigma}_q = \boldsymbol{\sigma}\mathbf{q}/|\mathbf{q}|$ is the vector of stresses on the pad with a normal directed along the displacement vector.

In the mechanics of deformed bodies, the physical value of this vector is connected with non-coaxiality of the stress and deformation tensors, which leads to localization of deformation [149, 150, 325]. In the general case, we have a possibility to find $\vec{\mathcal{E}}$ from (5.7.5), by using known solutions for $\boldsymbol{\sigma}$ in the case of dynamic propagation [101, 238].

It is well known that an important difference between a discrete and continuum model is the existence of radiation from the tip of the moving crack, due to existence of periodic modulation in the velocity in the presence of an underlying lattice [98]. From this viewpoint, the considered model of a crackon (§5.3) is a discretization of the classic continual model of the medium.

In real systems, the energy introduced into the volume by external loads is localized by the crack through various mechanisms. These can be, for example, acoustic emission [248], stress waves [318], or surface waves [281]. From the structure of the fissured body (the presence of free surfaces), the surface waves in particular principally change the energy relaxation picture in this case. In the case of geological structures, an essential role belongs also to Love waves, and the share of the energy taken away by all types of waves makes up to 5% of the overall deformation energy [309]. In this case, microcracks are emitting insignificant energy. According to experimental data, for sandstone, the average dimension of a microcrack lies within the limits of (1.4–28.4) $\times$ $10^{-6}$m, and the total energy of acoustic signals at formation of a submicroscopic crack is 0.03–58.25nJ [352]. A considerable share of the energy 'pumped' into the volume is absorbed by inter-block motions, and by elastic and plastic deformation of individual blocks.

### 5.7.2 *Ray path and energy storage in geomechanics*

It follows from (5.7.4) that to study the energy flow field we need to build a field of normals to the preset surface. In the general case, a normal is defined locally in the given point of state space with radius-vector **r**. In principle, we can consider the values included in equation (5.7.5) as functions of generalized coordinates, which do not mandatorily coincide with spatial coordinates $x, y, z$. In this case, definition of the normal to the surface requires clarification. It follows from the geometrical definition of a gradient that in the fibered Euclidean space (2.1.48) a normal can be defined as a gradient to the layer.

A normal in the space of any nature should be defined in a similar way. Let a layer of vector field (for example, an equipotential surface) be set as $F(\xi_i)$, $i = 1, 2 \ldots, n$. Then:

$$\mathbf{n} = \text{grad } F(\xi_i) \equiv \nabla F(\xi_i).$$

For the main fissured space $\mathcal{P}(\mathcal{M}, G)$ (p. 48), we need to build a horizontal subspace of manifold $\mathcal{M}$.

A similar problem—to build a field of normals to a certain surface—arises in building a beam trajectory in an inhomogeneous medium [40, 356]. This problem is of great importance in geologic exploration in analysis of seismic data. V. Červený [356] considers the Finsler space, in which distance $s$ is measured by travel time $\tau$, so that:

$$d\tau^2 = g_{ij}\, dx^i dx^j. \tag{5.7.6}$$

The Fermat functional there reads:

$$\mathcal{I}(l) = \int\limits_S^R \left(g_{ij}(x^{i'})(x^{j'})\right)^{1/2} du, \quad (x^{i'}) = \frac{dx^i}{du}, \tag{5.7.7}$$

in agreement with the general definition (2.4.13). The Fermat functional satisfies the Euler–Lagrange equation (3.2.14) and determines the geodesic line, according to §2.2.1. In this formulation, the problem of wave propagation in an inhomogeneous medium is mathematically equivalent to the problem of crack propagation. In this case, the equation of flow surface (2.5.1) corresponds to the slowness dynamics equation. Phase-slowness vector $\mathbf{p}$ corresponds to the Christoffel equation [40, 94]

$$\left(\Gamma^{ij}(\mathbf{x}, \mathbf{p}) - \delta^{ij}\right) A_j(\mathbf{x}) = \mathbf{0}, \quad i, j = 1, 2, 3, \tag{5.7.8}$$

where $\mathbf{A}(\mathbf{x})$ is the wavefront amplitude and $\Gamma(\mathbf{x}, \mathbf{p})$ is the Christoffel matrix:

$$\Gamma^{ij}(\mathbf{x}, \mathbf{p}) = \sum_{k=1}^{3} \sum_{l=1}^{3} \frac{c^{ikjl}(\mathbf{x})}{\rho(\mathbf{x})} p_k p_l, \quad i, j = 1, 2, 3, \tag{5.7.9}$$

with $\rho(\mathbf{x})$ being mass density, and $c^{ijkl}(\mathbf{x})$ being the elasticity tensor.

It has been shown in [40] that the eigenvalue of the Christoffel matrix generates the Hamiltonian of a system of rays

$$H(\mathbf{x}, \mathbf{p}) := \frac{1}{2} G(\mathbf{x}, \mathbf{p}),$$

where $G(\mathbf{x}, \mathbf{p})$ is one of the eigenvalues. For a crack, the Hamiltonian can be built, for example, with account of the link (2.4.10), definition (2.4.14) and metric tensor in the form of (3.2.13). In this case, the Hamilton orthogonality condition [40] is equivalent to the orthogonality in the Finsler space.

The energy balance of the isolated space of the medium with account of (5.7.2) can be presented as the difference of the incoming $W^{in}$ and outgoing $W^{out}$ energy flows in the form of:

$$\Delta W = W^{in} - W^{out} = \int\limits_{\partial\Omega_1} \vec{\mathcal{E}}\mathbf{n}^+ d\partial\Omega - \int\limits_{\partial\Omega_2} \vec{\mathcal{E}}\mathbf{n}^- d\partial\Omega. \tag{5.7.10}$$

With account of integration by one and the same surface, which is covering the volume, we have:

$$\Delta W = W^{in} - W^{out} = \int_{\partial\Omega} \vec{\mathcal{E}}(\mathbf{n}^{+} - \mathbf{n}^{-}) d\partial\Omega. \tag{5.7.11}$$

Because of the orthogonality asymmetry [303] in the Finsler space (2.4.34) the energy accumulated in the volume is different from zero.

In this way the stored energy can be used for the prediction field of damage by an earthquake because the energy is a leading parameter for seismic incidents [298].

**Remark 5.7.1.** With account of a possibility to regulate the crack trajectory, a practical problem arises to take preventive measures to control the direction of energy emission of an earthquake through preliminary creation of the field of defects of a definite structure. Naturally, this idea requires a detailed additional investigation.

## 5.8 Chaotic hierarchical dynamical systems and application of non-standard analysis for its description*

Studying the chaotic behavior of multiple dynamical systems and definition of the sequence of transition from order to chaos is a complicated mathematic problem. One of its solutions can be to calculate the entropy of dynamical systems. As noted above in § 5.6.3, information and dual value—entropy—are the control parameters for dynamical systems. However, the problem of calculating the entropy for dynamical systems is rather complex, and its general solution is unknown. The problem becomes especially complicated for the dynamical systems set on fractal functions, or for hierarchical systems [215]. However, for a rather important special case—for Markovian hierarchical systems—it seems possible. In this case, we can find the respective entropy of Kolmogorov–Sinai, define the entropy of Rényi and assess the chaotic behavior of the system. For development of the respective theory, let us recall certain definitions of the functional analysis.

*Hölder space*

**Definition 5.8.1.** The Banachian space of limited and continuous functions $f(x) = f\left(x^1, x^2, \ldots, x^n\right)$, defined on manifold $\mathcal{E}$ $n$-dimensional Euclidean space and satisfying the Hölder condition is the Hölder space.

**Definition 5.8.2.** The Hölder conditions are inequations, in which the function increments are evaluated through the increment of its argument.

For random metric spaces, the inequations of the following form are valid:

$$\rho_E\left(f(x),\ f(x_0)\right) \leq A(x_0)\, \rho_X^{\alpha}(x,\ x_0),$$

where $\rho_E,\ \rho_X$ are metrics of spaces *X, E*. In the plasticity theory, a classic example of a Hölder inequation is the fundamental inequation of the mathematic plasticity theory:

$$|(\sigma\varepsilon)| \leq \|\sigma\|_{(P)}\, \|\varepsilon\|_{(d)},$$

where $(\cdot)$ is the internal product and $(P)$ and $(d)$ are the straight and dual norms [383].

### 5.8.1 Iterated function system

According to the simplest definition, a dynamical system (DS) refers to the system whose state depends on time. From this point of view, any mechanical or geological system which realizes movement or evolution development is a dynamical system. Analysis of a DS evolution from initial time $t_0$ to finite time $t_f$ with sufficient generality can be reduced to the analysis of a set of differential equations of the form:

$$\dot{\vec{y}} = \vec{Q}(\vec{y}, t, \mu),$$

where $\vec{y} = [y_1, y_2, \ldots, y_k]^T$ is a vector of $k$ degrees of freedom and $\mu$ is an additional parameter on which the system depends. The system movement geometrically can be represented as a line, linked sequenced state of the system in $K$-dimensional space with the special geometrical structure. This space is named 'phase space'. Assume that the system has Markov character and the possibility of state can be determined by the probability $P_i$. Let us investigate the possibility to determine the dynamical entropy of the system $\vec{y}$. We can investigate not the present system but the *approximation* of DS by the system of functions which have the same entropy.

Let us investigate an iterated function system (IFS) [19, 335]. An IFS consists of a certain number of $k$ functions $F_i$, $i = 1, 2, \ldots, k$, which act randomly with given probabilities $p_i$, $i = 1, 2, \ldots, k$. IFS can be considered as a combination of deterministic and stochastic dynamics. This IFS reduces system evolution to the discrete consequences of stationary states. It can be shown [19] that IFS generates unique invariant measures, in the general case fractal. This is valid for the IFS $F = \{F_i, p_i, i = 1, 2, \ldots, k\}$, following some additional assumptions:

1. $X$ is a compact metric space.
2. $F_i : X \to X$, $i = 1, 2, \ldots, k$ is the Lipschiz function with the Lipschiz constant $L_i > 0$.
3. $p_i : X \to [0, 1]$, $i = 1, 2, \ldots, k$ are Hölder continuous functions fulfilling $\sum_{i=1}^{k} p_i(x) = 1$ for each $\forall\, x \in X$.
4. $p_i(x) > 0$ for every $x \in X$ and $i = 1, 2, \ldots, k$.

Such IFS are called hyperbolic. We investigate only this sorts of IFS. It can be shown [19] that for every IFS a unique fixed point $A$ exists,

$$A = W(A) = \bigcup_{i=1}^{k} F_i(A),$$

where $A$ is a nonempty compact subset of $X$. The fixed point is called the attractor of IFS.

The IFS $F = \{F_i, p_i, i = 1, 2, \ldots, k\}$ generates a Markov operator $V$, acting on space $M(X)$:

$$(V_\nu)(B) = \sum_{i=1}^{k} \int\limits_{F_i^{-1}(B)} p_i(\lambda)\, d\nu(\lambda). \tag{5.8.1}$$

The space $M(X)$ is the space of all probabilities $X$, $\nu \in M(x)$ and $B$ is the measurable subset $X$ [335]. Markov operators of this form describe the evolution of probability measures under the action $F = \{F_i, p_i, i = 1, 2, \ldots, k\}$, and the IFS is the continuous random function [19, 335]. The Markov process linked with (5.8.1) can be built in the following way. Take the probability measure on $\Omega$:

$$\begin{aligned} P_x(i_1, i_2, \ldots, i_n) \\ := P_x\Big(\{\omega \in \Omega : \omega(j) = i_j, j = 1, \ldots, n\}\Big) \\ := p_{i_1}(x)\, p_{i_2}(F_{i_1}(x)) \ldots \times p_{i_n}\Big(F_{i_{n-1}}(\ldots(F_{i_1}(x)))\Big), \end{aligned} \tag{5.8.2}$$

where $x \in X$, $i_j = 1, \ldots, k$, $j = 1, \ldots, n$; $n \in N$. $\Omega$ produces the code space, $\Omega = \{1, \ldots, k\}^{\mathbb{N}}$. In this case the Markov process can be built as:

$$Z_n^x(\omega) = F_{\omega(n)}\left(F_{\omega(n-1)}\left(\ldots\left(F_{\omega(1)}(x)\right)\right)\right), \; Z_0^x(\omega) = x, \tag{5.8.3}$$

where $\omega \in \Omega$, $n \in \mathbb{N}$. As is shown in [335], for an IFS which fulfills assumptions (1)–(4), and expressions (5.8.1)–(5.8.3), there exists a unique invariant probability $\mu$ satisfying the equation:

$$V\mu = \mu.$$

This invariant measure is attractive, which means $V^n\nu$ converges weakly to $\mu$ for every $\nu \in M(x)$, if $n \to \infty$ [335]. In other words, $\int_X u dV^n\nu \to \int_X u d\mu \quad \forall u : X \to \mathbb{R}$. It means that the invariant measure is the non-standard (or hyperreal) number in the sense of Robinson [3, 73, 176].

### *5.8.2 Entropy of IFS*

Now we investigate the entropy of the IFS because the chaotic state of any dynamical system closely depends on the entropy properties. Let $\mu$ be the attractive invariant measure for the IFS $F = \{F_i, p_i, i = 1, 2, \ldots, k\}$. For the code space $\Omega = \{1, \ldots, k\}^{\mathbb{N}}$ we can determine [335] a measure of probability $P_\mu$ as:

$$\begin{aligned} P_\mu(i_1, \ldots, i_n) &:= P_\mu(\{\omega \in \Omega : \omega(j) = i_j, j = 1, \ldots, n\}) \\ &= \int_X P_x(i_1, \ldots, i_n)\, d\mu(x) \end{aligned} \tag{5.8.4}$$

for $i_j = 1, \ldots, k$, $j = 1, \ldots, n$; $n \in (N)$. For brevity we use only the index $i$ for enumeration. Thus taking into account the definition of the entropy and expression (5.8.4) we obtain for the partial entropy of a discrete set of variables:

$$H(n) = -\sum_{i_1, \ldots, i_n = 1}^{k} P_\mu(i_1, \ldots, i_n) \ln P_\mu(i_1, \ldots, i_n). \tag{5.8.5}$$

Relative entropy is given by the expression:

$$G(1) := H(1), \quad G(n) = H(n) - H(n-1) \text{ for } n > 1. \tag{5.8.6}$$

The KS entropy is the exact value of the non-standard number because the range of integration in expressions (5.8.4)–(5.8.6) is the non-standard number:

$$K_1 = \lim_{n\to\infty} G(n) = \lim_{n\to\infty} \frac{H(n)}{n}.$$

If we introduce the Rényi entropy in the standard manner [155] analogously to the Rényi dimension we obtain:

$$K_\beta = \frac{1}{1-\beta} \ln \sum_i (p_i)^\beta = \lim \sup_{n\to\infty} \frac{1}{1-\beta}$$

$$\times \ln \left[ \sum_{i_1,\ldots,i_n}^{k} P_\mu (i_1, \ldots, i_n) \left( \frac{P_\mu (i_1, \ldots, i_n)}{P_\mu (i_1, \ldots, i_{n-1})} \right)^{\beta-1} \right]. \tag{5.8.7}$$

In (5.8.7) $\beta$ is the free parameter. The Rényi entropy is equal to the topological entropy if $\beta = 0$. It means that topological entropy can be correlated with the Hausdorff–Bezikovich dimension and accommodating KS entropy is the limiting case of fractal information dimension:

$$K_1 = \lim_{\beta\to 1} K_\beta.$$

For the distribution (5.8.4) an escort distribution can be introduced [155, 335]. Let us present this distribution (5.8.4) in the form:

$$P_\mu (i_1, \ldots, i_n) = \exp\left(-b_\mu (i_1, \ldots, i_n)\right) \times \sum_{i_1,\ldots,i_n} P_\mu (i_1, \ldots, i_n)$$

$$= \sum_{i_1,\ldots,i_n} \exp\left(-b_\mu (i_1, \ldots, i_n)\right) = 1, \tag{5.8.8}$$

then the escort distribution takes the form:

$$p_\mu = \frac{(P_\mu)^\beta}{\sum\limits_{i_1,\ldots,i_n} (P_\mu)^\beta}, \quad \sum_{i_1,\ldots,i_n} (p_\mu) = 1.$$

If we use the representation (5.8.8), the escort distribution for the system is a Gibbs canonical distribution:

$$p_\mu = \exp\left[\beta \left(F(\beta) - b_\mu\right)\right],$$

$$F(\beta) = -\frac{1}{\beta} \ln \sum_{i_1,\ldots,i_n} (-\beta b_\mu), \tag{5.8.9}$$

where $b_i$ plays a role of effective Hamilton function, $F(\beta)$ is the corresponding free energy, and $1/\beta$ is the effective temperature.

From (5.8.9), taking into account expression (5.8.7), there follows a link between the free energy and the Rényi entropy:

$$F(\beta) = -\frac{1}{\beta}\ln\sum_i \exp(-\beta b_i)$$
$$= -\frac{1}{\beta}\ln\sum_i \exp(P_i)^\beta = -\frac{1-\beta}{\beta}K_\beta[P]. \tag{5.8.10}$$

From equation (5.8.10) we obtain that the free energy is equal to zero for $\beta = 1$:

$$F(\beta = 1) = -\ln\sum_i P_i = 0,$$

and additionally $p_i = P_i$.

Thus the Rényi entropy has statistic nature and characterizes a non-equilibrium state coinciding with the chaos state. It means that the Rényi entropy can be taken as a possible measure of the chaotic state of the system. From the physical meaning of chaos and according to the main principles of non-standard analysis [3, 73] distribution (5.8.8) forms the monad of standard number, and the standard number corresponds to the KS entropy. The standard part of the entropy is the function of fixed point $A$ and KS entropy is a characteristic of the equilibrium state because the measure $\mu$ is attractive. The halo of monad of non-standard numbers in one-dimensional code space can be estimated as:

$$\Gamma = \lim_{\beta\to 1}\left(K_\beta - K_1\right).$$

The value $\Gamma$ represents the internal indeterminacy of the appropriate dynamical system.

### *5.8.3 Entropy of hierarchical space*

Let us consider an application of the general theory to hierarchical space. In the general case, according to § 1.1, the motion of a system within a hierarchical space is described by the action of two hierarchical operators: $\rho^+$ and $\rho^\times$. Operator $\rho^+$ responds for the motion along the levels of the hierarchical system, and $\rho^\times$ for the motion within the bounds of one layer. Therefore, the space of indices $\Omega$ is two-dimensional (2-D). The metrical properties of the hierarchical fracture space were not studied deliberately, but its compactness follows from the definition of the multilevel hierarchical system. Strictly speaking, the action of operators $\rho^+$, $\rho^\times$ is not independent—the system cannot randomly change levels and position in these levels (for example, to jump over several cells as a result of a unit translation). However, in the local zone, in the proximity of the current state, a translation of an elementary system by a unit cell is an independent event.

Therefore, the action of operators $\rho^{+}$ and $\rho^{\times}$ generates a 2-D Markovian process. From the properties of independent Markovian processes, we assume:

$$\begin{aligned} P_{\mu}^{\nu}\binom{l_1,\ldots,l_m}{i_1,\ldots,i_n} &:= P_{\nu}(l_1,\ldots,l_m)P_{\mu}(i_1,\ldots,i_n) \\ &:= P_{\nu}\Big(\{\varpi\in\Omega:\varpi(k)=l_k,\ k=1,\ldots,m\}\Big) \\ &\quad\times P_{\mu}\Big(\{\omega\in\Omega:\omega(j)=i_j,\ j=1,\ldots,n\}\Big) \\ &= \iint\limits_X P_{\rho^{+}}(l_1,\ldots,l_m)\,P_{\rho^{x}}(i_1,\ldots,i_n)\,d\mu(x)d\nu(x), \end{aligned} \tag{5.8.11}$$

where $\mu$ is the invariant measure for operator $\rho^{+}$, and $\nu$ is the invariant measure for operator $\rho^{\times}$.

In this case, a halo of a non-standard number represents a hypersphere in the space of indices, and entropies can be introduced similar to the procedure described in [21, 155]:

$$\begin{aligned} H(n,m) &= \sum_{i_1,\ldots,i_n=1}^{k}\ \sum_{j_1,\ldots,j_n=1}^{l} P_{\mu}(i_1,\ldots,i_n)\ln P_{\mu}(i_1,\ldots,i_n) \\ &\quad\times P_{\nu}(j_1,\ldots,j_n)\ln P_{\nu}(j_1,\ldots,j_n) \\ &= P_{\mu}P_{\nu}\ln P_{\mu}+P_{\mu}P_{\nu}\ln P_{\nu}=P_{\nu}H_{\mu}+P_{\mu}H_{\nu}. \end{aligned}$$

Non-standard numbers (5.8.11) according to the general procedure of introducing a hierarchy of structural levels [296] fill the numerical axis more densely than the manifold of irrational numbers. The hypersphere can be presented as the basic element of the second hierarchical level:

$$\Gamma^{(2)}(x,\nu(\mu)) = \operatorname*{Lim}_{n\to\infty}\Gamma_n^{(2)}(x,\nu(\mu)),$$

where $\operatorname*{Lim}_{n\to\infty}$ means the limit of the sequence of functions.

## Certain conclusions

- The fractal dimension of fracture and deviation of metric, introduced by equations (5.6.45) and (5.6.46) in conformity with experimental data [74] and computer models [387], depend both on the properties of real crystals and on the conditions of fracture. In this case, it is thought that the crack trajectory does not depend clearly on distribution of dislocations, since a dislocation is associated with the torsion tensor [27, 109, 118], while the geodesics in the Finsler space do not depend on the curvature. The obtained result was postulated earlier for a special case of anti-flat deformation [22] on the basis of the continual consideration.
- The established fact opens the possibility to regulate the process of crack propagation. In consideration of the modes of the preset fracture of the ideal material (when the fracture of the material should take place in the preset place and, desirably, along the

preset trajectory), the trajectory of crack propagation is defined by equation (5.6.36). In this case, by setting the distribution of defects and regulating this distribution by technological means, we can obtain the required characteristics of the crack trajectory. In this case, the distribution function of dislocations acts in the capacity of a free parameter, which does not affect the fractality of the surface, but affects the mechanical parameters of the material (for example, the Lame parameters).

- Amorphous materials, for example, can be considered in the capacity of such ideal material. This is connected with the fact that, for amorphous materials, the structural borders (the borders of grains and metallographic phases) are missing, and equation (4.2.5) will most precisely describe the crack trajectory. Besides, due to well-known fragility of amorphous materials, the effects related to plasticity will be insignificant, and the difference of the Finsler metric from the Riemannian one is weak.
- For dynamical systems, entropy can be introduced on the basis of the mechanism of non-standard analysis. It has been shown that entropy of Kolmogorov–Sinai is a precise value of a non-standard number, the monad of which is defined by the Rényi entropy. The Rényi entropy can be used for definition of the halo of a non-standard number. The two-dimensional probability measure (5.8.11) can be summarized in the multi-measure case, which will allow in future considering the criteria of appearance of the hierarchical haze in the hierarchical space.

# Appendix A
# Spaces: Some Definitions

**Topological space.** If $S$ is a set, and $\mathbf{T}$ is a set of subsets of $\{U_i\}$ such that (i) every union of elements of $\mathbf{T}$ is an element of $\mathbf{T}$, (ii) the intersection of two elements of $\mathbf{T}$ is in $\mathbf{T}$, and (iii) both $\{U_i\}$ and the empty set are in $\mathbf{T}$, then $(\{U\}, \mathbf{T})$ is a *topological space* $\mathbb{X}$ with *topology*, $\mathbf{T}$. The elements of $\mathbf{T}$ are open sets. A (open) neighborhood of a point is an open set containing that point. The complement (in $U_i$) of an open set is a *closed* set. If $A \subset U_i$ then $s \in U_i$ is a limit point of $A$ if for all neighborhoods, $\mathfrak{U}_a$ of $s$ $(\mathfrak{U})_a - s) \cap A \neq \varnothing$. The closure, $\bar{A}$, of $A \subset U_i$ is the union of $A$ and all its limit points. $(\{U_i\}, \mathbf{T})$ is a *Hausdorff space* if distinct elements of $S$ are contained in disjoint elements of $\mathbf{T}$.

*Axioms of linear vector space*

Provided a number of axioms have been observed, a set of elements is called a linear vector space. Besides, a rule should be set of summing up two elements of the space, as a result of which an element of that very space appears. An operation should be defined of multiplying a vector by a real number. It is required, in this case, to observe the usual properties of addition and multiplication operations.

Let objects $\mathbf{a}$, $\mathbf{b}$, $\mathbf{c}$ be the vectors belonging to the linear vector space; $k$ and $m$ be real numbers. Then:

- Operations of adding and multiplying by a number should meet the requirements:

$$\mathbf{a} + (\mathbf{b} + \mathbf{c}) = (\mathbf{a} + \mathbf{b}) + \mathbf{c}; \quad \mathbf{a} + \mathbf{b} = \mathbf{b} + \mathbf{a}.$$

- There should exist a zero vector: $\mathbf{a} + \mathbf{0} = \mathbf{a}$.
- For $\forall$ $\mathbf{a}$ $\exists$ $(-\mathbf{a})$, which means:

$$\mathbf{a} + (-\mathbf{a}) = 0.$$

- Multiplication by a number should meet the following requirements:

$$(km)\mathbf{a} = k(m\mathbf{a}),$$
$$(k + m)\mathbf{a} = k\mathbf{a} + m\mathbf{a},$$
$$k(\mathbf{a} + \mathbf{b}) = k\mathbf{a} + k\mathbf{b}.$$

- Multiplication by one should not change the vector:

$$1\mathbf{a} = \mathbf{a}.$$

For a vector space, the notion of convexity can be defined. Set $C$ of vector space $E$ is called convex if for vectors **a** and **b** all the elements $\lambda\mathbf{a} + \mu\mathbf{b}$, where $\lambda$, $\mu$ are positive numbers, and $\lambda + \mu = 1$, also belong to $E$.

**Definition A.1.** Hyperplane $H$ is called the reference hyperplane of convex set $C$, if $C$ is located completely in one of half-subspaces defined by $H$, and $H$ has at least one common point with closure $\overline{C}$ of set $C$.

**Definition A.2.** Given vector spaces $V_1, V_2, \ldots, V_p, W$. A mapping $\phi : V_1 \times \cdots \times V_p, \rightarrow W$ from the Cartesian product $V_1 \times \cdots \times V_p$ to $W$ is multilinear if it is linear in each argument.

**Example A.1.** We get an important example of a bilinear function ($p = 2$, $W = \mathbb{R}$) if we choose $V_1 = \mathbb{R}^n$, $V_2 = \mathbb{R}^m$. Then, given any $a_{ij}$, $i = 1, \ldots, n$, $j = 1, \ldots, m$, the map $\phi : \mathbb{R}^n \times \mathbb{R}^m \rightarrow \mathbb{R}$ defined by

$$((v^1, \ldots, v^n), (w^1, \ldots, w^m)) \rightarrow a_{ij} v^i w^j$$

is bilinear. Moreover, every bilinear function $\phi : \mathbb{R}^n \times \mathbb{R}^m \rightarrow \mathbb{R}$ can be written in this form.

# Appendix B
# Certain Relations of Vector Analysis

Let us recall for reference certain correlations of vector analysis.

Tensor rotation and vortex $\tau$:

$$\langle \tau \rangle = e_{ijk} \tau_{ij} \mathbf{e}_k, \tag{B.1}$$

$$\nabla \times \tau = \mathbf{e}_i e_{ijk} \partial_j \tau_{kl} \mathbf{e}_l. \tag{B.2}$$

Here, $e_{ijk}$ is the Levi–Civita symbol, and $\mathbf{e}_i$ are unitary vectors of the Cartesian basis.

The associated rotation vector is defined as:

$$\vec{\tau} = {}^1\!/_2 \langle \tau \rangle. \tag{B.3}$$

In index notation:

$$\tau_k = {}^1\!/_2 e_{ijk} \tau_{ij}. \tag{B.4}$$

The antisymmetric part of tensor $\tau_{ij}$:

$$\tau_{[ij]} = e_{ijk} \tau_k, \tag{B.5}$$

$$\begin{aligned} \nabla \times \tau_{[.]} &= \mathbf{e}_i e_{ijk} \partial_j e_{klm} \tau_m e_l = \left( \delta_{il} \delta_{jm} - \delta_{im} \delta_{jl} \right) \mathbf{e}_i \partial_j \tau \mathbf{e}_l \\ &= \mathbf{e}_i \partial_j \tau_j \mathbf{e}_i - \mathbf{e}_i \partial_j \tau_i \mathbf{e}_j \\ &= \left( \nabla \cdot \vec{\tau} \right) \delta - \vec{\tau} \nabla, \end{aligned} \tag{B.6}$$

where $\delta$ is the Kronecker symbol:

$$\begin{aligned} \langle \nabla \times \tau \rangle &= e_{ilm} e_{ijk} \partial_j \tau_{kl} e_m = \left( \delta_{il} \delta_{mk} - \delta_{lk} \delta_{mj} \right) \partial_j \tau_{kl} \mathbf{e}_m \\ &= \partial_j \tau_{kj} \mathbf{e}_k - \partial_j \tau_{kk} \mathbf{e}_j = \tau \cdot \nabla - \nabla \left( \mathrm{tr} \tau \right). \end{aligned} \tag{B.7}$$

# Appendix C
# Groups: Basic Definitions and Properties

Set $\Im$ of elements $G_1, G_2, G_3, \ldots$ is called a group, if for the elements of the plurality, a law of multiplying elements has been determined, which meets the respective requirements:

$$G_a G_b = G_d \; \forall G \; \in \Im. \tag{C.1}$$

There exists a unit element:

$$E G_a = G_a \; \forall G \; \in \Im. \tag{C.2}$$

There exists a inverse element:

$$G_a G_a^{-1} = E \; \forall G \; \in \Im. \tag{C.3}$$

For multiplication, the law of transitivity is observed:

$$G_a \left( G_b G_c \right) = \left( G_a G_b \right) G_c \; \forall G \; \in \Im. \tag{C.4}$$

An additional requirement is commuting of the group elements (in this case the group is Abelian):

$$G_a G_b = G_b G_a \; \forall \; G \in \Im. \tag{C.5}$$

Real applications allow superimposing a range of additional requirements on the elements of the group. Thus, for most physical processes the requirement of continuity is observed. Therefore, elements $\Im$ should be continuous. Since according to definition (C.1), a product of two elements of the group is also an element of the group, we can write down:

$$G_a G_b \equiv G(a_1, a_2, \ldots, a_r) G(b_1, b_1, \ldots, b_r) = G(d_1, d_2, \ldots, d_r), \tag{C.6}$$

where $(a_i, b_i, d_i)$ are the parameters defining the elements of the group. All the $r$ parameters of the group are significant in the sense that when using fewer parameters, the elements of the group cannot be differentiated among each other ($r$ is the dimension of the group). Then, parameters $d_q$ are the functions of the initial parameters:

$$d_q = \phi_q(a_1, a_2, \ldots, a_r; b_1, b_1, \ldots, b_r). \tag{C.7}$$

To meet the group conditions (C.4–C.6) a number of additional conditions should be imposed on function $\phi$ [86]. For most of the processes realized by the mechanics of solids, it is sufficient to represent that functions $\phi$ are differentiated functions of the parameters. Groups of this type are called Lie groups[1].

A group of dimension $r$ with parameters $a_1, a_2, \ldots, a_r$ has an endless number of elements, but almost all the properties of the group are defined by a finite number of operators called infinitesimal.

**Definition C.1.** If in vector space $L$ we can find set $\mathbf{T}$ of linear operators $\mathbf{T}(G_a)$, corresponding to the elements of group $\Im$ in the sense that

$$\mathbf{T}(G_a)\mathbf{T}(G_b) = \mathbf{T}(G_aG_b), \quad \mathbf{T}(E) = 1,$$

such set $\mathbf{T}$ is called a *representation* of group $\Im$ in space $L$.

If all the parameters of the group are sufficiently small, then with the precision down to the members of the first order, the representation of the group can be written down as [86]:

$$\mathbf{T}(\mathbf{a}) \approx 1 + \sum_{q=1}^{r} a_q \mathbf{X}_q, \tag{C.8}$$

where $\mathbf{a} \equiv (a_1, a_2, \ldots, a_r)$, $\mathbf{X}_q$ are certain fixed linear operators, which do not depend on parameters $a_q$.

These operators $\mathbf{X}_q$ act as infinitesimal operators of representation $\mathbf{T}$. It follows from expression (C.8) that they are defined as partial derivatives of the representation:

$$\mathbf{X}_q = \lim_{a_q \to 0} \left[\mathbf{T}(0, 0, \ldots a_q, \ldots, 0) - 1\right]/a_q = \left[\frac{\partial}{\partial a_q}\mathbf{T}(\mathbf{a})\right]_{a=0}. \tag{C.9}$$

It is adopted here that for a unit element all the parameters $a_q = 0$, that is: $\mathbf{T}(0, 0, \ldots, 0) = 1$. The whole group can be restored by the infinitesimal operators. Let us consider this procedure in a special case of a single-parameter group $\Im$ with the multiplication law $G(c) = G(a)G(b)$, where $c = a + b$. In other words, the parameter of the group is additive. Then, on the basis of the rules of action with linear operators, we can write down operator $\mathbf{T}(a)$ in the following form: $\mathbf{T}(a) = \{\mathbf{T}(a/n)\}^n$, where $n$ is an integer. At large $n$, parameter $a/n$ is small, and in the limit, at $n \to \infty$, in formula (C.8) it is enough to preserve the members of the first and zero orders only. Then, on the basis of definition of exponential function as an endless power series, we have:

$$\mathbf{T}(a) = \lim_{n \to \infty} \{1 + (a/n)\mathbf{X}\}^n = \exp\{a\mathbf{X}\}. \tag{C.10}$$

Similar considerations can be made also for a multi-parameter group.

There are several theorems referred to infinitesimal operators. We formulate them without proof.

---

[1] Sophus Lie (1842–1899) was a Norwegian mathematician, who studied systemically endless groups, which was perceived by his contemporaries to be mere academic curiosity.

**Theorem C.1.** If two representations of group $\Im$ have identical infinitesimal operators, then these representations are equivalent.

**Theorem C.2.** For any representation **T** of group $\Im$, the set of infinitesimal operators $\mathbf{X}_q$ satisfies the commutation relations:

$$[\mathbf{X}_q, \mathbf{X}_p] = \sum_t c_{qp}^t \mathbf{X}_t, \tag{C.11}$$

where digital coefficients $c_{qp}^t$, called *structural constants*, are the same for all representations **T** of group $\Im$.

**Theorem C.3.** Any set of operators $\mathbf{X}_q$, defined in space $L$, forms a set of infinitesimal operators of representation **T** of group $\Im$ in space $L$, if these operators satisfy the commutation relations (C.11).

In the group theory, these theorems play the role of additional 'integrability conditions', which are quite similar to those in mechanics [375, p. 222]. Theorem 1 gives the 'multiplication law' for infinitesimal operators. In the general case, an infinitesimal operator, corresponding to a product of two elements of the group, coincides with the sum of the infinitesimal operators for the multipliers:

$$\mathbf{T}(\mathbf{a})\mathbf{T}(\mathbf{b}) \approx \left(1 + \sum_q a_q \mathbf{X}_q\right)\left(1 + \sum_p b_p \mathbf{X}_p\right) \approx 1 + \sum_q \left(a_q + b_q\right)\mathbf{X}_q,$$

at small parameters **a** and **b**.

Studying of Lie groups is somewhat simpler than studying finite groups, as we can consider only the algebra of infinitesimal operators [5], and the multiplication table can be replaced by a set of structural constants.

*Certain standard groups*

In the fracture mechanics, a number of standard groups are used; the mathematical properties of them are rather well studied.

**Group of translations** $T$**.** This group has an endless number of elements, corresponding to displacements along axis $X$: $\mathbf{P}(\xi) = x + \xi$, where $-\infty < \xi < \infty$. This is an Abelian group with the multiplication law $\mathbf{P}(\xi_1)\mathbf{P}(\xi_2) = \mathbf{P}(\xi_1 + \xi_2)$. The infinitesimal operator has the form of: $\mathbf{P}_x = -\partial/\partial x$.

The generalization to the 3-D case is trivial because of commutability of all the displacements. We obtain that a 3-D group of translations is a direct product of three one-dimensional translation groups along axes $x$, $y$ and $z$. The common element is designated as $\mathbf{P}(\boldsymbol{\rho})$ and defined in the trivial way:

$$\mathbf{P}(\boldsymbol{\rho})\mathbf{r} = \mathbf{r} + \boldsymbol{\rho}, \tag{C.12}$$

where **r** is the radius-vector of a point in the 3-D space. For a 3-D group of translations, there are three infinitesimal operators: $\mathbf{P}_x = -\partial/\partial x$, $\mathbf{P}_y = -\partial/\partial y$, $\mathbf{P}_z = -\partial/\partial z$.

**Group of rotations** $\mathcal{R}_3$**.** In a 3-D space, rotation is usually designated as $\mathbf{R}_k(a)$, $0 \le a \le \pi/2$ is the rotation angle, $0 \le a \le \pi/2$, **k** is a unit vector directed along the rotation axis. In fact, rotation depends on three parameters: rotation angle $a$ and two spherical angle vectors $\mathbf{k}$. However, it is easier to make use of three other parameters, namely: $\mathbf{a} = (ak_x, ak_y, ak_z) = (a_x, a_y, a_z)$.

Sometimes, to designate a group of rotations $\mathcal{R}_3$ one uses notation $O_3^+$. In this notation, '+' means that the group determinant is equal to +1. For a group of all orthogonal transformations (including inversions), one uses notations $O_3$.

**Euclidean group** $\mathcal{E}_3$**.** The Euclidean group in three dimensions is generated by a product of proper rotations $\mathbf{R}(\mathbf{a})$ and translations $\mathbf{P}(\boldsymbol{\rho})$. Consequently, with account of (C.12), the common element of the group can be written down as $\mathbf{R}(\boldsymbol{\rho})\mathbf{R}(\mathbf{a})$, and its action on vector $r$ is given by expression:

$$\mathbf{R}(\boldsymbol{\rho})\mathbf{R}(\mathbf{a})\mathbf{r} = \mathbf{R}(\mathbf{a})\mathbf{r} + \boldsymbol{\rho}. \tag{C.13}$$

Since it is geometrically obvious that displacements and rotations do not commutate, then group $\mathcal{E}_3$ cannot be presented as a direct product of the groups.

Thus, for the elasticity theory, initial group $G_0$ is a semidirect product $G_0 = SO(3)_0 \rhd T(3)_0$, respectively, of the group of real-valued orthogonal rotations $SO(3)_0$ and translations $T(3)_0$.

The linear transformations $\mathbb{R}^n$, which preserve the Euclidean structure, form an *orthogonal group* $O(2n)$, and the ones preserving the complex structure form a *complex linear group* $GL(n, C)$.

# Appendix D
# Dimensions

We recall that a system of subsets $\{U_i\}$ of topological space $\mathbb{X}$ is called its cover, if every point $x \in \mathbb{X}$ $\forall x \in \mathbb{X}$ belongs to any of the sets (at least one) $U_i$, i.e., $\forall x \in \mathbb{X} \exists U_i \in \{U_i\} | x \in U_i$. For simplicity, we shall consider finite covers only.

A subcovering is a system of sets $\{V_i\}$, if any $V_i$ belongs to at least one $U_i \in \{U_i\}$.

The covering multiplicity $\{U_i\}$ is the biggest of such numbers $n \in \mathbb{Z} > 0$, $\mathbb{Z}$ is the set of integers, that there exist $n$ elements of covering $\{U_i\}$, having a non-empty intersection (i.e., there always exists (at least one) point belonging to $n$ elements covering all these $\{U_j\}$, $j = 1, \ldots, n$ simultaneously).

**Topological dimension.** For simplicity, let us consider a compact, i.e., a closed limited set. Every compact at $\forall \varepsilon > 0$ allows $\varepsilon$-covering, i.e., can be presented in the form of unification of a finite number of closed sets, each of them having diameter $d < \varepsilon$. Or from any of its open covers, a finite subcover can always be chosen.

**Definition D.1.** A topological dimension $d_t$ or dim of compact $\mathbb{X}$ is called the smallest of such integers $n$, that into any open covering of space $\mathbb{X}$ we can write-in a closed subcovering of order $\leq n + 1$. If there are no such numbers, then we assume $\dim \mathbb{X} \overset{\text{def}}{=} \infty$.

The visual sense of this definition is rather simple. For example, at $n = 2$ it asserts that any 2-D 'platform' can be paved no matter how with small stones (closed sets) in such a way that the stones are adjacent to each other no more than by three. At the same time, this platform cannot be paved with stones no matter how small having only two abutments. When filling in a certain 3-D volume with sufficiently small stones (for example, with brickwork), necessarily abutments by four appear.

**Hausdorff–Besikovich dimension.** It is obvious that a point has dimension equal to zero; a section, circle, generally any usual curve in the plane or in the space is of dimension 1; a sphere is 2-D; and bodies are 3-D. In all the above cases, the dimension is equal to the number of independent variables required to set a point in the considered object. However, the sense of the notion *dimension* is broader. It characterizes more 'delicate' topological properties of objects and coincides with the number of independent variables required to describe an object, only in special cases. We relate the notion of length with one-dimensional objects, the notion of area with two-dimensional ones, etc. But how can we represent a set with dimension 3/2? For this purpose, probably, we need something intermediate between the length and area, and if we call conventionally the length to be 1-dimensional, and area 2-dimensional, then we need some (3/2)-dimension. In 1919, Felix Hausdorff defined this $\alpha$-dimension for any $\alpha > 0$, $\alpha \in \mathbb{R}$ and on this basis, he correlated each subset in the Euclidean space with a number called a metric dimension.

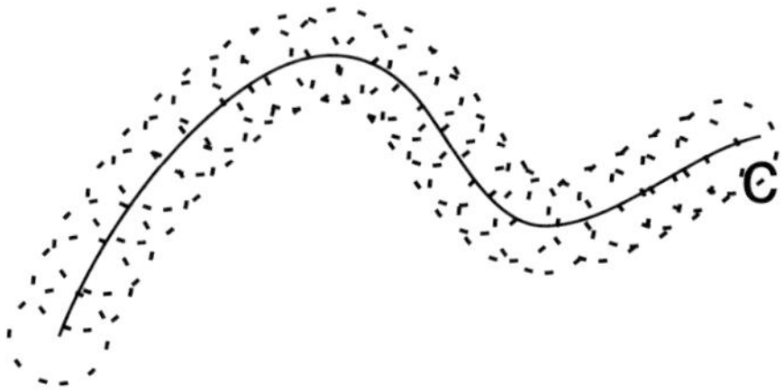


Fig. D.1. Coverage of the curve.

Let us consider known expressions for the length (this means, diameter $r$), area and volume of a sphere in the Euclidean space. The respective formulas in the Euclidean space of any integer number of measures are well known:

$$V_d = \gamma(d) r^d, \quad d = 1, 2, 3, \ldots, \quad \gamma(d) = \frac{\Gamma\left(\frac{1}{2}\right)^d}{\Gamma\left(1 + \frac{1}{2}\right)} \tag{D.1}$$

where $\Gamma(x)$ is the gamma-function of Euler:

$$\Gamma(x) = \int_0^\infty e^{-t} t^{x-1} dt, \quad x > 0.$$

We can define the $d$-th dimension of a sphere of radius $r$ in $\mathbb{E}^n$, where $d$ is any non-negative real number. It is achieved through expansion of formula (D.1) on all real $d > 0$. For example, the measure (dimension) of the sphere in the 3/2-dimension space is defined as $\gamma(3/2) r^{3/2}$. The next step implies a transfer of the notion of $d$-dimension from the sphere to the random set $A \subset \mathbb{R}^n$. For this purpose, let us build a covering of set $A$ (in this case, of chosen two-dimension curve $C$) by set $B_\varepsilon(x_i)$ of spheres of radius $\varepsilon$ (Fig. D.1).

The total volume of the $2 - d$ spheres, which cover our curve, can be calculated as

$$\sum_{i=1}^{M} \gamma(d) \varepsilon^d.$$

Analogously we can extend the coverage procedure to space of higher dimension.

**Definition D.2. (fractal measure)** $\varepsilon$-fractal $d$-dimension of the set is the number

$$\mu(A, D, \varepsilon) \stackrel{\text{def}}{=} \min\{M\} \cdot \varepsilon^d \equiv N(\varepsilon) \varepsilon^d.$$

For example, if $A_1 = [0, 1] \in \mathbb{R}^1$, then $N(\varepsilon) = \frac{1}{2\varepsilon} + 1$. This means if $\varepsilon = \infty$, it is enough to have one sphere to cover set (section [0,1]). If $\varepsilon = 1/4$, we need one and a half spheres to cover the set.

**Definition D.3. (Hausdorff dimension)** A fractal $d$-dimension of Hausdorff is called the number

$$\mu_F(A, d) = \limsup_{\varepsilon \to 0} \mu(A, d, \varepsilon) = \limsup_{\varepsilon \to 0^+} (\varepsilon^d \cdot N(\varepsilon)) \equiv \mu_F(A, d). \tag{D.2}$$

Besikovich showed that for each $\mathbb{X}$, there exists number $d_h \in \mathbb{R}$, that the $d$-dimension measure of Hausdorff of the compact $\mathbb{X}$ is endless at $d < d_H$, and, in contrast, is equal to 0 at $d > d_H$. Thus, for compact $A_1$

$$\mu_F(A_1, 1) = \lim_{\varepsilon \to 0^+} \varepsilon^d \cdot N(\varepsilon) = \lim_{\varepsilon \to 0^+} \varepsilon^1 \cdot \left( \left[ \frac{1}{2\varepsilon} \right] + 1 \right) = \frac{1}{2}; \text{ by } d = 1;$$

$$\mu_F(A_1, 1) = \lim_{\varepsilon \to 0^+} \varepsilon^d \cdot \left( \left[ \frac{1}{2\varepsilon} \right] + 1 \right) = 0 \text{ by } d > 1;$$

$$\mu_F(A_1, 1) = \lim_{\varepsilon \to 0^+} \varepsilon^d \cdot \left( \left[ \frac{1}{2\varepsilon} \right] + 1 \right) = \infty \text{ by } d < 1.$$

**Definition D.4. (dimension of Hausdorff–Besikovich)** Number $d_H$ satisfying the relation

$$d_H = \inf\{d | \mu_F(A, d) = 0\}$$

is the dimension of Hausdorff–Besikovich (metric or fractal dimension of set $A$). It is defined as $d_H$, $d_F$, or $d$.

In the previous example

$$\mu_F(A_1, d) = \begin{cases} 0 & d > 0; \\ +\infty & d < 0; \\ \dfrac{1}{2} & d = 1. \end{cases}$$

It follows from (D.2)

$$\mu(A, d, \varepsilon) = N(\varepsilon) \cdot \varepsilon^d \quad \Rightarrow \quad N(\varepsilon) = \frac{\mu}{\varepsilon^d}. \tag{D.3}$$

Let us take the logarithm of both parts:

$$\log N(\varepsilon) = \log \mu - \log \varepsilon^d \quad \Rightarrow \quad d = -\frac{\ln N(\varepsilon)}{\ln \varepsilon}$$

$$d = d_H = \lim_{\varepsilon \to 0} \frac{\ln N(\varepsilon)}{\ln \dfrac{1}{\varepsilon}}.$$

For the majority of 'good' objects, spaces and sets, the values dim and $d_H$ coincide. However, there are objects for which dim $< d_H$. These are fractals.

**Dimension of Minkowski.** The dimension of Minkowski can serve as an analogue of the dimension of Hausdorff–Besikovich, comfortable for use in application problems. These dimensions coincide as a rule, but the algorithm for definition of the Minkowski dimension is somewhat simpler.

Definition of the Minkowski dimension $d_M$ for a curve (fractal or smooth) lies in the following. Let the centre of a small Euclidean sphere (circle) of radius $r$ move along the curve covering the *Minkowski area*, that is area $S(r)$, arising at the motion of the sphere. Let us divide the area by $2r$, and tend $r$ to zero. In the case of a smooth curve we could have obtained in the limit the length of the curve, but for a fractal curve the result is endless. Indeed, the ratio $S(r)/2r \sim r^{1-d_M}$. In its turn, $\lim_{r\to 0} r^{1-d_M} = \infty | d_M > 0$. The value $d_M$ is the dimension of the divergence speed and is called the *dimension of Minkowski–Bouligand*. It can be calculated by formula:

$$d_M = \lim_{r\to 0} \frac{\ln S(r)}{\ln\left(\dfrac{1}{r}\right)} + 2. \tag{D.4}$$

In the case of a smooth curve $S(r) \sim r$ and $d_M = -1 + 2 = 1$, as should have been expected.

For all strictly self-similar fractals, dimension $d_M = d_H$. If these dimensions do not coincide, then $d_M > d_H$. This indicates that the Minkowski dimension is somewhat 'coarser' than the Hausdorff–Besikovich dimension, since it does not account for certain delicate structures of the object.

**Rényi dimension.** One of the most general definitions of the dimension for fractal object description was proposed by the Hungarian mathematician Alfred Rényi. The Rényi dimension is the continual set of dimensions:

$$D_q = \begin{cases} \lim\limits_{\varepsilon\to 0} \left[ \dfrac{\ln \sum\limits_{i=1}^{M} P_i^q(\varepsilon)}{(1-q)\ln\varepsilon} \right] & q \neq 1, \\[2em] \lim\limits_{\varepsilon\to 0} \left[ \dfrac{\ln \sum\limits_{i=1}^{M} P_i^q(\varepsilon) \ln P_i^q(\varepsilon)}{\ln\varepsilon} \right] & q = 1, \end{cases} \tag{D.5}$$

where $P(\varepsilon)$ is the probability that a given structure point belongs to the $i$-th element of the structure coverage by $d$-dimensional cubes of face $\varepsilon$. For uniform Euclidean objects, the identity $d_q \equiv d_T \equiv d$ holds, where $q_T$ is the topological dimension. For the regular fractals $d_T < d_q \leq d$.

By $q = 0$ dimension Rényi is congruent to the dimension of Hausdorff–Besikovich:

$$D_q = \lim_{\varepsilon \to 0} \frac{1}{-1} \frac{\log \sum_{i=1}^{N} p_i^0}{\log \varepsilon} = \frac{\log \sum_{i=1}^{N} 1}{-\log \varepsilon} = \lim_{\varepsilon \to 0} \frac{\log \sum_{i=1}^{N} N(\varepsilon)}{-\log \frac{1}{\varepsilon}}. \tag{D.6}$$

The dimension of Hausdorff–Besikovich does not depend on $q$; the numerator in equation (D.6) is the number of elements in limited subcovering of the initial compact set by the sphere of radius $\varepsilon$, and the denominator is the number which shows how many times the radius of the sphere $\varepsilon$ is put into the unit length.

For a self-similar fractal with equiprobable components, the Rényi dimension of appropriate power $q$ is congruent with the Hausdorff–Besikovich dimension.

**Correlation dimension.** The correlation dimension for discrete systems can be set by the expression:

$$C(r, M) = \lim_{N \to \infty} \left[ N^{-2} \sum_{i=j}^{N} \sum_{i=1}^{N} \Theta\big(r - |P_i - P_j|\big) \right], \tag{D.7}$$

where $C(r, M)$ is the number of neighborhoods in the distance $r$ from the given point, $\Theta$ is the Heaviside function, and $|\cdot|$ is the distance between points in $M$-dimensional Euclidean space. $M$ is defined as an *embedding dimension* [249, 326] of the Euclidean space. The embedding dimension can be defined in the following way. If we know certain temporal characteristics of the process (a time series exists), the whole interval is divided into identical intervals, and the vector of the state can be then built:

$$\mathbf{P}_M(x) = \big(y(x), y(x + x_0), y(x + 2x_0), \ldots, y(x + [M - 1]x_0)\big).$$

If we take $M$ big enough, the vector $\mathbf{P}_M(x)$ will be a set, equal to the attractor of the system. The correlation dimension is the Rényi dimension by $q = 2$.

**Iterated function system and attractors.** According to the definition of hierarchical operators $\rho^+$, $\rho^*$ if we investigate the limited case of one variable the successive action of operators is equal to the well known Hutchinson operator $W(A)$, which describes the set of affine transformation $w_1, w_2, \ldots, w_N$ (Heinz-Otto Peitgen, Harmut Jürgens, Dietmar Saupe *Chaos and Fractals: New Frontier of Science*, Springer,Verlag, N.Y, 1992). For a given initial state (image) small affine copies $w_1(A), w_2(A), \ldots, w_N(A)$ are produced. Finally the process overlays all these copies into one new state, the output

$$W(A) = w_1(A) \cup w_2(A) \cup \ldots \cup w_N(A).$$

Running the process in feedback mode this corresponds to iterating the operator $W$. The Hutchinson operator turns repeated acts into a dynamical system: an IFS. Let $a_0$ be an initial aed (state). Then we obtain

$$a_{k+1} = W(a_k), \quad k = 0, 1, 2, \ldots$$

a sequence of sets (states), by repeating applying $W$. An IFS generates a sequence which tends toward a final state $a_\infty$ which can be called the attractor of the IFS and which is left invariant by the IFS. It means

$$W(a_\infty) = a_\infty.$$

**Coarse Hölder exponent.** Taking into account the Hausdorff measurement procedure we can introduce the coarse Hólder exponent

$$\alpha \frac{\log \mu}{\log \varepsilon}.$$

After the coarse definition we can build the frequency distribution of $\alpha$ in the following way. For each value $\alpha$ one evaluates the number $N_\varepsilon(\alpha)$ of boxes size $\varepsilon$ having coarse Hölder exponent equal to $\alpha$. We may take the weighted logarithm

$$f_\varepsilon(\alpha) = -\frac{\log N_\varepsilon(\alpha)}{\log \varepsilon}$$

as the probability distribution to hitting the value $\alpha$. As $\varepsilon \to 0$ the function $f_\varepsilon(\alpha)$ tends to well defined limit $f(\alpha)$.

The definition $f(\alpha)$ means that, for each $\alpha$ the number of boxes increased for decreasing $\varepsilon$ as $N_\varepsilon(\alpha) \sim \varepsilon^{-f(\alpha)}$. The exponent $f(\alpha)$ is a continuous function of $\alpha$ and a graph of function looks like as the mathematical symbol $\bigcap$. The value $f(\alpha)$ could be interpreted as a fractal dimension of the subset of boxes of size $\varepsilon$ having $\alpha$ in the limit $\varepsilon \to 0$. As $\varepsilon \to 0$, there is an increasing number of subsets, each of them characterized by its own $\alpha$ and the fractal dimension $f(\alpha)$. Between other reasons it causes the term multifractal.

# BIBLIOGRAPHY


1. M. Adda-Bedia, R. Arias, M. Ben Amar, and F. Lund. Dynamic instability of brittle fracture. *Phys. Rev. Lett.*, 82(11):2314–2317, 1999.
2. E. S. Aifantis. Gradient deformation models at nano, micro, and macro scales. *Trans. ASME: J. Eng. Mater. Technol.*, 121(2):189–122, 1999.
3. S. Albeverio, J. E. Fenstad, R. Høegh-Krohn, and T. Lindstrøm. *Nonstandard methods in stochastic analysis and mathematical physics*. Academic Press, Orlando/San Diego, 1986.
4. T. L. Anderson. Application of fractal geometry to damage development and brittle fracture in materials. *Scripta Metall.*, 23(1):97–102, 1989.
5. B. D. Annin, V. O. Bytev, and S. I. Senashov. *Group properties of equation of elasticity and plasticity*. Nauka, Novosibirsk, 1985. In Russian.
6. K.-H. Anthony and A. Azirhi. Lagrangian field theory of plasticity and dislocation dynamics. Attempts towards unification with thermodynamics of irreversible processes. *Arch. Mech.*, 50(3):345–365, 1998.
7. P. L. Antonelli, A. Bóna, and M. A. Slawinski. Seismic rays as Finsler geodesic. *Nonlinear Anal. Ser. A, Real World Application*, 4(5):711–722, 2003.
8. R. W. Armstrong. Cleavage crack propagation within crystals by the Griffith mechanism versus a dislocation mechanism. *Mater. Sci. Eng.*, 1(4):251–254, 1966.
9. V. I. Arnold. *Mathematical methods of classical mechanics*. Springer, New York, 1989.
10. G. S. Asanov. *Finsler geometry, relativity and gauge theories*. Reidel, Dordrecht, 1985.
11. V. Ya. Asanovich and I. A. Miklashevich. MT effective medium approximation in **r**-representation for the electron structure of 3d metals. *Phys. Stat. Sol. (b)*, 164:445–451, 1991.
12. Ph. Avouirs, T. Hertel, R. Martel, T. Schmidt, H. R. Shea, and R. E. Walkup. Carbon nanotubes: nanomechanics, manipulations, and electronic devices. *Appl. Surf. Sci.*, 141:201–209, 1999.
13. L. B. Babkin and E. M. Morozov. *Fizika i mechanika deformacii i razrushenija*, chapter Zakriticheskaja skorost' rasprostranenija teshchiny v rastianutoj poposie. Atomizdat, Moscow, 1979, p. 9–17. In Russian.
14. V. F. Badaeva, P. P. Kaminskij, and Yu. A. Khon. To theory of macroscopic plastic deformation localisation. *Phys. Mesomechanics*, 4(1):43–48, 2001. In Russian.
15. A. S. Balankin. Fractal mechanics of deformable media and failure topology of solids. *Trans. USSR Acad. Sci. Earth Sci. Sections*, 323(2):29, 1994.
16. A. S. Balankin. Physics of fracture and mechanics of self-affine cracks. *Enging. Fract. Mech.*, 57(2/3): 135–203, 1997.
17. M. V. Baninchuk. To one of variational principles in crack theory. *Izvestija AN. Mechanica tverdogo tela*, 4:197–199, 1969. In Russian.
18. V. V. Barkaline and I. A. Miklashevich. To develop the programme system of hierarchical modeling the properties of materials, construction and processes of nanotechnology, nanoelectronics and chemical techonology for the cluster high-performance multiprocessors systems. Technical Report 2222, Belarusian National Technical University, Minsk, Belarus, 2006.
19. M. F. Barnsley, S. G. Demko, J. H. Elton, and J. S. Geronimo. Invariant measures for Markov processes arising from iterated function systems with place-dependent probabilities. *Ann. Ins. Henri Poincare, Sect. A.*, 24:367–394, 1988.
20. S. J. Basinski and Z. S. Basinski. Plastic deformation and work hardening. In *Dislocations in Solids*, Ed. by F. Nabarro. North Holland, Amstrdam, 1979, pp. 261–362.
21. C. Beck and F. Schlogl. *Thermodynamics of chaotic systems*. Cambridge University Press, Cambridge, 1993.
22. G. E. Beltz and L. L. Fisher. Effect of finite crack length and blunting on dislocation nucleation in mode III. *Philos. Mag. A*, 79(6):1367–1378, 1999.

23. J. J. Benbow. Cone cracks in fused silica. *Proc. Phys. Soc. (London)*, 75(5):697–699, 1960.
24. B. Bhushan and A. Majumdar. Fractal model of elastic-plastic contact between rough surfaces. *J. Trib.*, 113:1–11, 1990.
25. B. Bhushan and A. Majumdar. Role of fractal geometry in roughness characterisation and contact mechanics of surfaces. *J. Trib.*, 112:205–216, 1990.
26. B. A. Bilby, A. H. Cottrell, and K. H. Swinden. The spread of plastic yield from a notch. *Proc. Roy. Soc.*, A272(1350):304–314, 1963.
27. B. A. Bilby, R. Bullough, and E. Smith. Continuous distribution of dislocations: a new application of the methods of non-Riemanninian geometry. *Proc. Roy. Soc.*, A231(1185):263–273, 1955.
28. X. Blanc, C. Le Bris, and P.-L. Lions. From moleuclar models to continuum mechanics. *Arch. Rational Mech. Anal.*, 164:341–381, 2002.
29. E. M. Blanter and M. G. Scnirman. Simple hierarchical systems: Stability, self-organisation criticality, and catastrophic behavior. *Phys. Rev. E*, 55(6):6397–6403, 1997.
30. D. I. Blochincev. *Acoustic of non-homogeneous moving media (Akustika neidnorodnoj dvizhushchejsia sredy)*. Nauka, Moscow.
31. N. M. Bogatov and H. P. Kryjniaia. Gauge field theory of structural defects and thermal tensions. *Phys. Stat. Sol. (b)*, 207(2):317–321, 1998.
32. N. N. Bogoliubov and D. V. Shirkov. *Introduction to the theory of quantized fields*. Wiley, New York, 3rd edition, 1980.
33. A. A. Bogush. *Introduction to the field theory of elementary particles*. Nauka i Technika, Minsk, 1981. In Russian.
34. L. A. Borisoglebskii. *Quantum mechanics*. Belarusian State University, Minsk, 1981. In Russian.
35. Luca Bortoloni and Paolo Cermelli. Rate-independent plasticity and statistically stored dislocations. *Z. Angew. Math. Phys. (ZAMP)*, 55(1):105–120, 2004.
36. E. Bouchand. Scaling properties of cracks. *J. Phys: Condens. Matter*, 9(21):4319–4344, 1997.
37. E. Bouchand, J.-P. Bouchand, J. Peans, and G. Lapasset. The statistics of crack branching during fast crack propagation. *Fractals*, 4:1051–1058, 1993.
38. W. F. Brown and J. E. Srawley. Plane strain crack toughness testing of high strength metallic materials. *ASTM Special Technical Publication*, 410, 1966.
39. L. Bruno. Inelastic material and fracture mechanics: a variational approach. *Arch. Appl. Mech.*, 69(6): 393–405, 1999.
40. I. Bucatatu and M. A. Slawinski. Generalized orthogonality between rays and wavefronts in anisotropic inhomogeneous media. *Nonlinear Anal. Ser. A, Real World Application*, 6(1):111–121, 2005.
41. V. Bulatov, F. Abraham, L. Kubin, B. Devincte, and S. Yip. Connecting atomistic and mesoscale simulations of crystal plasticity. *Nature (GB)*, 391(6668):669–672, 1998.
42. V. S. Buldyrev. Asymptotics of solution wave equation concentrated near the axis of plane waveguide in inhomogeneous media. *Problemy matematicheskoj fiziki (Problems of mathematical physics)*, 3(5):5–30, 1968. Leningrad, in Russian.
43. W. L. Burke. *Spacetime, geometry, cosmology*. University Science Books, Mill Valley, CA, 1980.
44. J. W. Cahn. *Interfacial segregation*, volume 3. American Society for Metals, Metals Park, OH, 1979.
45. R. W. Cahn and P. Haasen, editors. *Physical metallurgy*, volume 1. North-Holland Physics Publishing, Amsterdam, 3rd edition, 1983.
46. G. Capriz. Sploshnyje sredy s substructuroj. Chast' I. *Fizicheskaja mesomechanica*, 3(4):5–14, 2000. In Russian.
47. G. Capriz. Sploshnyje sredy s substructuroj. Chast' II. *Fizicheskaja mesomechanica*, 3(6):37–50, 2000. In Russian.
48. G. Capriza and C. Davini. A geometric classification of inhomogeneities in continua with dislocations. *Meccanica*, 19(4):309–314, 1984.
49. E. Charkaluk, M. Bigerelle, and A. Iost. Fractals and fracture. *Enging. Fract. Mech.*, 61(1):119–139, 1998.
50. A. V. Chechkin. *Mathamatical informatics*. Nauka, Moscow, 1991. In Russian.
51. V. F. Chekurin. Inverse problem of nondestructive control of the level of hardening of sheet glass. *Mech. Solids*, 33(3):68, 1998.
52. G. P. Cherepanov. *Mechanics of brittle fracture*. McGraw-Hill, New York, 1979.

53. G. P. Cherepanov. *Mechanika razruchenija kompozicionnych materialov (Fracture mechanics of composite materials)*. Nauka, Moscow, 1983. In Russian.
54. G. P. Cherepanov. General theory of failure. *Mater. Sci.*, 22(1):32–39, 1986.
55. G. P. Cherepanov. Quantum fracture mechanics. *Strength Mater.*, 22(2):155–163, 1990.
56. G. P. Cherepanov, A. S. Balankin, and V. S. Ivanova. Fractal fracture mechanics: a review. *Engng. Fract. Mech.*, 51(6):997–1033, 1995.
57. G. N. Chernyshov, A. P. Popov, V. M. Kozincev, and I. I. Ponomarev. *Ostatochnyje napriazhaenija v deformirujemysh tverdych telach (Residual stress in deformed solids)*. Nauka, Moscow, 1996. In Russian.
58. A. V. Chigarev. *Stochastic and regular dynamics of inhomogeneous media*. UP Tekhnoprint, Minsk, 2000. In Russian.
59. A. V. Chigarev, Yu. V. Chigarev, and I. A. Miklashevich. Some models of crack trajectory stochastisation in composite materials. *Problems of mechanics non-elastic deformation*, pp. 371–381. Fizmatlit, Moscow, 2001. In Russian.
60. A. V. Chigarev and Yu. V. Chigarev. About possibility of arising stochastic instability in inhomogeneous media. *Akusticheskij jurnal (Acoustic Journal)*, XXIV(5):765–771, 1978. In Russian.
61. A. V. Chigarev and I. A. Miklashevich. Statistical properties of stochasically distributed dislocation by the plane deformation. In *Materialy Sympozjum Mechaniki Zniszczenia Materialow i Konstrukcji*, Awgustow, 23–26 Maja, 2001. Bialystok, Poland, 2001, pp. 57–62. In Russian.
62. A. Chudnovsky, B. Kunin, and M. Gorelik. Modeling of britle fracture based on the concept of crack trajectory ensemble. *Eng. Fract. Mech.*, 58(5-6):437–457, 1997.
63. Ching-Hwei Chue and Shih-Ming Weng. Fracture analysis of piezoelectric materials with an arbitrarily oriented crack using energy density theory. *Comput. Struct.*, 83:1251–1265, 2005.
64. F. Costanzo and D. H. Allen. A continuum mechanics approach to some problems in subcritical crack propagation. *Int. J. Fracture*, 63(1):27–57, 1993.
65. B. Cottrell and J. R. Rice. Slightly curved or kinked cracks. *Int. J. Fracture*, 16(2):155–169, 1980.
66. J. G. Mc Crea, F. W. Hehl, and E. W. Mielke. Mapping Noeter identities into Bianchi identities in the general relativistic theories of gravity and in the field theory of static lattice defects. *Int. J. Theor. Phys.*, 29(11):1185–1205, 1990.
67. E. S. Şuhubi and A. Bakkaloĝlu. Symmetry groups for arbitrary motions of hyperplastic solids. *Int. J. Engng. Sci*, 35(7):637–657, 1997.
68. Shu-Ho Dai and J. C. M. Li. Dislocation-free zone at the crack tip. *Scripta Metall.*, 16(2):183–188, 1982.
69. D. L. Davidson. Fracture surface roughness as a gauge of fracture toughness: aluminium particulate SiC composites. *J. Mater. Sci.*, 24(2):681–687, 1989.
70. R. de Wit. *Linear theory of static disclinations*, volume 1 of *Fundamental aspects of dislocation*. National Bureau of Standards, 1970, pp. 651–673.
71. S. V. Derezin and L. M. Zubov. Equation of nonlinear elastic continuum with continuously distributed dislocations and disclinations. *Doklady AN*, 366(6):762–765, 1999. In Russian.
72. B. V. Derjaguin, N. V. Churaev, and V. M. Muller. *Surface forces*. Consultants Bureau, New York, 1987.
73. M. Devis. *Applied nonstandard analysis*. Wiley-Interscience, New York/London/Sydney, 1977.
74. E. Doege, B. Lackman, and B. Kischnick. Fractal geometry used for the characterization of sheet surfaces. *Ann. CIRP*, 44(1):197–200, 1995.
75. G. A. Domrachev and A. I. Lazarev. Application of the theory of algebraic systems for creating a hierarchy of solid structures formed under equilibrium and nonequilibrium conditions. *Phys. Solid State*, 41(5):720–724, 1999.
76. L. K. Dong, H. Y. Zhang, and C. W. Lung. Application of gauge theory of defects to fracture. *Int. J. Solids Structure*, 25(7):707–713, 1989.
77. G. Drazer, H. Auradou, J. Koplik, and J. P. Hulin. Self-affine fronts in self-affine fractures: Large and small scale structure. *Phys. Rev. Lett.*, 92(1):014501, 2004.
78. F. Ducastelle. Electronic structure: from bonding to mechanical properties. In V. Pontikis J. Lépinoux, D. Mazière and G. Saada, editors, *Multiscale phenomena in plasticity. From experiments to phenomenology, modelling and materials engineering*, volume 367 of *NATO Science Series, Series E: Applied Science*. Kluwer Academic Publisers; Dordrecht/Boston/London, 2000, pp. 239–257.
79. J. Dundurs. *Elastic interactions of dislocations with inhomogeneities*. Mathematics theory of dislocations. American Society of Mechanical Engineering, New York, 1968, pp. 70–115.

80. J. Dundurs. Edge-bonded dissimilar orthogonal elastic wedges. *J. Appl. Mech.*, 36(4):650–652, 1969.
81. D. Dutta, P. Barat, and S. N. Ganguly. Acoustic characterisation of fractal crack surface. *Physica D*, 122(1-4):155–160, 1998.
82. D. G. B. Edelen. Aspects of variational arguments in the theory of elasticity: fact and folklore. *Int. J. Solids Structure*, 17:729–740, 1981.
83. A. V. Eletskii. Carbon nanotubes. *Physics—Uspekhi*, 40(9):899–924, 1997.
84. E. Cartan. *Lecons sur la geometrie des espaces de Riemann*. Gauthier-Villars, Paris, 1951. Russian edition: Moscow, 1960, 307pp. from first edition.
85. E. Cartan. *Théorie des groupes finis et continus et la géométrie differentielle traitées par la methode du repere mobile : leçons professées à la Sorbonne*. Gauthier-Villars, Paris, 1951. Russian edition: Moscow, 1963, 367pp. from Lection, in Moscow.
86. J. P. Elliott and P. G. Dawber. *Symmetry in physics*, volume 1 of *Principles and simple applications*. MacMillan, London, 1979.
87. D. Ertaş and M. Kardar. Dynamic roughening of directed lines. *Phys. Rev. Lett.*, 69(6):929, 1992.
88. D. Ertaş and M. Kardar. Anisotrope scaling in depinning in a flux line. *Phys. Rev. Lett.*, 73(12):1703, 1994.
89. D. Ertaş and M. Kardar. Anisotrope scaling in threshold critical dynamics of driven directed lines. *Phys. Rev. B*, 53(6):3520–3542, 1996.
90. J. D. Eshelby. The elastic field of a crack extending non-uniformly under general anti-plane loading. *J. Mech. Phys. Solids*, 17:177, 1969.
91. V. I. Fabricant. *Application of potential theory in mechanics: a selection to new results*. Kluwer Academic, The Netherlands, 1989.
92. K. Falconer. *Fractal geometry*. John Wiley and Sons, Chichester/New York, 1990.
93. J. Feder. *Fractals*. Plenum Press, New York, 1991.
94. F. I. Fedorov. *Theory of elastic waves in crystals*. Plenum Press, New York, 1968.
95. L. A. Fil'shtinskii and S. P. Shapovalov. Comparison of statistical and deterministic approaches to the determination of mechanical properties of fiber composites. *Mech. Solids*, (5):55–61, 1985. Translation from Izvestija AN SSSR. *Mechanika tverdogo tela.*
96. J. Fineberg and M. Marder. Instability in dynamic fracture. *Physics Reports*, 313(1–2):1–108, 1999.
97. E. S. Flejshner. *Osnovy sistemologii (Basis of systemology)*. Radio i sviaz, Moscow, 1982. In Russian.
98. S. Fratini, O. Pla, P. González, F. Guinea, and E. Louis. Energy radiation of moving cracks. *Phys. Rev. B*, 66(10):104104, 2002.
99. L. B. Freund. Crack propagation in an elastic solid subject to general loading. I. Constant rate of extension. *J. Mech. Phys. Solids*, 20:129–140, 1972.
100. L. B. Freund. Crack propagation in an elastic solid subject to general loading. II. Non-uniform rate of extension. *J. Mech. Phys. Solids*, 20:141–152, 1972.
101. L. B. Freund. *Dynamic fracture mechanics*. Cambrigde University Press, Cambridge, 1998.
102. Ya. B. Fridman. Diagramma otnositelnoy structurnoj neodnorodnosti materialov (diagram of relative structural nonhomogeneity of materials). *Doklady Akademii Nauk SSSR*, 106(2):258–261, 1956. In Russian.
103. T. Fukuhara and H. Nakanishi. Dynamical stability of crack front line. *J. Phys. Soc. Japan*, 67(12): 4064–4067, 1998.
104. J. F. Ganghoffer. New concepts in nonlocal continuum mechanics and new materials obeying a generalised continuum behaviour. *Int. J. Eng. Sci.*, 41:291–304, 2003.
105. G. Gao, T. Çagin, and W. A. Goddart III. Energetics, structure, mechanical and vibrational properties of single-walled carbon nanotubes. *Nanotechnology*, 9(3):184–191, 1998.
106. I. Goldhirsch and C. Goldenberg. On the microscopic foundations of elasticity. *Eur. Phys. J. E*, 9:245–251, 2002.
107. R. V. Goldshtein and A. B. Mosolov. Treshchiny s fractalnoj poverchnost'jy (cracks with fractal surface). *Doklady Akademii Nauk SSSR*, 319(4):840–844, 1991. In Russian.
108. K. Golos and F. Ellyin. Total strain energy density as a fatigue damage parameter. In *Adv. Fatigue Sci and Technol.*, Proc. NATO Adv. Study Inst. Conf. Alvor, April 4–15, 1988.
109. A. V. Grachev, A. I. Nesterov, and S. G. Ovchinnicov. The gauge theory of point defects. *Phys. Stat. Sol. (b)*, 156(2):403–410, 1989.
110. A. E. Green and P. M. Naghdi. A unified procedure for construction of theories of deformable media. I. Classical continuum physics. *Proc. Roy. Soc. London A*, 448(1934):335–356, 1995.

111. A. E. Green and P. M. Naghdi. A unified procedure for construction of theories of deformable media. II. Generalized continua. *Proc. Roy. Soc. London A*, 448(1934):357–377, 1995.
112. A. E. Green and P. M. Naghdi. A unified procedure for construction of theories of deformable media. III. Mixtures of interacting continua. *Proc. Roy. Soc. London A*, 448(1934):379–389, 1995.
113. M. Grinfeld. *Thermodynamic methods in the theory of heterogeneous systems*. Longman, Harlow, 1991.
114. M. G. Gringauz, G. P. Starikovskii, L. A. Fil'shtinskii, S. P. Shapovalov A. M. Kharchenko, and V. T. Shcherbakov. Averaging of the properties of hybrid fibrous composites. *Mech. Compos. Mater.*, 22(6):698–702, 1987.
115. Yu. V. Grinyaev and N. V. Chertova. Mechanical properties of materials and the object of description in gauge theories. *Tech. Phys.*, 43(7):809–813, 1998.
116. Yu. V. Grinyaev and N. V. Chertova. Physical content of a gauge model describing media with structure and defects. *J. Appl. Mech. Tech. Phys.*, 40(6):1143–1147, 1999.
117. F. Guiu, M. Alguero, and M. J. Reece. Crack extension force and rate of mechanical work of fracture in linear dielectrics and piezoelectrics. *Philos. Mag.*, 87(7):8873–888, 2003.
118. M. A. Gusev and V. P. Myasnikov. A thermomechanical model of an elastoplastic material with structural defects. *Mech. Solids*, 33(4):131, 1998. Translation from *Izvestija AN. Mechanika tverdogo tela*.
119. P. Hähner and H. Stamm. A dislocation dynamical theory of the ductile-to-brittle transition. *Acta Metall. Mater*, 43(7):2797–2805, 1995.
120. T. C. Halsey, M. N. Jensen, L. P. Kadanoff, I. Procaccia, and B. I. Shrailtlan. Fractal measures and their singularities: the characterization of strange sets. *Phys. Rev. A.*, 33(2):1141–1151, 1986.
121. A. Hansen and E. L. Hinrichsen. Roughness of crack interfaces. *Phys. Rev. Lett.*, 66(19):2476–2479, 1991.
122. M. H. He and J. W. Hutchinson. Crack deflection at an interface between dissimilar elastic materials. *Int. J. Solids Structure*, 25(9):1053–1067, 1989.
123. K. P. Herrmann and V. V. Loboda. Special approach for the determination of fracture mechanical parameters at an interface crack tip. *Arch. Appl. Mech.*, 68(3/4):227–236, 1998.
124. T. Hertel, R. E. Walkup, and P. Avouirs. Deformation of carbon nanotubes by surface van der Waals forces. *Phys. Rev. E*, 58(20):13870–13873, 1998.
125. D. W. Hess and N. Bernstein. The effects of quantum-mechanics on fracture in silicon. *APS Meeting Abstracts*, p. 17003, March 2001.
126. J. P. Hirth and J. Lote. *Theory of dislocation*. Krieger Publishing, Malabar, FA, 1992.
127. P. J. Holmes. A nonlinear oscillator with a strange attractor. *Phil. Trans. R. Soc.*, 292(1394):419–448, 1979.
128. V. R. Hoves and S. Tolansky. Pressure crack-figures on diamond faces. I. The octahedral face. *Proc. Roy. Soc*, A230(1182):287–294, 1955.
129. J. W. Hutchinson, M. Y. He, and A. G. Evans. The influence of imperfections on the nucleation and propagation of buckling driven delaminations. *J. Mech. Phys. Solids*, 48:709–734, 2000.
130. A. A. Iljushin. *Solids mehanics*. Moscow University Publishing, Moscow, 1971. In Russian.
131. V. N. Ionov and V. V. Selivanov. *Dinamika razrushenija deformirujemogo tela*. Mashinostrojenije, Moscow, 1987. In Russian.
132. A. Yu. Ishlinsky and D. D. Ivlev. *Mathematical theory of plasticity*. Fizmatlit, Moscow, 2001. In Russian.
133. A. G. Ivanov. Lokalnyj i integralnyj energeticheskija podchody v probleme rezrushenija (Local and integral approximation in problem of fracture). In *Prikladnyje problemy prochnosti i plastichnosti, chislennoje modelirovanije fiziko–mechanicheskich processov*. Gorkovskij gosudarstvennyj universitet, Gor'kij, 1990, pp. 4–20. In Russian.
134. V. N. Ivanov. Sinergetika deformirovanija i razrushenija structurno-neonorodnych tverdych tel. Matematicheskij formalism (Synergetics of deformation and fracture of structural-inhomogeneous solids. Mathamatical formalism). In *Deformirovanije i razrushenije structurno-neonorodnych materialov i konstrukcij: Sbornik nauchnych trudov*. UrO AN SSSR, Sverdlovsk, 1989, pp. 3–13. In Russian.
135. V. S. Ivanova, A. S. Balankin, I. M. Bunin, and A. A. Oksochev. *Synergetica and fractals in theory of metals*. Nauka, Moscow, 1994. In Russian.
136. V. S. Ivanova and G. V. Vstovsky. Mesomechanics of stability of fractal structures of solids in non-linear conditions. *Phys. Mesomech.*, 2(5):19–25, 1999.

137. D. D. Ivlev. To general equations of theory of ideal plasticity and statics of dry substances. *Prikladnaja Matematika Mechanika*, 22(1):90–96, 1958. In Russian. Reprinted in *Mechanics of plastic media*, Volume 1, Fizmatlit, Moscow, 2001, pp. 34–37.
138. D. D. Ivlev. To relation which are determined plastic flow by the tresca condition and its generalisation. *Doklady Ademii Nauk SSSR*, 124(3):546–549, 1959. In Russian. Reprinted in *Mechanics of plastic media*, volume 1, Fizmatlit, Moscow, 2001, pp. 15–19.
139. B. Jin, A. K. Soh, and Z. Zhong. Propagation of an anti-plane moving crack in a functionally graded piezoelectric strip. *Arch. Appl. Mech.*, 73(3–4):252–260, 2003.
140. C. H. Johannson and P. A. Persson. *Detonic of high explosives*. Swedish Detonic Research Foundation Vinterviken, Stokholm, Sweden, 1970.
141. B. E. Johnson and A. Hoger. To use of strain energy to quantify the effect of residual stress on mechanical behavior. *Mathem. Mech. Solids*, 3(4):447–470, 1998.
142. Aida Kadič and Dominic G. B. Edelen. *A gauge theory of dislocations and disclinations*, volume 174 of *Lecture notes in physics*. Springer, Berlin, 1983.
143. M. O. Katanaev. Geometric theory of defects. *Physics—Uspekhi*, 48(7), 2005.
144. A. Kelly. The strengthening of metals by dispersed particles. *Proc. Roy. Soc.*, A282(1388):63, 1964.
145. A. Kelly. Interface effects and the work of fracture of a fibrous composite. *Proc. Roy. Soc.*, A319(1536):95, 1970.
146. A. P. Kfouri and K. J. Miller. Crack separation energy-rates for the dbcs model under biaxial modes of loading. *J. Mech. Phys. Solids*, 27:135–150, 1979.
147. A. P. Kfouri and J. R. Rice. Elastic-plastic separation energy rate for crack advance in finite growth steps. In D. M.R. Taplin, editor, *Fracture 1977. Advances in research on the strength and fracture of materials. Proc., 4th Int. Conf. on Fracture, ICF4, Waterloo, June 1977*, volume 1, Univ. of Waterloo, Ontario, Canada. Pergamon Press, 1978, pp. 43–59.
148. A. M. Khludnev and J. Sokolowski. Griffith formulae for elasticity systems with unilateral conditions in domains with cracks. *Eur. J. Mech. A/Solids*, 19:105–119, 2000.
149. S. A. Khristianovich and E. I. Shemiakin. K teorii idealnoj plastichnosti. *Izvestija AN. Mechanika tverdogo tela*, (2):63–69, 1967. In Russian.
150. S. A. Khristianovich and E. I. Shemiakin. O ploskoj deformacii plasticheskogo materiala pri slozhnom nagrushenii. *Izvestija AN. Mechanika tverdogo tela*, (5):138–149, 1969. In Russian.
151. R. Kienzler and G. Herrmann. Conservation laws in nonhomogeneous plane elastostatics. *J. Mech. Phys. Solids*, 45(5):789–805, 1997.
152. R. Kienzler and G. Herrmann. On four-dimensional formalism in continuum mechanics. *Acta Mechanica*, 161(1-2):103–125, 2003.
153. R. Kienzler and G. Herrmann. On conservation laws in elastodynamics. *Int. J. Solids Structures*, 41: 3595–3606, 2004.
154. E. N. Kilchevkaya and N. A. Kilchevsky. Function of kinetical stresses and space geometry in deformed continua. In N. N. Bogolubov et al., editors, *Mechanika sploshnoj sredy i rodstvennyje problemy analiza: k vosmidesiatiletiyu akademika N. I. Muschelishvili (Mechnics of continua and related problems of analysis: to 80 anniversary of acedician N. I. Muschelishvili)*, Nauka: Redaction of physical and mathematical literature. Nauka Moscow, 1972, pp. 243–250. In Russian.
155. Yu. L. Klimontovich. *Statistical theory of open systems*. Kluwer, Dordrecht, 1995.
156. G. Kluge. Zur Dynamik der allgemeinen Verzetzungs-Theorie bei berücksichtigung von Momentenspannungen. *Int. J. Engng. Sci.*, 7(2):169–182, 1969.
157. V. D. Klushnikov. *Physical and mathematical basis of strength and plasticity*. Moscow State University, Moscow, 1994. In Russian.
158. Hideo Kobajashi. Micromechanics of crack growth. *J. Japn. Soc. Proces. Eng.*, 54(6):1035–1039, 1988.
159. S. Kobayashi and S. Ohr. In situ fracture exeriments in b. c.c. metals. *Philos.*, 42(6):763–772, 1980.
160. S. Kobayashi and S. M. Ohr. *37th Ann. Proc. Electron Microscopy Soc. Amer.*, San Antonio, TX, 1979, pp. 424–425.
161. S. Kobayashi and K. Nomizu. *Foundations of differential geometry*, volume 1. Interscience, New York/ London, 1963.
162. K. Kondo. On the geometrical and physical foundations of the theory of yielding. In *Proc. 2nd Japan Nat. Congr. Appl. Mech.*, Tokyo, 1952, p. 41.

163. K. Kondo. On the fundamental equations of the macroscopic mechanical behaviour of microscopically non-uniform materials. *RAAG Memoirs*, DII:470–483, 1955.
164. K. Kondo. Non-Riemannian and Finslerian approaches to the theory of yielding. *Int. J. Engng. Sci.*, 1(1):71–88, 1963.
165. K. Kondo. On the analytical and physical foundations of a theory of dislocations and yelding by the differential geometry of continua. *Int. J. Engng. Sci*, 2:219–251, 1964.
166. V. M. Kornev and I. V. Yakovlev. Model of undulation in explosive welding. *Combust. Explosion Shock Waves*, 20(2):204–207, 1984.
167. B. V. Kostrov. Unsteady propagation of longitudial shear cracks. *Appl. Math. Mech.*, 30(6):1241–1248, 1966. English translation from *Prikl. Mat. Mek.*
168. B. V. Kostrov and L. V. Nikitin. Some general problems of mechanics of brittle fracture. *Archivum Mech. Stosowanej*, 22(6):749–776, 1970.
169. G. V. Kozlov, V. U. Novikov, and O. Yu. Bur'jan. Fractal variant of Griffith equation by the fracture of inter-phase layer in polymer composites. *Materialovedenije (Material Science)*, (11):13–16, 1999. In Russian.
170. E. Kröner. Interrelations between various branches of continuum mechanics. In E. Kröner, editor, *IUTAM Symposium, Mechanics of Generalised Continua*. Springer Verlag, 1968, p. 330.
171. E. Kröner. *Conitnuum theory of defects*. Physics of defects. Ed. by R. Ballian, M. Kleman and J. P. Poirier. North-Holland Amsterdam, 1981, pp. 216–315.
172. E. Kröner. Dislocation in crystals and in continua: a confrontation. *Int. J. Engng. Sci.*, 33(15):2127– 2135, 1995.
173. I. A. Kunin. *Elastic media with microstructure*, volume 1. Springer, Berlin, 1982.
174. I. A. Kunin. Kinematics of media with continuously changing topology. *Int. J. Theor. Phys.*, 29(11): 1167–1176, 1990.
175. M. V. Kurlenia, V. N. Oparin, and V. I. Vostrikov. About possible description to mountain shock prognosis. *J. Mining Sci.*, 24(6), 1988. Translation from Russian *FTPRPI*, 6:3–15, 1998.
176. A. G. Kusraev and S. S. Kutateladze. *Non-standard methods of analysis*. Nauka, Novosibirsk, 1990. In Russian.
177. L. D. Landau and E. M. Lifshic. *Theory of elasticity*, volume VII of *Course of theoretical physics*. Pergamon Press, Oxford, 1959.
178. L. D. Landau and E. M. Lifshic. *The classical theory of fields*, volume 2 of *Course of theoretical physics*. Pergamon Press, 1976.
179. Kin Tak Lau, Mei Lu, and David Hui. Coiled carbon nanotubes: synthesis and their potential applications in advanced composite structures. *Composites B*, 37(6):437–448, 2006.
180. M. Lazar. Dislocation theory as a 3-dimensional translation gauge theory. *Ann. Phys. (Leipzig)*, 9(6): 461–473, 2000.
181. J.-B. Leblond. Basis results for elastic fracture mechanics with frictionless contact between the crack lips. *Eur. J. Mech. A/Solids*, 19:633–647, 2000.
182. N. I. Lebovka and V. V. Mank. Phase diagram and kinetics of inhomogeneous square lattice brittle fracture. *Physika A*, 181:346–363, 1992.
183. N. I. Lebovka, V. V. Mank, and N. S. Pivovarova. Modeling of fracture of inhomogeneous system by the nonstationary deformation. *Phys. Solid State*, 34(7):2007–2015, 1992.
184. E. H. Lee. Elastic-plastic deformation at finite strain. In Wei H. Yang, editor, *Topic in plasticity*. AM Press, Boston/Ann Arbor, MI, 1991, pp. 1–30. (Anniversary volume in honor of Profesor E. H. Lee.)
185. Weisheng Lei and Bingsen Chen. Fractal characterization of some fracture phenomena. *Enging. Fract. Mech.*, 50(2):149–155, 1995.
186. X.-F. Li. Griffith crack moving in a piezoelectric strip. *Arch. Appl. Mech.*, 72(10):745–758, 2002.
187. V. A. Lichachiov, V. E. Panin, E. E. Zasimchuk, et al. *Kooperativnyje deformacionnyje processy i lokalizacija deformacii*. Naukova Dumka, Kiev, 1989. In Russian.
188. A. J. Lichtenberg and M. A. Liberman. *Regular and stochastic motion*. Springer-Verlag, Berlin/Heidelberg/ New York, 1983.
189. Ch. -Bo Lin and Ch. -Sh. Yeh. The magnetoelastic problem of a crack in a soft ferromagnetic solid. *Int. J. Solids Structures*, 39:1–17, 2002.
190. Ch. -Sh. Liu. Lie symmetries of finite strain elastic-perfectly plastic models and exactly consistent schemes for numerical integrations. *Int. J. Solids Structures*, 41(7):1823–1853, 2004.

191. V. V. Loboda and A. E. Sheveliova. Determining prefracture zones at a crack tip between two elastic orthotropic bodies. *Int. Appl. Mech.*, 39(5):566–572, 2003.
192. V. V. Loboda. The quasi-invariant in the theory of interface cracks. *Engng. Fracture Mech.*, 44:573–580, 1993.
193. X. Lu and M. Comninou. The sinusoidal crack. *Engng. Fracture Mech.*, 34(3):649–656, 1989.
194. C. W. Lung and L. Y. Xiong. The dislocation distribution in the plastic zone at the crack tip. *Phys. Stat. Sol. A.*, 77:81–86, 1983.
195. N. M. Macmillan. The ideal strength of solids. In R. Latanison and J. R. Pickens, editors. *Atomistics of fracture*. Plenum Press, 1983, pp. 95–164.
196. B. Mandelbrot. *The fractal geometry of nature*. Freeman, New York, 1984.
197. B. Mandelbrot. Self-affine fractal sets. I. The basic fractal dimensions; II. Length and area measurements; III. Hausdorff dimension anomalies and their implications. In Luciano Pietronero and Erio Tosatti, editors, *Fractals in Physics. Proceedings of VI Int. Symp., 9–12 July*, ICTP, Trieste, Italy. North-Holland, Amsterdam, 1985, pp. 3–28.
198. M. Marder. New dynamical equation for cracks. *Phys. Rev. Lett.*, 66(19):2484–2487, 1991.
199. V. M. Markochev. Accident theory and fracture mechanics. *Strength Mater.*, 17(7):929–934, 1985.
200. G. Maughin. *Continuum mechanics of electromagnetics solids*. North-Holland, Amsterdam, 1988.
201. J. K. McCoy and A. J. Markworth. Effect of the discrete/continuum boundary on atomistic simulation of a crack tip and its dependency on model size. *Scripta Metall.*, 21(9):1247–1251, 1987.
202. Robert M. McMeeking. The energy release rate for a Griffith crack in a piezoelectric material. *Eng. Fracture Mech.*, 71:1149–1163, 2004.
203. P. Meakin. The growth of rough surfaces and interfaces. *Physics Reports*, 235(4 & 5):189–289, 1993.
204. V. K. Melnikov. Lines of force of a magnetic field. *Soviet Physics—Doklady*, 7(6):502–504, 1962.
205. V. K. Melnikov. Kachestvennoje opisanije silnogo resonansa v nelinejnoj sisteme (quality description of strong resonance in non-linear system). *Doklady Akademii Nauk SSSR*, 148(6):1257–1260, 1963. In Russian. English translation *Soviet Physics—Doklady*.
206. Michael B. Mensky. *Continuous quantum measurements and path integrals*. IOP Publishing, Bristol, 1993.
207. M. D. Mesarovic, D. Masko, and Y. Takahara. *Theory of hierarchical multilevel systems*. Academic Press, New York and London, 1970.
208. M. D. Mesarovic and Y. Takahara. *General systems theory: mathematical foundation*. Academic Press, New York and London, 1975.
209. A. B. Migdal. *Search of the trustful*. Znanije, Moscow, 1978. In Russian.
210. I. A. Miklashevich. Isomorphism of processes of deformation and motion in space with special geometry. In *Fizika processov deformacii i razrushenija i prognosirovanije mechanicheskogo povedenija materialov*, Proceedings XXXVI international seminar, 26–29 Sept. 2000. Vitebsk, 2000. In Russian.
211. I. A. Miklashevich. The trajectory of a crack in inhomogeneous media subjected to plane loading. *J. Compos. Mater. Design*, 6(3):408–418, 2000. In Russian.
212. I. A. Miklashevich. The crack front equation and the generalized theory of the discontinuity. In *Mechanica–2002*, Proceedings of the international conference April 4–5, 2002, Kaunas University of Technology, Lithuania. Technologija, Kaunas, 2002, pp. 195–200.
213. I. A. Miklashevich. Fractal dimensionality of deterioration based on medium properties. *Doklady Natsionalnoi Akademii Nauk Belarusi*, 46(1):48–50, 2002. In Russian.
214. I. A. Miklashevich. Group properties of deformation and procedure of introduction of gauge fields. *Materialy. Tekhnologii. Instrumenty (Materials, Technologies, Tools)*, 7(3):16–19, 2002. In Russian.
215. I. A. Miklashevich. Ob ierarchicheskom podchodie v mechanike deformirujemogo tverdogo tela (To hierarchical approximation in solid mechanis). *Vestnik BGPA*, (3):58–63, 2002. In Russian.
216. I. A. Miklashevich. The structure boundary influence on the crack propagation by the plane loading. *Mekhanika Kompozitsionnykh Materialov i Konstruktsii*, 8(2):255–260, 2002. In Russian.
217. I. A. Miklashevich. Crack propagation in a solid containing an interface with discontinuously changing Young's modulus. *Int. J. Fracture*, 116(3):49–54, 2003.
218. I. A. Miklashevich. The effect of influence zone shape near tip on the crack trajectory bifurcation. In *Mechanica–2003*, Proceedings of the international conference April 3–3, 2003, Kaunas University of Technology, Lithuania. Technologija, Kaunas, 2003, pp. 144–148.
219. I. A. Miklashevich, V. Ya. Asanovich, and B. P. Burylev. Mikroskopicheskije aspekty obrazovanija sojedinenij pri svarke vryvom. *Adgezija rasplavov i pajka materialov*, 25:69–74, 1991. In Russian.

220. I. A. Miklashevich, L. N. Bialiatskaja, and A. V. Chigarev. Nonlinear effects at the crack propagation. In V. Kuvshinov and L. Babichev, editors, *Nonlinear phenomena in complex systems: fractals, chaos, phase transitions, self-organization*, Proceedings of IX Annual Seminar NPCS'2000. Minsk, 2000, pp. 206–214.
221. I. A. Miklashevich and A. V. Chigarev. Stochastisation of crack growth direction in heterogenous media. In *8 International conference of fracture. Ukraine 93. Part 1.* Kiev, 1993, p. 227.
222. I. A. Miklashevich, A. V. Chigarev, and A. M. Korsunsky. Variational determination of the crack trajectory in inhomogeneous media. *Int. J. Fracture*, 111(2):L29–L34, 2001.
223. I. A. Miklashevich. Designing of the materials with established mechanical properties. In A. G. Gorshkov, editor, *Dynamic and technological problems of the mechanics of construction and media*, Proc. VI Int. Symp. Yaropolec, Moscow, 14–18 Feb. 2000. Grafros, Moscow, 2000, pp. 115–120. In Russian.
224. I. A. Miklashevich. Influence of the defect structure of the material on the fracture and crack motion. *Russian Phys. J.*, 45(12):1165–1169, 2002.
225. I. A. Miklashevich. Crack trajectory instability: propagation in inhomogeneous medium. *Theor. Appl. Fracture Mech.*, 43(3):360–368, 2003.
226. I. A. Miklashevich. Delamination of composites along the interface as a buckling failure of the stressed layer. *Mech. Compos. Mater.*, 40(4):441–450, 2004.
227. I. A. Miklashevich and A. V. Chigarev. The crack trajectory calculation in the composite materials by the linear approximation. *Doklady Akademii Nauk of Belarus (Report of Acad. Sci. Belarus)*, 39(2):114–121, 1995. In Russian.
228. I. A. Miklashevich and A. V. Chigarev. Equation of the crack front with allowance for the metric properties of the material. *Russian Phys. J.*, 45(12):1159–1164, 2002.
229. I. A. Miklashevich and A. V. Chigarev. Stability of a crack trajectory in heterogeneous media. *Mech. Solids*, 37(4):93–37, 2002.
230. S. Minagava. A trial of gauge theory for the stress-function space. *Int. J. Theor. Phys.*, 29(11):1271–1276, 1990.
231. R. Miron. Introduction to the theory of Finsler space. In *Proceeedings of the national seminar on Finsler spaces,* February, 1980, Timisoara, 1981, pp. 131–183.
232. K. L. Molski. Problemy modelowania rozwoju peknesc miedzyfazowysch. *Zeszyty Naukowe politechniki biaiostockiej. Nauki technicznie*, 134(Mecnanika), 2000, z. 22, pp. 143–159.
233. F. C. Moon and P. J. Holmes. A magnitoelastic strange attractor. *J. Sound Vibr.*, 65(2):275–296, 1979.
234. A. I. Moroz. *Curs of system theory (Kurs teorii sistem).* Vysshaja shkola, Moscow, 1987. In Russian.
235. F. M. Morozov. *Nelinejnyje problemy mechaniki i fiziki deformirijemogo tverdogo tela*, volume 1 of *Trudy nauchnoj shkoly akademika V. V. Novozhilova.* St. Petersburg State University, St. Petersburg, 1998, pp. 198–212. In Russian.
236. E. M. Morozov and Ya. B. Fridman. Trajectories of the brittle-fracture cracks as a geodesic lines on the surface of a body. *Soviet Physics—Doklady*, 6(7):619–621, 1962.
237. E. M. Morozov, L. S. Polak, and Ya. B. Fridman. On the variational principles of crack development in hard substances. *Doklady Akademii Nauk SSSR*, 156(3):537–540, 1964. In Russian, English translation *Soviet Physics—Doklady*.
238. N. Morozov and Y. Petrov. *Dynamics of fracture*. Foundation of Engineering Mechanics. Springer, 2000.
239. N. F. Morozov and Y. V. Petrov. Quantum nature and dual character of fracture dynamics in solids. *Physics—Doklady*, 47:85–88, 2002.
240. J. W. Morrisey and J. R. Rice. Crack front waves. *J. Mech. Phys. Solids*, 46(3):467–487, 1998.
241. J. W. Morrisey and J. R. Rice. Perturbative simulations of crack front waves. *J. Mech. Phys. Solids*, 48:1229–1251, 2000.
242. A. B. Movchan, H. Gao, and J. R. Willis. On perturbations of plane cracks. *Int. J. Solids Structures*, 35(26-27):3419–3453, 1998.
243. J. Muto and H. Nagahama. Dielectric anisotropy and deformation of crustal rocks: physical interaction theory and dielectric mylonites. *Phys. Earth Planet. Inter.*, 141:2735, 2004.
244. V. P. Myasnikov and M. A. Gusev. A geometrical model of the defect structure of an elastoplastic continuous medium. *J. Appl. Mech. Tech. Phys.*, 40(2):331–340, 1999.
245. V. P. Myasnikov and M. A. Gusev. Non-euclidean model of material deformation in diffferent hierarchical levels. *Physical Mesomechanics*, 3(1):5–16, 2000. In Russian.

246. O. B. Najmark, M. M. Davydoff, O. A. Plekhov, and S. V. Uvaroff. Experimental and theoretical investigation of dynamical stochastization and scaling by the crack propagation. *Fizicheskaja mesomechanika*, 2(3):47–58, 1999. In Russian. English translation *Physical Mesomechanics*.
247. K. Nakasa. Application of fracture mechanics to crack branching and crack curving phenomena. *Bull. Japn. Inst. Metals*, 28(9):753–759, 1989.
248. K. Nam. Acoustic emission from surface fatigue cracks in SS41 steel. *Fatigue Fract. Engng. Mater. Struct.*, 22:1103–1109, 1999.
249. A. H. Nayfeh and B. Balachandran. *Applied nonlinear dynamics*. Wiley Series in Nonlinear Science. Wiley, 1995.
250. V. B. Nemtsov. *Non-equlibrium statistical mechanics systems with orientational order*. Technalohija, Minsk, 1997. In Russian.
251. A. I. Nesterov and S. G. Ovchinnikov. Geometric approach to dislocation and disclination theory. *Russian Phys. J.*, 30(11):937–940, 1987.
252. R. Niefanger, V.-B. Pham, G. A. Schneider, H.-A. Bahr, H. Balke, and U. Bahr. Quasi-static straight and oscillatory crack propagation in ferroelectric ceramics due to moving electric field: experiments and theory. *Acta Mater.*, 52(1):117–127, 2004.
253. S. M. Nikolskij. *Course of mathematical analysis*, volume 2. Nauka, Moscow, 1973. In Russian.
254. W. D. Nix and H. Gao. An atomistic interpretation of interface stress. *Scripta Mater.*, 39(12):1653–1661, 1998.
255. E. Noether. Invariante Variationsprobleme. *Nachrichten der Königlichen Gesellshaft Wissenschaften. Der Göttingen, Mathematisch—Physikalische Klasse*, 2:235–257, 1918. English translation *Transport Theory and Stat. Phys.* 1:186–207.
256. W. Noll. A mathematical theory of mechanical behaviour of contionuous media. *Arch. Rational Mech. Anal.*, 2(3):197–226, 1958.
257. W. Noll. Material uniform simple bodies with inhomogeneities. *Arch. Rational Mech. Anal.*, 27(1):1–32, 1967/1968.
258. S. Novikava, G. Ananich, and K. Miatliuk. The structure and the dynamics of information in design systems, volume 2 of *Proceedings of 7th Conference on Engineering Design, ICED'90*, Dubrovnik, 1990, pp. 946–953.
259. S. Novikava, S. Gancharova, and A. Zhybul et al. The statute of hierarchical mathematics and its cybernatical maintenance. Preprints of Eleventh International Conference on Mathematical and Computer Modelling and Scientific Computing, Washington, DC, 1997, p. 149.
260. V. U. Novikov and G. V. Kozlov. Fractial parameterization of the structure of filled polymers. *Mech. Compos. Mater.*, 35(3):285–200, 1999.
261. V. V. Novozhilov. On a necessary and sufficient criterion for brittle strength. *Prikladnaya Matematika i Mekhanika*, 33(2):212–222, 1969. In Russian. English translation *J. Appl. Math. Mech.*
262. V. V. Novozhilov. On the foundations of a theory of equilibrium cracks in elastic solids. *PMM, J. Appl. Math. Mech.*, 33:777–790, 1970. Translation from *Prikl. Mat. Mekh.*, 33:797–812 (1969).
263. W. Nowik, V. Barkalin, et al. Demographical units in hierarchical coordinates: construction, activity and goals. In *Proceedings of the IFAC Conference on Supplementary Ways for Improving International Stability, SWISS'98*, May 14–16, Sinaia, Romania, 1998. pp. 95–100.
264. J. F. Nye. Some geometrical relations in dislocated crystals. *Acta Metall.*, 1(2):153–162, 1953.
265. J. W. Obreimoff. The splitting strength of mica. *Proc. Roy. Soc.*, 127A(805):290–297, 1930.
266. I. V. Obreimov and E. S. Trekhov. Optical contact of polished glass surfaces. *Soviet Physics—JETP USSR*, 5(2):235–242, 1957.
267. O. Obrezanova, A. B. Movchan, and J. R. Willis. Stability of an advancing crack to small perturbation of its path. *J. Mech. Phys. Solids*, 50(1):57–80, 2002.
268. S. F. Ohr and J. Narayan. Electron microscope observation of shear cracks in strain steel single crystals. *Philos. Mag.*, 41(1):81–89, 1980.
269. P. J. Olver. *Application of Lie groups to differential equations*, volume 107 of *Graduate texts in mathematics*. Springer-Verlag, New York, 1986.
270. M. Onami, S. Ivasimadzu, K. Genka, K. Siodzava, and K. Tanaka. *Vvedenije v mikromechaniku*. Metallurgija, Moscow, 1987. Russian translation from Japanese.
271. A. N. Orlov, Yu. M. Plishkin, and I. M. Shevelova. *Fizika metallov i metallovedenije*, 32:185, 1957.

272. L. V. Ovsiannikov. *Gruppovyje svojstva differentsialnysh uravnienij (Group properties of dofferential equations)*. Siberean Branch of Academy of Science, Nowosibirsk, 1962. In Russian.
273. L. V. Ovsiannikov. *Group analysis of differential equations*. Academic Press, New York, 1982.
274. K. L. Pan and J. Fang. Nonlocal interaction of a dislocation with a crack. *Arch. Appl. Mech.*, 64(1):44–51, 1993.
275. V. V. Panasjuk, M. P. Savruk, and A. P. Dacyshyn. *Raspredelenie naprjazenij okolo treshchin v plastinach i obolockach*. Naukova Dumka, Kiev, 1976. In Russian.
276. V. E. Panin, V. E. Egorushkin, and P. V. Makarov. Physical basis of mesomechanics of plastic deformation and fracture of solid-state materials. *Physical mesomechanics and computer-aided design of materials*, volume 1. Nauka, Novosibirsk, 1995, pp. 7–49. In Russian.
277. V. E. Panin, V. A. Lichachov, and Yu. B. Grinajev. *Structural levels of solid body deformation*. Nauka, Moscow, 1985. In Russian.
278. V. E. Panin, V. E. Yegorushkin, P. V. Makarov, et al. *Physical mesomechanics and computer-aided design of materials*, volume 1,2. Nauka, Novosibirsk, 1995.
279. V. E. Panin. *Structure level of plastic deformation and fracture*. Nauka, 1990.
280. Ya. G. Panovko. *Vvedenije v teoriju mechanicheskich kolebanij*. Mauka, Moscow, 1980. In Russian.
281. A. Parisi and R. C. Ball. Role of surface waves on the relation between crack speed and the work of fracture. *Phys. Rev. B*, 66(16):165432, 2002.
282. A. Parisi, G. Galdarelli, and L. Pietronero. Roughness of fracture surfaces. *Europhys. Lett.*, 52(3):304–310, 2000.
283. V. Z. Parton and B. A. Kudriavcev. *Electromagnetoelasticity of piezoelectric and electroconducting bodies*. Nauka, Moscow, 1988. In Russian.
284. V. Z. Parton and E. M. Morozov. *Mechanics of elasto-plastic fracture*. Hemisphere, New York, 1989.
285. R. B. Pecherski. Modelling of large plastic deformation based on the mechanism of micro-shear banding. Physical foundations and theoretical description in plane strain. *Arch. Mech.*, 44(5-6):563–584, 1992.
286. J. Peddieson, G. R. Buchanan, and R. P. McNitt. Application of nonlocal continuum models to nanotechnology. *Int. J. Eng. Sci.*, 41(3–5):305–312, 2003.
287. Volker Perlick. Fermat principle in Finsler spacetimes. ArXiv:gr-qc/0508029 v1, August 2005.
288. Y. V. Petrov. Quantum analogy in the mechanics of fracture of solids. *Phys. Solid State*, 38:1846–1850, 1996.
289. V. S. Polenov and A. V. Chigarev. Propagation of waves in an inhomogeneous viscoelastic medium with initial stresses. *J. Appl. Math. Mech.*, 58(3):563–568, 1994.
290. H. Primas. *Chemistry, quantum mechanics and reductionism*. Springer-Verlag, Berlin, 1981.
291. Yu. N. Radaev. Canonical invariants of equations theory of closed plasticity and damage. *Vestnik Samarskogo Gosudarstvennogo universiteta*, 4(14):70–92, 1999. In Russian.
292. S. Ramanathan, E. Deniz, and D. S. Fisher. Quasistatic crack propagation in heterogeneous media. *Phys. Rev. Lett.*, 79(5):873–875, 1997.
293. S. Ramanathan and D. S. Fisher. Dynamics and instabililties of planar tensile crack in heterogeneous media. *Phys. Rev. Lett.*, 79(5):877, 1997.
294. P. K. Raschewski. *Elementare Einfuhrung in die Tensorrechnung*. Deutscher Verlag, Berlin, 1966. Russian original: *Rimanova geometrija i tenzornyj analiz*, Moscow, 1967.
295. V. R. Regel. Present state of solid bodies strength physics. *Rheologica Acta*, 14(1):19–26, 1975. Extended variant in Russian: V. R. Regel, A. I. Slucker and E. E. Tomashevskij. *Kineticheskaja pripoda prochnosti tverdych tel*. Nauka, Moscow, 1974.
296. A. F. Revuzhenko. *Mekhanika uprugo-plasticheskikh sred i nestandartnyi analiz*. Izd-vo Novosibirskogo Un-ta, Novosibirsk, 2000. In Russian.
297. J. R. Rice. A path independent integral and the approximative analysis of strain concentration by notches and cracks. *Trans. ASME, E: J. Appl. Mech.*, 35(2):379–386, 1968.
298. J. R. Rice. The mechanics of earthquake rupture. In A. M. Dzewonski and E. Boshi, editors, *Physics of the Earth interior*, volume 78 of *Proc. International School of Physics Enrice Fermi*. Italian Physical Society and North-Holland, 1980, pp. 555–649.
299. N. B. Romalis and V. P. Tamuzh. Vlijanije mikrodefectov na treshchinostojkost materialov. In *Mechanika i nauchno-technicheskij progress*, volume 3. Nauka, Moscow, 1988, pp. 104–102. In Russian.
300. A. E. Romanov and E. S. Aifantis. Defect kinetics in crack instabilities. *Scripta Metall. Mater.*, 30(10):1293–1298, 1994.

301. A. R. Rosenfield. Fractal mechanics. *Scripta Metall.*, 21(10):1359–1361, 1987.
302. S. Roux, J. Schmittbuhl, J.-P. Vilotte, and A. Hansen. Some physical properties of self-affine rough surfaces. *Europhys. Lett.*, 23(4):277–282, 1993.
303. H. Rund. *The differential geometry of Finsler spaces.* Springer Verlag, Berlin, 1959.
304. M. Rybaczuk and P. Stoppel. The fractal growth of fatigue defects in materials. *Int. J. Fracture*, 103(1): 63–86, 2000.
305. T. Nishioka S. Shen. Fracture of piezoelastic materials: energy density criterion. *Theor. Appl. Fracture Mech.*, 33(3):57–65, 2000.
306. Jan Saczuk. Finslerian foundations of solid mechanics. *Zeshyty naukowe instytutu maszyn przeplywowych PAN w Gdansku. Studia i Materialy*, 472(1427):1–97, 1996. Gdansk: Wydawnictwo IMP PAN.
307. J. Saczuk. Mechanics of solids with microstructure modelled by Finslerian geometry. *Zeshyty naukowe instytutu maszyn przeplywowych PAN w Gdansku. Studia i Materialy*, 493(1440):1–156, 1998. Gdansk: Wydawnictwo IMP PAN.
308. J. Saczuk. Continua with microstructure modelled by the geometry of higher-order contact. *Int. J. Solids Structures*, 38(38), 2001.
309. M. A. Sadovsky. *Selected works: geophysics and physics of explosion.* Nauka, Moscow, 2004. In Russian.
310. R. Z. Sagdeev, D. A. Usikov, and G. M. Zaslavsky. *Nonlinear physics: from the pendulum to turbulence and chaos.* Harwood Academic, 1990.
311. C. Sansour. On the geometric structure of the stress and strain tensors, dual variables and objective rates in continuum mechanics. *Arch. Mechanics*, 44(5-6):527–556, 1992.
312. J. Schmittbuhl and K. J. Måaløy. Direct observation of a self-affine crack propagation. *Phys. Rev. Lett.*, 78(20):3888–3891, 1997.
313. J. Schmittbuhl, S. Roux, and Y. Berthaud. Development of roughness in crack propagation. *Europhys. Lett.*, 28(8):585–590, 1994.
314. K. Schmittbuhl and K. J. Måløy. Direct observation of a self-affine crack propagation. *Phys. Rev. Lett.*, 78(20):3888–3891, 1997.
315. J. A Schouten. *Tensor analysis for physicists.* Clarendon Press, Oxford, 1954.
316. L. I. Sedov. *A course in continuum mechanics.* Wolters-Noordhoff, Groningen, 1971.
317. L. I. Sedov and V. L. Berditchevski. A dynamic theory of continual dislocations. Mech. general. *Continua, Proc. IUTAM-sympos.* Freundenstadt and Stuttgart, 1967, pp. 214–238.
318. Th. Seelig and D. Gross. On stress wave induced curving of fast running cracks—a numerical study by a time-domain boundary element method. *Acta Mechanica*, 132(1):47–61, 1999.
319. A. A. Shaniavskij. The fractal nature of the cinetics of fatigue cracks in metals. I. Consideration of the winding crack trajectory in fracture mechanics. *Metalls*, (5):80–86, 1999. In Russian. English translation *Russian Metallurgy (Metally).*
320. A. A. Shaniavskij. The fractal nature of the kinetics of fatigue cracks in metals. II. About the crack growth. *Metalls*, (1):112–119, 2000. In Russian. English translation *Russian Metallurgy (Metally).*
321. A. A. Shaniavskij. Rotacionnaja neustojchivost deformacii i razrushenija metallov pri rasprostranenii treshchin na mesoskopicheskom masshtabnom urovnie. I. Mechanismy rasrushenija. *Fizicheskaja mesomechanika*, 4(1):81–95, 2001. In Russian. English translation *Phys. Mesomechanics.*
322. A. A. Shaniavskij. Rotacionnaja neustojchivost deformacii i razrushenija metallov pri rasprostranenii treshchin na mesoskopicheskom masshtabnom urovnie. I. Processy plasticheskoj deformacii v vershinie treshchiny. *Fizicheskaja mesomechanika*, 4(1):73–80, 2001. In Russian. English translation *Phys. Mesomechanics.*
323. E. Sharon, G. Cohen, and J. Fineberg. Crack front waves and the dynamics of a rapidly moving crack. *arXiv:cond-mat*, 0107470 v1 23 Jul, 2001.
324. E. Sharon, G. Cohen, and J. Fineberg. Crack front waves: localized solitary waves in dynamic fracture. *Nature*, 410(6834):68, 2001.
325. E. I. Shemiakin. Sinteticheskaja teorija prochnosti. Chast' I (Synthetical theory of fracture. Part I). *Fizicheskaja mesomechanika*, 2(6):63–69, 1999.
326. Dov Sherman and Ilan Be'ery. Nonlinear analysis of the fracture surface of a single-crystal brittle solid. *Physica D*, 119(3/4):424–432, 1998.
327. T. D. Shermegor. *Theory of elasticity of microinhomogeneous media (Teoriya uprugosti mikroneodnorodnykh sred).* Nauka, Moscow, 1977. In Russian.

328. M. G. Shnirman and Y. A. Tyurina. Generalised hierarchical model of defect develoment and self-organised criticality. *Phys. Rev. E*, 57(4):3804–3813, 1998.
329. M. A. Shtremel. Non-local interactions of many cracks. *Phys. Metals Metallogr.*, 91(3):221, 2001.
330. G. C. Sih. Strain energy density factor applied to mixed mode crack problems. *Int. J. Fract. Mech.*, 10(3):305–321, 1974.
331. G. C. Sih. *Mechanics of fracture initiation and propagation*, volume 11 of *Engineering application of fracture mechanics*. Kluwer, Dordrecht/Boston/London, 1991.
332. V. V. Silberschmidt. On the multifractal character of load distribution near the cracks. *Europhys. Lett.*, 23(8):593–598, 1993.
333. I. Simonsen, D. Vandembroucq, and S. Roux. Wave scattering from self-affine surfaces. *Phys. Rev. E*, 61:5914, 2000.
334. M. T. Siniawski, S. J. Harris, Quan Wang, and Shuangbiao Liu. Wear initiation of 52100 steel sliding against a thin boron carbide coating. *Tribol. Lett.*, 15(1):29–41, 2003.
335. W. Slomczynski, J. Kwapien, and K. Zyczkowski. Entropy computing via integration over fractal measures. *Chaos*, 10(1):180–188, 2000.
336. K. Sobczyk, J. Trebicki, and JR B. F. Spenser. Modelling of curvilinear random fatigue crack growth. *Engin. Fract. Mech.*, 52(4):703–715, 1995.
337. A. K. Soh, J.-X. Liu, K. L. Lee, and D.-N. Fang. On a moving Griffith crack in anisotropic piezoelectric solids. *Arch. Appl. Mech.*, 72(6–7):458–715, 2002.
338. I. M. Sokolov. Dimensions and other geometric critical exponents in percolation theory. *Uspekhi Fiz. Nauk*, 150(2):221–248, 1986.
339. N. M. Sokolova. The new approach to the great Fermat's theorem. *Vestnik BNTU*, (1):65–68, 2005. In Russian.
340. C. Soutis and Ihar A. Guz. Predicting fracture of layered composites caused by internal instability. *Composites A*, 32(9):1243–1253, 2001.
341. P. Ståhle. On the small crack fracture mechanics. *Int. J. Fracture*, 22(3):203–216, 1983.
342. P. Steinmann. Views on multicative elastolasticity and the continuum theory of dislocations. *Int. J. Engng. Sci.*, 34(15):1717–1735, 1996.
343. V. P. Stojan. Self-organisation of microfracture and crack localisation in brittle bodies by general three dimensional loading. *Phys. Mesomechanics*, 3(5):65–76, 2000. In Russian.
344. R. L. Stratonovich. *Theory of information*. Sovietskoe Radio, Moscow, 1975. In Russian.
345. Zhigang Suo. Singularities interacting with interfaces and cracks. *Int. J. Solids Structure*, 25(10): 1133–1142, 1989.
346. F. Takayashi and N. Hiizu. Dynamical stability of the crack front life. *J. Phys. Soc. Japan*, 67(12): 4064–4067, 1998.
347. H. Tanaka and Akira Tsuri. Random propagation of a semi-elliptical surface crack as a bivariate stochastic process. *Engng. Fract. Mech.*, 33(5):787–800, 1989.
348. T. Y. Thomas. *Plastic flow and fracture in solids*, volume 2 of *Mathematics in science and engineering*. Academic Press, New York, 1961.
349. R. Thomson. Brittle-fracture in a ductile material with application to hydrogen embrittlement. *J. Mater. Sci.*, 13(1):128–142, 1978.
350. R. Thomson. Dislocation–crack interaction. *Scripta Metall.*, 20(11):1473–1476, 1986.
351. C. Truesdell and W. Noll. *The non-linear field theories of mechanics*, volume III/3 of *Encyclopedia of physics*. Springer, Berlin/Heidelberg/New York, 1965.
352. B. N. Tsai. Physical aspects of mechanism of rock failure. *J. Mining Sci.*, 40(1):67–73, 2004. Translation of *Fiziko-Tekhnicheskie Problemy Razrabotki Poleznykh Iskopaemykh*.
353. P. E. Turchi, A. Gonis, and L. Colombo. Tight binding approach to computational material science. In *High speed computer and algorithm organization*, volume 491 of *MRS Symposium Proceedings*. MRS, Warrendale, 1998.
354. T. Yajima and H. Nagahama. Kawaguchi space, Zermelo's condition and seismic ray path. *Nonlinear Anal.: Real World Applications*, 8:130–135, 2007.
355. V. Červenỳ. *Seismic ray theory*. Cambridge University Press, Cambridge, 2001.
356. V. Červenỳ. Fermat variational principle for anisotropic inhomogeneous media. *Stud. Geophys. Geodet.*, 46(6):568–588, 2002.

357. A. A. Vakulenko and M. L. Kachanov. Kontinualnaja teorija sredy s treshchinami. *Izvestija AN. Mechanika Tverdogo Tela*, (4):159–166, 1971. In Russian. English translation *Mechanics of Solids*.
358. T. Vegge and K. W. Jacobsen. Atomistic simulation of dislocation processes in copper. *J. Phys.: Condens. Matter*, 14(11):2929–2956, 2002.
359. F. Verhulst. *Nonlinear differential equations and dynamical systems*. Springer Verlag, Berlin, 1990.
360. A. I. Vesnitsky. *Waves in systems with moving boundaries and loads*. Fizmatlit, Moscow, 2001. In Russian.
361. V. Z. Vlasov. *Thin-walled elastic beams*. National Science Foundation and Department of Commerce, 1961.
362. S. D. Volkov and V. P. Stavrov. *Statistical mechanics of composite materials*. Belarusian State University, Minsk, 1978. In Russian.
363. A. S. Volmir. *Non-linear dynamics of plates and shells*. Nauka, Moscow, 1972. In Russian.
364. I. P. Volobuev and Yu. A. Kubyshkin. *Differentsialnaja geometrija i algebry Li b ich prilozhenija v teorii pola*. Editorial URSS, Moscow, 1998. In Russian.
365. L. von Bertalanffy, W. Beier, and R. Lane. *Biophysik des Fließgleichgewichts*. Akademie Verlag, Berlin, 1977.
366. G. V. Vstovskii. Model of the fractal profile of a fatigue crack. *J. Appl. Mech. Tech. Phys.*, (2):130–137, 1992.
367. W. Wack and A. Tourabi. Some remarks on macroscopic observations and related microscopic phenomenon of the mechanical behaviour of metallic materials. *Arch. Mech.*, 44(5-6):621–662, 1992.
368. C. C. Wang. On the geometric structure of simple bodies, a mathematical foundation for the theory of continuous distributions of dislocations. *Arch. Rat. Mech. Anal.*, 27(1):33–94, 1967.
369. X. D. Wang and L. Y. Jiang. Fracture behaviour of crack in piezoelectric media with electromechanically coupled boundary conditions. *Proc. R. Soc. London A*, 458(2026):2545–2560, 2002.
370. R. E. Warren, A. Majumdar, and D. Krajcinovich. A fractal model for the rigid-perfectly plastic contact of rough surfaces. *Trans. ASME: J. Appl. Mech.*, 63:47–54, 1996.
371. R. H. Wasserman. *Tensors and manifolds with applications to mechanics and relativity*. Oxford University Press, New York/Oxford, 1992.
372. K. Weerstraß. *Abhandlungen aus der Functionenlehre*. Springer, Berlin, 1886.
373. J. Weertman. Fracture mechanics—unified view for Griffith–Irwin–Orowan cracks. *Acta Metall.*, 26(11):1731–1738, 1978.
374. W. H. Weibull. *A Statistical theory of the strength of materials*. Generalstabens Litografiska Anstalts Förlag, Stockholm, 1939.
375. H. Weil. *The theory of groups and quantum mechanics*. Dover Publication, 1931.
376. H. Weyl. *The classical groups: their invariants and representations*. Princeton University press, Princeton, NJ, 1946.
377. E. T. Whittaker. *A treatise of the analytical dynamics of particles and rigid bodies*. University Press, Cambridge, 1927.
378. R. E. Williford. Multifractal fracture. *Scripta Metall.*, 22(11):1749–1754, 1988.
379. J. M. Wills and W. A. Harrison. Further studies on interactions in simple metals and transition metals. *Phys. Rev. B*, 29(10):5486–5490, 1984.
380. E. M. Wright. The asymptotic expansion of the generalized hypergeometric function. *Proc. London Math. Soc. Second Series*, 46:389–408, 1940.
381. Yunping Xi and Zd. Bazant. Random growth of crack with r-curve: Markov process model. *Engng. Fract. Mech.*, 57(6):593–608, 1997.
382. Y. Xu and L. M. Keer. Crack curving in anisotropic elastic solids. *Engin. Fracture Mech.*, 44(1):63–73, 1993.
383. Wei H. Yang. On generalized Hölder inequality. *Nonlinear anal. meth. applications*, 16(5):489–498, 1991.
384. W. H. Yang, editor. *Topic in plasticity*. AM Press, Ann Arbor, MI, 1991. (Anniversary volume in honor of Profesor E. H. Lee.)
385. Ryoiji Yuuki and Jin-Quan Xu. Fracture criteria on kinking of a crack out the interface in dissimilar materials. *Trans. Japan Soc. Mech. Eng.*, 1945–1951, 1990.
386. G. M. Zaslavskij. *Statistical irreversibility in non-linear systems*. Nauka, Moscow, 1970. In Russian.
387. Xiaodong Zhang, M. A. Knackstedt, D. Y. C. Chan, and L. Paterson. On the universality of fracture surfaces roughness. *Europhys. Lett.*, 34(2):121–126, 1996.

388. S. N. Zhurkov. Kinetic concept of the strength of solids. *Int. J. Fracture*, 26(4):295–307, 1984.
389. S. N. Zhurkov. Physical basis of the strength of materials. *Soviet Physics—Solid State (US)*, 22(11): 1958–1961, 1984. Translation from *Fizika Tverdogo Tela*.
390. S. N. Zhurkov, V. S. Kuksenko, V. A. Petrov, et al. Koncentratcionnyj kriterij objomnogo razrushenija tverdych tel. *Fizicheskije processy v ochagach zemletriasenij*. Nauka, Moscow, 1979. In Russian.
391. J. Z. Zuo and G. C. Sih. Energy density theory formulation and interpretation of cracking behavior for piezoelectric ceramics. *Theoret. Appl. Fract. Mech.*, 34:17–33, 2000.

# INDEX